Attribute-Based Encryption and Access Control

Data-Enabled Engineering

Series Editor

Nong Ye
Arizona State University, Phoenix, USA

Published Titles

Data Mining
Theories, Algorithms, and Examples
Nong Ye

Convolutional Neural Networks in Visual Computing
A Concise Guide
Ragav Venkatesan and Baoxin Li

Software-Defined Networking and Security
From Theory to Practice
Dijiang Huang, Ankur Chowdhary, and Sandeep Pisharody

Social Media Analytics for User Behavior Modeling
A Task Heterogeneity Perspective
Arun Reddy Nelakurthi and Jingrui He

Attribute-Based Encryption and Access Control
Dijiang Huang, Qiuxiang Dong, and Yan Zhu

For more information about this series, please visit: www.crcpress.com/
Data-Enabled-Engineering/book-series/CRCDATENAENG

Attribute-Based Encryption and Access Control

Dijiang Huang
Qiuxiang Dong
Yan Zhu

CRC Press
Taylor & Francis Group
Boca Raton London New York

CRC Press is an imprint of the
Taylor & Francis Group, an **informa** business

CRC Press
Taylor & Francis Group
6000 Broken Sound Parkway NW, Suite 300
Boca Raton, FL 33487-2742

International Standard Book Number-13: 978-0-8153-8135-8 (Hardback)

Library of Congress Cataloging-in-Publication Data

Names: Huang, Dijiang, author. | Dong, Qiuxiang, author. | Zhu, Yan (Professor of computer security), author.
Title: Attribute-based encryption and access control / by Dijiang Huang, Qiuxiang Dong, Yan Zhu.
Description: First edition. | Boca Raton, FL : CRC Press, 2020. | Series: Data-enabled engineering | Includes bibliographical references and index.
Identifiers: LCCN 2019053725 | ISBN 9780815381358 (hardback) | ISBN 9781351210607 (ebook)
Subjects: LCSH: Computer networks--Security measures. | Computer security. | Identification.
Classification: LCC TK5105.59 .H77 2020 | DDC 005.8/24--dc23
LC record available at https://lccn.loc.gov/2019053725

Visit the Taylor & Francis Web site at
http://www.taylorandfrancis.com

and the CRC Press Web site at
http://www.crcpress.com

Any attribute that we have does not possess the power to define itself as 'good' or 'bad'. Rather, any such definition is based solely on what we 'attribute' to that attribute.

— Craig D. Lounsbrough

Contents

PART I Foundations of Attribute-Based Encryption for Attribute-Based Access Control

PART II Applications of Attribute-Based Encryption

Preface

WHY THIS BOOK?

Very few existing books focus on using Attribute-Based Encryption (ABE) to realize Attribute-Based Access Control (ABAC). This book serves as a bridge to attract more research and development attention from both academia and industry to implement secure data access control based on ABE approaches. First, we need to understand the difference between ABAC and Role-Based Access Control (RBAC), and then we will discuss why ABE approaches are promising solutions to realize ABAC for practical security systems, especially in today's booming data-centric science, IoT, and mobile applications.

ABAC vs. RBAC

Gartner, a leading research and advisory company, predicted that by 2020, 70% of enterprises will use ABAC as the dominant mechanism to protect critical assets, up from less than 5% since 2014 [86]. However, it is still very difficult to shake the uses of dominating solutions such as RBAC today. When contemplating using ABAC, one must take into consideration both the good and bad. On one hand, ABAC can be designed to support more complex policies, in which we call it fine-grained access control compared to its counterpart RBAC; however, on the other hand, it is also inherently more complex to operate, in which there are many more attributes than roles, and thus it increases management complexity. Some basic concepts are summarized here between ABAC and RBAC:

- RBAC is for coarse-grain access control and ABAC is for fine-grain access control. RBAC is usually realized by first creating an access control group (i.e., defined as a role) with pre-setup privileges, and then latter users can be added into the group for their desired access privileges. For ABAC, users are assigned attributes to describe their job functions or properties, and the access control system just needs to focus on required access control policies that are described by a set of attributes to check users' privileges to decide if the access should be granted or not. From this view, we can also call ABAC a Policy-Based Access Control (PBAC) and we use these terms interchangeably in this book.
- An access control system is usually formed by two system-level functions: identity and credential management and access control enforcement. Both ABAC and RBAC can be implemented by using either centralized or decentralized credential issuing parties, i.e., Identity Management (IdM). Access control enforcement is also done in a centralized fashion for both ABAC and RBAC, which requires one or multiple dedicated access control policy enforcement parties, a.k.a., Policy Enforcement Point (PEP).

- In access control application scenarios where access decisions have broad strokes, RBAC is preferred. On the contrary, ABAC is suitable for more complex access control environments, where the control is usually driven by dynamic changing access policies. Although ABAC is powerful, the rule *less is more* always holds. Creating complex ABAC filters may lead to system management disaster or uncontrollable access control scenarios. Thus, getting to the right level is the key to making ABAC practical.
- RBAC and ABAC are compatible. A hierarchical approach can be used to use RBAC and ABAC together; for instance, using RBAC to control who can access one or a set of modules, and then using ABAC to control access to what is inside a module.

Why ABE-Enabled ABAC?

Using rigid and dedicated computer infrastructures as the ABAC Policy Enforcement Point (PEP) does not fully enjoy the concept of attribute-based solution. Using Attribute-Based Encryption (ABE), the access control policies can be incorporated into the ciphertext, and thus the Access Control (AC) is *mobile*. This means ABE-based ABAC is suitable for Data-Centric Access Control (DCAC), in which it provides two unique and critical capabilities for emerging data-centric applications:

- The ABE-based ABAC solution does not need to change existing identity management (IdM) solutions. In this way, we will not dramatically change existing security infrastructures.
- Since the access control policies are incorporated into the ciphertext, the PEP can be mobile, decentralized, or even distributed, which means each data hosting party can serve as a PEP. As a result, there is no need for a dedicated access control enforcement infrastructure to define the boundary between trusted and untrusted parties for data access.

ABE-enabled ABAC can serve as a fundamental trust infrastructure that can greatly improve the data-centric applications to support a scalable data access and sharing paradigm. We must note that ABE solutions fit into the ABAC concept naturally due to its "attribute" nature. However, ABE can support the RBAC when considering attributes are required to validate its group-based roles. This book focuses on ABE-enabled ABAC, and how to use ABE to support RBAC is left for future investigation.

AUDIENCE

Our goal has been to create a book that can be used by a diverse set of audiences, especially for computer network, system, and software researchers and engineers who may not have strong cryptography backgrounds. To this end, except a few critical schemes, this book does not provide in-depth proofs for many schemes presented in Part II (ABE applications). This book can serve as a reference for college students, instructors, researchers, and software developers who are interested in developing ABE-based Attribute-Based Access Control (ABAC) solutions. It can serve as a good reference book for undergraduate and graduate courses focusing on applied cryptography and computer security. Specifically, we present several new research outcomes to set out to create a book that can be used by professionals, as well as students and researchers. In general, this is intended as a self-study. We assume that the reader already has some basic knowledge of applied cryptography. Although we expect readers have some level of advanced background in pairing-based cryptographic constructions, this book will be sufficiently self-learned without referring to additional resources. Among professionals, the intent has been to cover two broad groups: developers of secure data access controls and secure communication/networking protocols, with the overall goal to bring out issues that one group might want to understand about what the other group faces. For students, this book is intended to help learn about both pairing-applied cryptography (especially in ABE) and ABAC in depth, along with lessons from security aspects, and operational and implementation experience. For researchers, who want to know what has been done so far and what are critical issues to address for next generation access control, this is intended as a helpful reference.

ORGANIZATION AND APPROACH

The organization of the book is summarized as follows:

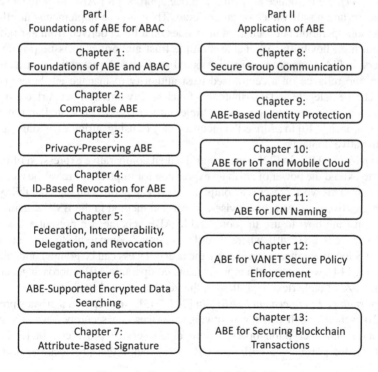

Figure 1: Organization of the book.

The organization of this book includes two parts: Part I: Foundations of Attribute-Based Encryption for Attribute-Based Access Control, which contains basic constructions of ABE schemes and a few important ABE features such as comparable attributes, attributes revocation, and ABE-based keywords search, etc.; and Part II: Applications of Attribute-Based Encryption, which contains several secure systems built on ABE-based data access control. The overall organization and chapter layout are presented in Figure 1. A succinct chapter description is provided at the beginning of each chapter, for which it provides a quick overview for the chapter and gives readers an overview of the presented materials.

Most of the presented solutions are based on previously published research work, which are summarized as follows:

Revocable ABE: We have done recent work in this area. For example, the solution in [216] presented a basic ID-based revocation approach by incorporating users' unique IDs into private keys corresponding to each attribute. This approach provides both individual and multiple users' revocation by identifying the revoked users' IDs in the given ABE access policy tree. Recent work [75, 77] provides a hierarchical delegation framework and ID management strategy; [76, 151] provides a federation protocol allowing multiple delegators to collaboratively generate private keys for attributes managed by different delegators.

ABE-Based Policy Management: To address the conversion from RBAC to ABAC, in [248, 247], we introduce an attribute lattice approach for ABE to define a seniority relation among all values of an attribute. This scheme implements an efficient comparison operation between attribute values on a Poset derived from attribute lattice with high flexibility. To enable multiple trust authorities to issue private keys for users, in [141], we proposed a distributed ABE-based trust authority framework to relax the reliance on a centralized trust authority to manage attributes. In this approach, we delegate and distribute this functionality amongst network entities. To manage the attributes managed by multiple authorities, we presented an ontology-based approach [140] to address the inconsistency of using attributes among multiple administrative domains.

Comparable ABE: In [226], we propose a dual comparative expression of integer ranges to extend the power of attribute expression for implementing various temporal constraints. This work presents a comparative ABE scheme with high flexibility, and it is an improvement over the bitwise-comparison method in the BSW scheme [25]. To demonstrate how to use the comparable ABE scheme, we present a Location-Based Service (LBS) [249], where according to users' comparable location attributes (e.g., GPS location range information), the users' access can be granted or denied. In [219, 113, 114], we presented a more efficient comparable ABE approach generating constant size of ciphertext regardless of the number of used attributes.

Performance Enhancement of ABE: In [242, 243], we proposed a privacy preserving ABE scheme such that the lightweight devices can securely outsource heavy encryption and decryption operations to cloud service providers. In [241, 240, 245, 116], we presented an efficient CP-ABE scheme that provides the following features: (1) the ciphertext is constant regardless of the number of involved attributes; (2) it provides a unique bit-assignment approach to achieve efficient secure group communication. In this scheme, a bit-assignment is a position in a user's ID representing three possible statuses: A, A^-, and A^* (i.e., Yes, No, and Don't Care) of an attribute. Using the bit-assignment and corresponding attributes, we can establish secure group communication groups efficiently.

Privacy-Preserving for ABE: To anonymize users' IDs, in [110], we proposed a variant of IBE wherein the PKG (Private Key Generator) is removed and the anonymous users can derive their own private keys based on public parameters and pseudonyms. In [245, 240], we proposed a new construction of CP-ABE, named Privacy Preserving Constant CP-ABE, which leverages a hidden policy construction. To provide privacy protection of attributes in the attribute policy tree, in [117, 244], we present an attribute gradual exposure approach. Based on the available attributes, users can reveal attributes one by one from the attribute policy tree structure. A powerful user, who owns more privileged attributes, can reveal all the attributes in a secure policy tree for ABE. In [218], we presented an ABE-based privacy-preserving mobile access control approach, and in [111], an efficient pseudonymous authentication-based conditional privacy protocol is presented for Vehicular Ad-hoc Networks (VANETs).

Searchable Encryption: In [73], we presented a solution for fuzzy keyword searches in the public key setting. In [74], based on symmetric key-based encryption, we presented a solution to encrypt users' profiles with a new searchable encryption scheme to provide privacy preserving for location-based services. In [72], we presented new techniques for attribute-based encryption approaches that split the computation for the keyword encryption and trapdoor/token generation into two phases. First, there's a preparation phase that does the vast majority of the work to encrypt a keyword or create a token before it knows the keyword or the attribute list/access control policy that will be used, and second phase then rapidly assembles an intermediate ciphertext or trapdoor when the specifics become known.

ABE-Based Application Scenarios: In the past, we have focused on various ABE-based applications. Here, we just enumerate a few. Besides the above-mentioned research, we have also investigated various application scenarios using ABE solutions, such as secure mobile cloud data storage [242], vehicular networking [112, 111, 53], healthcare [219], etc. In [243], we presented a solution to secure mobile cloud data based on ABE approaches. In [110], a pseudonym-based cryptography for anonymous communications in mobile ad hoc networks was presented. In [117, 244], we arrange the policies in a skewed policy tree structure. In this way, a user can reveal the security policies by decrypting the policy-tree level by level to protect the policy information. In [140, 139], we presented a privacy-preserving access control scheme for Information-Centric Networking (ICN) and its corresponding attribute management approach, in which the solution is compatible with flat-name–based ICN architectures.

WHAT IMPORTANT WORK IS NOT COVERED IN THIS BOOK ABOUT ABE?

Attribute-Based Encryption (ABE) research has been very active for more than 10 years. Recent development trends show ABE-based research is still very active and many new solutions addressing application needs have been proposed; see surveys in [135, 21, 148, 133, 16, 78].

We must note that this book does not intend to cover all significant research outcomes in the area of ABE-related research. However, we do notice that researchers have been inspired to construct lattice-based ABE schemes due to the fact that the hardness of the lattice problem is able to resist quantum attacks. Lattice-based ABE research is one of main research areas omitted by this book. Here, we provide a glimpse of this research direction, and we encourage both researchers and security engineers in their further investigations.

Zhang et al. [237] presented ciphertext policy ABE from lattices. In 2013, Boyen [43] proposed an efficient key-policy ABE scheme on lattice. He introduced a broad lattice manipulation technique for expressive cryptography, and realized functional encryption with access structures on post-quantum hardness assumptions. In 2014, Han et al. [98] proposed a general transformation from ABE to attribute-based encryption with keyword search and a concrete attribute private key-policy ABE scheme. Zhao et al. [238] proposed a new ABE scheme for circuits on lattice. Shraddha et al. [200] gave an enhancing flexibility CP-ABE scheme with multiple mediators. In 2016, Li et al. [142] constructed a concrete KP-ABE outsourcing scheme. Karati et al. [126] proposed a threshold-based ABE scheme without a *bilinear* map and pointed out the new scheme was much more efficient and flexible than others. In 2019, Liu et al. [149] constructed an ABE scheme for a large-scale distributed environment on lattices from decisional LWE assumptions. It is secure against CPA under the selective attribute model.

ACKNOWLEDGMENTS

This book could not have been done without many days and nights of hard work from our three main authors, who have collaboratively contributed the materials and presentations for each chapter. Moreover, the materials for each chapter also have significant contributions from previous graduates from Dr. Dijiang Huang's research group. As mentioned in the contributor section, some book chapters are partially based on several PhD dissertations and their related publications during their study at Arizona State University. Moreover, Dr. Weijia Wang visited ASU during the years 2011 and 2015-2016 and focused his research on ABE. Here we summarize their contributions for this book.

Contributors	Chapters
Zhibin Zhou	3, 8, 9, 10
Zhijie Wang	2
Weijia Wang	4
Bing Li	11
E. Chen	7

Moreover, the MS thesis by Ashwin Narayan Prabhu Verleker had been focused on ABE solution testing and prototype establishment and Mayank Verma's thesis contributes to the ABE solutions for VANETs.

Dr. Dijiang Huang leads the Secure Networking and Computing (SNAC) group at ASU. SNAC is formed by graduate students who are working on various research and development projects in the areas of computer networking security, cloud computing security, applied cryptography, and IoT security, etc. SNAC hosts regular meetings for group members to share research results and discuss research issues. Many SDN security-related work has been studied through SNAC meetings. The authors also gratefully thank current SNAC members: Adel Alshamrani, Chun-Jen Chung, Yuli Deng, Jiayue Li, Fanjie Lin, Duo Lu, Sowmya Myneni, Abdulhakim Sabur, and Zeng Zhen, who inspired them through discussions, seminars, and project collaborations.

Special thanks is given to the Naval Research Lab (NRL) project for sponsoring the ABE-based research. Particularly, authors from ASU thank Jim Luo and Myong Kang from NRL for their discussions and guidance to lead our ABE research to a more meaningful and thoughtful direction to address real-world problems.

Another special thanks to the National Science Foundation of China's (NSFC) support and for allowing Professor Dijiang Huang to travel to China and collaborate with Professor Yan Zhu on the ABE research. Our collaborations have produced fruitful research outcomes and have contributed many resources to this book.

Our immediate family members suffered the most during our long hours of being glued to our laptops. Throughout the entire duration, they provided all sorts of support, entertainment, and "distractions." Dijiang would like to thank his wife Lu, and their son Alexander and daughter Sarah, for love and patience, and for enduring this route. He would also like to thank his father Biyao and mother Zhenfen, and his brother Dihang, sister-in-law Cuishan, and nephew Yicheng for taking care of his parents in Beijing, and allowing him to focus on his research. Finally, he would like to thank his many friends for an outpouring of support.

Qiuxiang would like to thank her parents Yanbin and Shuqin for their unconditional love and constant support, her sister Chunxiang for taking care of her parents during her study at Arizona State University, her husband Liang for all the encouragement and understanding, her grandma Guiqing for her unconditional love, sister Junfeng for her encouragement and support during her study, all the families who gave her constant love, and her ginger, fluffy roommate Frankie for making her happy. She also would like to thank all the people who worked together with her, including her labmates Yuli, Duo, Yang, Jiawei, and Chun-Jen; advisors Dijiang, Zhong, Zhi, Chunqi, Zhonqiang, and Guiying; industry mentors Gansha, Yue-Hsun, and Callen; and friends Fred and Rio for all the inspiring discussions and help in daily life.

Yan Zhu would like to thank his students, Huaixi Wang, Shanbiao Wang, Ruyun Yu, and E. Chen, for the ABE/ABS and cryptography research, and the National Key Technologies R&D Programs of China under Grant 2018YFB1402702, in part by the National Natural Science Foundation of China under Grant 61972032. He also would like to thank all the people who worked together with him at Peking University and University of Science and Technology Beijing.

Dijiang Huang
Tempe, Arizona, USA
Dijiang.Huang@asu.edu

Qiuxiang Dong
Tempe, Arizona, USA
qiuxiang.dong@asu.edu

Yan Zhu
Beijing, China
zhuyan@ustb.edu.cn

Authors

LinkedIn QR Code

Dr. Dijiang Huang graduated in telecommunications from Beijing University of Posts and Telecommunications (China) with a bachelor's degree in 1995; his first job was a network engineer in the computer center of Civil Aviation Administration of China (CAAC); with four years of industry working experience, he then came to the University of Missouri-Kansas City (UMKC) in the United States to pursue his graduate study in the joint computer networking and telecommunication networking program in Computer Science; and he earned a master's of science and PhD in computer science in 2001 and 2004, respectively. During his study at UMKC, he became interested in the research areas of mobile computing and security, and focused his research on network security and mobile networks.

After graduating with his PhD, Dr. Huang joined Arizona State University (ASU), and now he is an associate professor at the School of Computing Decision Systems Engineering (CIDSE) of ASU. One of his early research areas focused on securing MANET communication and networking protocols, and pairing-based cryptography. Later, he realized that the cross-layer approach is extremely important to design secure computer networking systems based on advanced cryptographic approaches. Gradually, he looked into the research problem of how to design and develop attribute-based cryptographic solutions to solve the real-world issues in various application domains, such as the vehicular network, cloud storage, content delivery network, blockchains, etc.

In 2010, Dr. Huang was awarded the Office of Naval Research (ONR) Young Investigator Program (YIP) award working on a research project to establish a secure mobile cloud computing system. The main task of the award is to develop a secure and robust mobile cloud and networking system to support trustworthy mission-critical operations and resource management considering communication, networking, storage, computation, and security requirements and constraints. With the boom of Software-Defined Networking (SDN), it has changed the playground of computer network security, which has become more dynamic, automatic, and intelligent in the past few years. His research has been focused on a more intelligent and Moving Target Defense (MTD) by incorporating dynamic learning models in security analysis and decisions. This book can share his past research and development outcomes and provides a starting point to ride on the next research and development wave for software-defined security, which can benefit both research communities and practitioners.

Dr. Huang is currently a Fulton Entrepreneur Professor in the School of Computing Informatics Decision Systems Engineering (CIDSE) of ASU, and he has published six US patents and is a co-founder of two start-up companies: Athena Network Solutions LLC and CYNET LLC. He is currently leading the Secure Networking and Computing (SNAC) research group. Most of his current and previous research is supported by federal agencies such as National Science Foundation (NSF), ONR, Army Research Office (ARO), Naval Research Lab (NRL), National Science Foundation of China (NSFC), and North Atlantic Treaty Organization (NATO); and industries such as the Consortium of Embedded Systems (CES), Hewlett-Packard, and China Mobile. In addition to ONR Young Investigator Award, he was also a recipient of HP Innovation Research Program (IRP) Award, and a JSPS Fellowship. He is a senior member of IEEE, ACM, and the National Academy of Inventors (NAI).

Dr. Qiuxiang Dong is currently a graduate research associate at Arizona State University. Before that, she obtained her PhD (2016) from Peking University and BE (2011) from Tongji University. Her research interests include applied cryptography, network security, data privacy, and security. She did research on advanced cryptographic algorithms, including searchable encryption, homomorphic encryption, identity-based encryption, attribute-based encryption, etc. She also did research on how to utilize cryptographic algorithms to protect data security in multiple applications, such as cloud computing, healthcare, blockchain, etc.

LinkedIn QR Code

Dr. Dong is also interested in privacy-preserving data mining by way of secure multiple computation and trusted computing.

She gained industrial experience in data protection, such as how to implement data encryption/decryption in an efficient and user-friendly way, how to implement authorization systems based on ACL (Access Control List) within a large organization, how to perform data mining or machine learning algorithms over data from multiple parties in a secure way, etc.

She is also interested in how to utilize machine learning algorithms to detect malware or other malicious activities. For more information, please refer to her LinkedIn profile.

Home Page QR Code

Dr. Yan (Vicent) Zhu received a PhD in computer science from Harbin Engineering University, China, in 2005. He is currently a full professor of computer science at the University of Science and Technology Beijing (USTB), China. He was an associate professor at Peking University, China, from 2007–2012. He was a visiting associate professor at Arizona State University, from 2008–2009, and a visiting research investigator at the University of Michigan-Dearborn in 2012. He has authored over 100 journal and conference papers in computer and network security. His research interests include cryptography, secure computation, and access control.

Contributors

In addition to the authors' previous works, some book chapters are partially based on research outcomes of several PhD dissertations supervised by Professor Dijiang Huang at Arizona State University. Several chapters are based on partial PhD dissertations and peer-reviewed research publications from the following contributors.

Chun-Jen Chung
Cofounder and CTO of Athena Network Solutions LLC
Chandler, Arizona

Bing Li
Software Engineer at Google
Mountain View, California

Zhijie Wang
Senior Software Engineer at GE Digital
San Francisco, California

Zhibin Zhou
Principal Cloud Architect at Huawei Technologies
Bellevue, Washington

At Arizona State University (ASU), we hosted visiting scholars from overseas. Here we list whose work has had significant contributions to this book during the visit.

Ashwin Narayan Prabhu Verleker's master's thesis contributed to the ABE solutions' testing and prototype establishment, and **Mayank Verma**'s thesis contributed to the ABE solutions for VANETs.

Weijia Wang
Associate Professor of Beijing Jiaotong University
Beijing, China

Finally, from the University of Science Technology of Beijing (USTB), **E. Chen** has the main contribution to the draft of Chapter 7 on attribute-based signature schemes.

Part I

Foundations of Attribute-Based Encryption for Attribute-Based Access Control

The Attribute-Based Access Control (ABAC) concept [107] offers improvement in access control capabilities over Role-Based Access Control (RBAC). It provides more granularity, flexibility, and better handles complexity by using attributes compared to the confines of roles.

One way to deploy ABAC is based on Attribute-Based Encryption (ABE) [25, 95]. This *cryptographic* approach supports access control for a new data-centric computing and security paradigm. The major benefit of using ABE-based ABAC solutions is that the access control is incorporated into ciphertext. It does not rely on an access control infrastructure to protect data. The ABE cryptographic algorithm performs authentication, authorization, and enforcement. The data owner simply encrypts the data with access policies, and data users with the appropriate attributes will automatically be able to decrypt. In this way, data owners only need to focus on data access control policy and do not need to worry about how to establish an access group in advance. Moreover, ABE-protected data are mobile, which means they can be stored on untrusted servers or sent through untrusted intermediaries without compromising security.

ABE provides revolutionary capabilities that allow for new secure information-sharing paradigms that are difficult or impossible to achieve otherwise. It can significantly improve the usability of protected data by providing tremendous flexibility and decentralization. The main drawback of ABE-based ABAC approaches is that access control features are restricted by features of the underlying ABE cryptographic algorithms. As the result, many researchers put their significant efforts toward how to improve ABE algorithms to achieve better usability for various potential data access control applications.

In the first part of the book, we first present the foundation of ABE solutions as the basic ABE scheme. To extend new cryptographic features to the basic ABE scheme, we present a few important improvements of ABE algorithms including comparable ABE in Chapter 2, in which attributes are not restricted by deterministic terms and they can be a used to specify a comparable range; secondly, we present a privacy-preserving scheme to protect attributes (or policies) for ABE access control in Chapter 3; in Chapter 4, we present an ID-based ABE revocation solution by incorporating users' ID into their private keys, and in this way, we can effectively include user(s)' ID(s) into the cyphertexts' revocation list to prevent one or a group of users from decrypting the ciphertext; furthermore, in Chapter 5, we present an extended ID-based ABE revocation solution by considering the scalability of ABE revocation management capabilities such as ABE key federation, delegation, interoperability, etc.; in Chapter 6, we present a searchable encryption approach based on ABE schemes; finally, in Chapter 7, we describe an attribute-based signature solution to enforce secure access control policies.

1 Foundations of Attribute-Based Encryption

In public key encryption schemes, each user has a randomly generated public/private key pair. The private key is kept secret by the user, and the public key is published to the public. Whenever *Bob* sends a message to *Alice*, he will at first get *Alice*'s public key pk_A and then encrypt the message with the encryption algorithm with inputs of the message and pk_A. Only *Alice* who has the private key sk_A could decrypt the ciphertext. However, if a malicious *Carol* cheated *Bob* into using her public key pk_C to encrypt the message, then *Carol* is capable of decrypting the ciphertext while *Bob* still thinks he sent the message to *Alice*. Therefore, it is very important to get users' public keys in a trustworthy way. To deal with this issue, Public Key Infrastructure (PKI) [120] was proposed. Each user will publish its public key with a certificate, which includes both the user's identity and its public key. We could consider this as a trustworthy map from user's identity to its public key. It is computationally impossible for the malicious user *Carol* to create a certificate, which maps *Alice*'s identity to pk_C.

PKI management overhead is heavy and sometimes it is difficult to access the PKI services to validate users' certificates. It would be better if we use a user's well-known and unique identity, such as email address, driver license ID, etc., as the user's public key. This will remove the certificate distribution and validation during the run-time of an application. To this end, [195] proposed identity-based cryptosystems and signature schemes. The Identity-Based Encryption (IBE) is a special type of public key encryption with an extra twist. In IBE, the public key is not randomly generated; it could be a string that uniquely identifies this user. For example, it could be an email address, a social security number, telephone number, *etc.*, or their combinations.

For the use case described above, when *Bob* wants to send a secret message to *Alice*, there is no need to get pk_A anymore, while he could just encrypt the message with *Alice*'s identity ID_A, for example, *Alice*'s email. Only *Alice* whose identity is verified to be ID_A could decrypt the ciphertext. In this way, *Carol* could not make the fake binding of ID_A and pk_C anymore since in the IBE cryptosystem, $(ID_A = pk_A) \neq (ID_C = pk_C)$. Shamir proposed the first IBE cryptosystem [195] in 1984, while a practical IBE scheme based on pairing was constructed in 2001 [39].

Based on the IBE scheme in [39], there are more constructions based on different mathematical assumptions [221, 60, 37]. All these schemes have a common feature that identities are viewed as a string of characters.

Considering biometric identities, which inherently include some noises in each sample, [186] proposed fuzzy identity-based encryption, which provides error-tolerance property. In fuzzy IBE, an identity is viewed as a set of descriptive attributes. For example, to encrypt a message M with an identity ID, the ciphertext

3

is $C = ENC(M,ID)$. A user with identity ID' is capable of decrypting the ciphertext C if and only if ID and ID' are close to each other. The distance is measured by the "set overlap" distance metric. It is in [186] where the concept "attribute-based encryption" was proposed. However, in this paper, "attribute-based encryption" was just defined as a type of application where the data owner wishes to encrypt a file to all data users who were assigned a certain set of attributes.

In an "attribute-based encryption" system, both ciphertexts and users' keys are labeled with a set of attributes. A user could decrypt the ciphertext only if there is a match between the user's attributes (or access policy) and the ciphertext's access policy (or attributes). The decryption of the fuzzy IBE-based ABE succeeds only when at least a pre-defined threshold, say k, attributes overlapped. Although this is enough for the biometric identity applications, this type of ABE lacks expressibility, thus limiting its applications in other larger system [95]. To this end, Goyal *et. al.* proposed a new crypto-system, which used the same name, Attribute-Based Encryption.

The booming of Attribute-Based Encryption (ABE) started in the seminar work in 2007's "Ciphertext-Policy Attribute-Based Encryption (CP-ABE)" [25], in which the data access policies are naturally incorporated into crypto key generations. This feature makes the ABE scheme easy to be used and managed in real data access control applications. Since then, many feature improvements and additions for CP-ABE schemes have been proposed in the literature. In this chapter, we focus on the foundations of ABE schemes. Many presented models are used as the basis to construct new schemes in later chapters.

This chapter is arranged based on the following contents: we first introduce Attribute-Based Access Control (ABAC), and then describe motivations of using ABE-based ABAC solutions and their facing issues in Section 1.1. In Section 1.2, mathematical backgrounds for ABE and proofs are presented. Section 1.3 presents basic construction components of ABE schemes. The frequently used symbols in this book are presented in Section 1.4. Finally, in Section 1.5, a summary of this chapter is provided.

1.1   ATTRIBUTE-BASED ACCESS CONTROL—AN ABE APPROACH

Attribute-Based Access Control (ABAC) is an emerging form of access control that is starting to garner interest in both recent academic literature and industry applications [86]. While there is currently no single agreed-upon model or standardization of ABAC, there are commonly accepted high-level definitions and descriptions of its functions. One such high-level description is given in National Institute of Standards and Technology (NIST)'s publication, *Guide to Attribute Based Access Control (ABAC) Definition and Considerations* [107].

> ***Attribute-Based Access Control:*** *An access control method where subject requests to perform operations on objects are granted or denied based on assigned attributes of the subject, assigned attributes of the object,*

environmental conditions, and a set of policies that are specified in terms of those attributes and conditions.

Unlike widely deployed existing access control models such as Role-Based Access Control (RBAC), ABAC allows for the creation of access policies based on existing attributes of users and objects in the system, rather than the manual assignment of roles, ownership, or security labels by a system administrator. Removing the need for manual intervention when authorizing users for certain roles or security levels, simplifying administration in complex systems with a large number of users, as well as creating the possibility of automating access control decisions for remote users from foreign systems [192].

To realize the ABAC concept, two major groups of ABAC solutions exist in current literature. Many existing approaches refer to eXtensible Access Control Markup Language (XACML) [92] as the practical realization of ABAC; although not strictly a part of the ABAC model, it is the access control policy language used to define access policy rules for a system. Similarly, Security Assertion Markup Language (SAML) [118] provides a standardized markup language and protocol for exchanging attribute-based authorization and authentication information among data service providers, identity/attribute providers, and users. Commercialization solutions include companies such as Axiomatics' ABAC solution [4], Jericho's EnterSpace [6], NextLabs' Dynamic Authorization [8], Xpressrules-PM funded by NIST SBIR Phase I [10], etc. The second group of ABAC solutions are based on ABE [25, 95], a cryptographic approach. To implement ABAC is an intuitive and unique approach, however, it depends on security properties enabled by selected ABE schemes. Zeutro [11] is one of main startup companies pushing in this direction. The major benefit of using ABE-based ABAC solutions is due to its mobility, where the access control is incorporated into ciphertext and it does not need to rely on traditional infrastructure-based access control solutions. In this way, protected data can be stored on any public and private storage providers' servers, in which way it can significantly improve the flexibility and usability of data sharing for protected data.

Generally speaking, an access control system is usually formed by two system-level functions: identity and credential management and access control enforcement. Both ABAC and RBAC can be implemented by using either centralized or decentralized credential issuing parties, i.e., Identity Management (IdM). Access control enforcement is also done in a centralized fashion for both ABAC and RBAC, which requires one or multiple dedicated access control policy enforcement parties, a.k.a., Policy Enforcement Point (PEP).

PEP-based access control approaches do not fully utilize the flexibility of ABAC's capabilities. Using Attribute-Based Encryption (ABE), the access control policies can be incorporated into the ciphertext, and thus the Access Control (AC) is *mobile*. This means ABE-based ABAC is suitable for Data-Centric Access Control (DCAC), in which it provides two unique and critical capabilities for emerging data-centric applications. First, the IdM can stay the same way we traditionally manage users' identities and credentials and corresponding key management. In this way, we will not dramatically change existing security infrastructures. Second, since the access control policies are incorporated into the ciphertext, the PEP can be mobile, decentralized or even distributed, which means each data hosting party can serve as a PEP. As a result, there is no need for a dedicated access control enforcement infrastructure to define the boundary between trusted and untrusted parties for data access.

1.1.1 MOTIVATION OF ABE-BASED ATTRIBUTE-BASED ACCESS CONTROL

We use an access control example used in a hospital environment to illustrate the salient features of ABE-based ABAC approaches. A congenital cardiac disease patient at a Phoenix hospital and his parents had a visit in California during a weekend. They had an emergent situation but could not have access to their previous diagnostic data that was hosted at the Phoenix hospital, but had to wait through the entire weekend to get the data due to lack of effective data access and sharing solutions among hospitals. However, according to HIPAA individual right [102], an individual has the right to access and manage his/her health and medical data. This cannot be easily achieved through existing *provider-centric* or *infrastructure-centric* data access control approaches, e.g., most of XACML solutions rely on a trusted access control service provider to manage users' data access. In this example, the Phoenix hospital wants to have a more *data-centric* access approach, which allows the patient, e.g., through a mobile phone, to easily access, manage, and share data with related parties so as to avoid data-sharing barriers among hospitals. Users' medical records can be stored on any storage services including public storage and personal devices, where it is highly desirable that data must be always encrypted. Additionally, security access control policies must be always enforced uniformly on both storage service providers and user-end devices. To satisfy these requirements, ABE-based ABAC is a suitable solution.

Besides the personal/home healthcare application scenario, which has a potential $350B market by 2020 [5], emerging decentralized storage services, Software Defined Storage (SDS), and blockchain-based applications, which share similar access control requirements, can benefit from ABE-based ABAC approaches. An appealing application example is Sia [9], a service storing data on distributed user-provided storage.

1.1.2 POTENTIALS AND ISSUES OF ABAC

Gartner predicted that by 2020, 70% of enterprises will use ABAC as the dominant mechanism to protect critical assets, up from less than 5% in 2014 [86]. However, as of today, ABAC still has not gained its momentum to take over RBAC solutions. According to the discussion on National Cybersecurity Center of Excellence (NCCoE) on the slow adoption of ABAC [3], "*one obstacle is lack of detailed guidance on how to integrate and configure ABAC components; hence the Practice Guide*", i.e., there is a lack of well-documented (preferably with some before-and-after metrics) "case studies" of how ABAC has delivered one or more business benefits. Moreover, organizations are usually reluctant to change their existing infrastructures, especially if the *infrastructure-centric* access control will have major impacts on their operations requiring newly established regulations, which usually take a longer time to adopt for such a solution. As a result, a practical and innovative access control solution must not only be technically sound but also address real system problems with fewer changes on institutions' operation procedures. ABE-based ABAC approaches can address these issues by moving the data access control from the infrastructure to users, and thus it can minimize regulation changes by simply granting users access their data; for example, a patient can keep his/her healthcare and medical data and provide it to corresponding parties during an emergency situation. However, existing ABE-based ABAC solutions face two major challenges, which prevent them from being widely deployed:

- *Delegation and Federation:* Existing solutions cannot provide an effective delegation solution, where an ABE key generation authority can grant full or partial of its key generation privileges to multiple key generation delegators. At the same time, users' private keys derived from multiple delegators can be used together without security issues, such as colluding issues for users sharing their private keys to gain unauthorized privileges. To achieve the desired delegation feature will not only reduce the workload of the key generation authority, but also can use the hierarchical key management framework to implement the trust management framework for the ABE-based ABAC approaches.
- *Revocation and Auditing:* Revoking an attribute can effectively revoke a group of entities, and usually composition of the group memberships may not be easily known in advance. If an attribute can be used uniquely to identify a group of users or an individual, we can treat a group ID or a user's ID as an attribute, and then we can simply implement the "NOT" logic on the attribute in the ABE scheme to revoke a known group of entities or individuals. However, this approach will significantly increase the size of the attribute set and make the attribute management extremely complicated. Similarly, auditing service demands the capability to identify a group or an individual, which requires an ABE-based challenge-response protocol to track users' access history.

To address these described issues and develop a practical ABE-based ABAC approach, in the following chapters, we present a few new ABE schemes.

1.2 MATHEMATICAL BACKGROUND

1.2.1 GROUP AND CYCLIC GROUP

In mathematical context, a group G is a set of elements equipped with a binary operator $\times$ that are related with each other according to the following four well-defined conditions called group axioms.

- Closure: for any two elements a and y, $x \times y$ is in the group G as well.
- Associativity: for any three elements x, y, and z, $(x \times y) \times z = x \times (y \times z)$.

- Identity element: there exists an element $e \in G$ such that for every element $x \in G, x \times e = e \times x = x$. The identity element e is denoted by ⯑.
- Inverse element: for each element $x \in G$, there exists an element y, such that $x \times y = y \times x = e$, where e is the identity element. y is the inverse element of x and is denoted by x^{-1}.

A group is called cyclic if there exists at least one element $g \in G$, such that all the elements in the group are powers of it. When using multiplication as the group binary operator shown as above, the elements of the group can be denoted by

$$\cdots, g^{-3}, g^{-2}, g^{-1}, g^0 = e, g, g^2, g^3, \cdots$$

1.2.2 PRIME-ORDER BILINEAR PAIRING

Prime-order pairing is a *bilinear* map function $e : \mathbb{G}_1 \times \mathbb{G}_2 \to \mathbb{G}_T$, where $\mathbb{G}_1, \mathbb{G}_2$, and $\mathbb{G}_T$ are three cyclic groups with large prime order p. The $\mathbb{G}_1$ and $\mathbb{G}_2$ are additive group, and $\mathbb{G}_T$ is multiplicative group. The discrete logarithm problem on $\mathbb{G}_1, \mathbb{G}_2$, and $\mathbb{G}_T$ are hard. Pairing has the following properties:

- *Bilinearity*:

$$e(g^a, h^b) = e(g, h)^{ab}, \ \forall g \in \mathbb{G}_1, h \in \mathbb{G}_2, a, b \in \mathbb{Z}_p^*.$$

- *Nondegeneracy*:
  $e(g_0, h_0) \neq 1$, where g_0 is the generator of $\mathbb{G}_1$ and h_0 is the generator of $\mathbb{G}_2$.
- *Computability*:
  There exists an efficient algorithm to compute the pairing.

1.2.3 COMPOSITE-ORDER BILINEAR PAIRING

Two types of composite-order *bilinear* pairing are used in this book. One is of order $N = p_1 p_2$, and the other is of order $N = p_1 p_2 p_3$. We present their definition below, respectively.

Type 1 [137] A three-prime composite pairing: the three-prime composite pairing is a *bilinear* map function $e : \mathbb{G} \times \mathbb{G} \to \mathbb{G}_T$, where $\mathbb{G}$ and $\mathbb{G}_T$ are cyclic groups of order $N = p_1 p_2 p_3$ and p_1, p_2, and p_3 are distinct primes. The map function satisfies the following conditions:

- Bilinear: $\forall g, h \in \mathbb{G}, a, b \in \mathbb{Z}_N, e(g^a, h^b) = e(g, h)^{ab}$.
- Nondegerate: $\exists g \in \mathbb{G}$ such that $e(g, g)$ has order n in $\mathbb{G}_T$.

Assume that the group operations in $\mathbb{G}$ and $\mathbb{G}_T$, as well as the *bilinear* map e are computable in polynomial time with respect to λ. $\mathbb{G}_{p_1}, \mathbb{G}_{p_2}$, and $\mathbb{G}_{p_3}$ denotes the subgroups of order p_1, p_2, and p_3 in G, respectively. Suppose $h_i \in \mathbb{G}_{p_i}$ and $h_j \in \mathbb{G}_{p_j}$ for $i \neq j$.

- Orthogonality: $e(h_i, h_j)$ is the identity element of G_T.

Type 2 [246] A *bilinear* map group system $S_N = (N = pq, G, G_T, e)$ used in [246] where $N = pq$ is the RSA modulus, p and q are two large primes. We have G and G_T as two cyclic groups with composite order n, where $n = sn' = s_1 s_2 p' q'$ and p, q, p', q', s_1, s_2 are all secret large primes. e denotes a computable *bilinear* map $e : G \times G \to G_T$ with the following properties:

- Bilinearity: $\forall g, h \in G, \forall a, b \in \mathbb{Z}, e(g^a, h^b) = e(g, h)^{ab}$;
- Nondegeneracy: g and h are the generators of G, $e(g, h) \neq 1$;
- Computability: $e(g, h)$ is efficiently computable.

We have G_s and $G_{n'}$ as the subgroups of order s and n' in G, respectively, and $e(g, h)$ becomes the identity element in G_T if $g \in G_s$, $h \in G_{n'}$. As an example, suppose w is the generator of G, then $w^{n'}$ is the generator of G_s, and w^s is the generator of $G_{n'}$. Assume $g = (w^{n'})^{\rho_1}$ and $h = (w^s)^{\rho_2}$ for some ρ_1, ρ_2, it holds that $e(g, h) = e(w^{\rho_1}, w^{\rho_2})^{sn'} = 1$. Our systems leverage the orthogonality between $G_{n'}$ and G_s and keep $N, n, s, p, q, , p', q'$ secret.

1.3 BASIC CONSTRUCTION COMPONENTS OF ABE

In this section, we presented basic components of building ABE and two well-known ABE solutions, KP-ABE and CP-ABE.

1.3.1 ACCESS STRUCTURE

Access Structure [95]. Let $\{P_1, P_2, \ldots, P_n\}$ be a set of parties. A collection $\mathbb{A} \subseteq 2^{\{P_1, P_2, \ldots, P_n\}}$ is monotone if $\forall B, C$: if $B \in \mathbb{A}$ and $B \subseteq C$ then $C \in \mathbb{A}$. An access structure is a collection $\mathbb{A}$ of non-empty subsets of $\{P_1, P_2, \ldots, P_n\}$, i.e., $\mathbb{A} \subseteq 2^{\{P_1, P_2, \ldots, P_n\}} \setminus \{\emptyset\}$. The sets in $\mathbb{A}$ are defined as authorized sets, and sets that do not belong to $\mathbb{A}$ are defined as unauthorized sets.

1.3.2 LINEAR SECRET-SHARING SCHEME

Linear Secret Sharing Schemes (LSSS) matrices can be utilized to express any monotonic access structure, which are the most commonly used access structures in most CP-ABE schemes. The algorithm of constructing an LSSS matrix is proposed by Lewko and Waters. LSSS is defined as follows.

Definition 1.1. Linear Secret Sharing Schemes (LSSS) [137]. A secret sharing scheme Π over a set of parties is called linear over $\mathbb{Z}_p$ if the following two conditions are satisfied:

- the shares for each party form a vector over $\mathbb{Z}_p$;

• a share-generating matrix for Π has ℓ rows and n columns. For all $i = 1,\ldots,\ell$, the ith row of M, we define $\rho(i)$ as the party labeling row i. For the column vector $v = (s, r_2, r_3, \ldots, r_n)$, where $s \in \mathbb{Z}_p$ is the shared secret and $r_2, r_3, \ldots, r_n \in \mathbb{Z}$ are randomly chosen numbers, then Mv is the vector of ℓ shares of the secret s according to Π, where the share $(Mv)_i$ belongs to party $\rho(i)$.

As shown in [23], every linear secret sharing-scheme according to the definition also enjoys the following **linear reconstruction** property:

Assume that Π is an LSSS for the access structure $\mathbb{A}$. Define $\mathbf{S} \in \mathbb{A}$ as an authorized set and $\mathbf{I} \subset [1, l]$ as $\mathbf{I} = \{i : \rho(i) \in \mathbf{S}\}$. Then, constants $\{w_i \in \mathbb{Z}_p\}_{i \in \mathbf{I}}$ can be derived in polynomial time, such that such that for valid shares $\{\lambda_i\}$ of any secret s we have $\sum_{i \in \mathbf{I}} w_i \lambda_i = s$.

1.3.3 CONVERSION ALGORITHM

A monotone access structure could be expressed by an access tree with AND, OR gates as interior nodes and attributes as leaf nodes. The Lewko-Waters' conversion algorithm works as follows.

Algorithm 1.1 The Lewko-Waters Algorithm

Input: An Access Tree T, $c = 1$
Output: The Corresponding LSSS Matrix
  1: **for** each level of the tree T
  2: **for** each node N in T
  3:     **if** the parent node is an OR gate labeled by the vector v
  4:     **then**
  5:         Label the left child of N by vector $\leftarrow v$
  6:         Label the right child of N by vector $\leftarrow v$
  7:     **else**
  8:         Pad N's vector with 0 at the end (if necessary) to make it of length c
  9:         Label the left child by vector $\leftarrow v\|1$
 10:         Label the right child by vector $\leftarrow (0, \cdots, 0)\| - 1$, where $(0, \cdots, 0)$ denotes the vector $\mathbf{0}$ of length c
 11:         $c \leftarrow c + 1$
 12:     **endif**
 13: **end loop**
 14: **end loop**

1.3.4 ACCESS STRUCTURE EXAMPLE

Here, we show an example of how to convert a Boolean formula to an access tree, and from an access tree to an access structure that is expressed by an LSSS matrix.

Denote the attribute by A_1, A_2, A_3, A_4, and A_5. The example Boolean formula is as follows

$$A_5 \wedge ((A_1 \wedge A_2) \vee (A_3 \wedge A_4))$$

The access tree with "AND" and "OR" gate is presented as in Figure 1.1. Each subset of the rows of matrix M includes $(1,0,0)$ in its span, if and only if the corresponding attributes satisfy the Boolean formula $A_5 \wedge ((A_1 \wedge A_2) \vee (A_3 \wedge A_4))$.

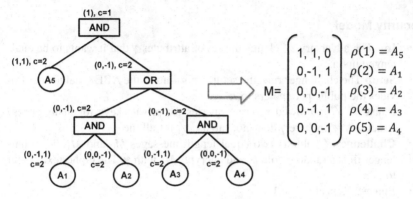

Figure 1.1: Access tree and LSSS matrix for Boolean formula $A_5 \wedge ((A_1 \wedge A_2) \vee (A_3 \wedge A_4))$.

If a user has attributes A_1, A_2, and A_5, then the corresponding row of the matrix will be $(1,1,0)$, $(0,-1,1)$, and $(0,0,-1)$. If we add these three vectors, we will get $(1,0,0)$. These are the most important features used to construct an attribute-based encryption scheme, i.e., only authorized users could recover $(1,0,0)$. If we created a random vector $v = (s,r1,r2)$ and changed M to be $M = Mv$, then the recovered vector would be $(s,0,0)$. The value s is used in the encryption scheme to hide the message.

1.3.5 KEY-POLICY ATTRIBUTE-BASED ENCRYPTION

In a Key-Policy Attribute-Based Encryption (KP-ABE) scheme, ciphertexts are labeled with a set of attributes and a user's private key is associated with an access policy, which will control which ciphertext this user is allowed to decrypt [95]. First, the proposed syntax of a KP-ABE scheme as well as a construction presented below.

KP-ABE Syntax

- **Setup** This is a randomized algorithm that takes no input other than the implicit security parameter. It outputs the public parameters PK and a master key MK.
- **Encryption** This is a randomized algorithm that takes as input a message m, a set of attributes γ, and the public parameters PK. It outputs the ciphertext E.

- **Key Generation** This is a randomized algorithm that takes as input an access structure $\mathbb{A}$, the master key MK, and the public parameters PK. It outputs a decryption key D.
- **Decryption** This algorithm takes as input the ciphertext E that was encrypted under the set $\mathbf{S}$ of attributes, the decryption key D for access control structure $\mathbb{A}$ and the public parameters PK. It outputs the message M if $\mathbf{S} \in \mathbb{A}$.

Security Model

- **Init** The adversary $\mathscr{A}$ claims the set of attributes γ that it wants to be challenged upon.
- **Setup** The challenger runs the setup algorithm of the ABE scheme and provides the public parameters to $\mathscr{A}$.
- **Phase 1** $\mathscr{A}$ is allowed to issue queries for private keys for many access structures $\mathscr{A}_j$, where γ does not satisfy $\mathscr{A}_j$ for all the j.
- **Challenge** $\mathscr{A}$ submits two equal length messages M_0 and M_1. The challenger flips a random coin b, and encrypts M_b with γ. The ciphertext is sent to $\mathscr{A}$.
- **Phase 2** Repeat Phase 1.
- **Guess** The adversary outputs a guess b' of b.

Access Tree

Definition 1.2. Access Tree [95]. Let T be a tree representing an access structure. Each non-leaf node of the tree represents a threshold gate, described by its children and a threshold value. If num_x is the number of children of a node x and k_x is its threshold value, then $0 < k_x \leq num_x$. When $k_x = 1$, the threshold gate is an OR gate and when $k_x = num_x$, it is an AND gate. Each leaf node x of the tree is described by an attribute and a threshold value $k_x = 1$.

The parent of the node x in the access tree is denoted by $parent(x)$. The leaf node x is denoted by $att(x)$. Children of a node is numbered from 1 to num. The function $index(x)$ returns this number for the node x. Let T denote an access tree with the root node r. T_x denotes the subtree of T rooted at the node x. $T_x(S) = 1$ denotes that a set of attributes S satisfies the access tree T_x.

KP-ABE Construction

The KP-ABE scheme is built upon a prime order *bilinear* map $e : \mathbb{G}_1 \times \mathbb{G}_1 \to \mathbb{G}_T$. the Lagrange coefficient Δ_{iS} for $i \in \mathbb{Z}_p$ and a set S of elements in $\mathbb{Z}_p$ is denoted by $\Delta_{i,S}(x) = \prod_{j \in S, j \neq i} \frac{x-j}{i-j}$.

- **Setup** Define the universe of attributes $U = \{1, 2, \cdots, n\}$. Each attribute $i \in U$, a random number $t_i \in \mathbb{Z}_p$ is selected. Chose a random $y \in \mathbb{Z}_p$. The generated public parameters PK are as follows.

$$PK = \{T_1 = g^{t_1}, \cdots, T_{|U|} = g^{t_{|U|}}, Y = e(g,g)^y\}$$

The master key MK is

$$MK = \{t_1, \cdots, t_{|U|}, y\}$$

- **Encryption** The encryption algorithm takes as inputs the message M, a set of attributes S, and the public parameters PK. Choose $s \in \mathbb{Z}_p$ randomly. Generate the ciphertext as follows:

$$E = (S, E' = MY^s, \{E_i = T_i^s\}_{i \in S}).$$

- **Key Generation** The algorithm takes as input a tree T and the master secret key MK. Choose a polynomial q_x for each node x in the tree T. The polynomials are chosen in a top-down manner from the root node r as follows. For each node x in the tree, the degree d_x of the polynomial q_x is defined to be $k_x - 1$, where k_x is the threshold value of node x. For the root node r, set $q_r(0) = y$ and d_r other points of the polynomial q_r randomly. For the other node x, set $q_x(0)$ to be $q_{parent(x)}(index(x))$ and choose d_x other points randomly. After all the polynomials have been decided, the decryption key D is given to the user.
  The following secret value for each leaf node x is given to the user.

$$D = \{D_x = g^{\frac{q_x(0)}{t_i}}, \text{ where } i = att(x)\}_{x \text{ is a leaf node}}$$

- **Decryption** Define a recursive algorithm $DecryptNode(E,D,x)$, where $E = (S, E' = MY^s, \{E_i = T_i^s\}_{i \in S})$, D is the private key, and x is a node in the tree. The function is defined to be

$$DecryptNode(E,D,x) = \begin{cases} e(D_x, E_i) & = & e(g,g)^{s \cdot q_x(0)} \text{ if } x \in S \\ \bot & & \text{otherwise} \end{cases} \quad (1.1)$$

if x is a leaf node and $i = att(x)$. If x is a non-leaf node, $DecryptNode(E,D,z)$ is called for all the nodes z, which are children of x; the output is stored as F_z. S_x denotes an arbitrary k_x-sized set of child nodes z, which satisfies $F_z \neq \bot$. If no such set could be found then the node cannot be satisfied and the function returns $\bot$. Otherwise, F_x is calculated in the following way:

$$\begin{aligned} F_x &= \prod_{z \in S_x} F_z^{\Delta_{i,S_x'}(0)}, \text{ where } i = index(z), S_x' = \{index(z) : z \in S_x\} \\ &= e(g,g)^{s \cdot q_x(0)} \text{ using polynomial interpolation} \end{aligned} \quad (1.2)$$

1.3.6 CIPHERTEXT-POLICY ATTRIBUTE-BASED ENCRYPTION

In [25], Ciphertext-Policy Attribute-Based Encryption (CP-ABE) scheme is presented. A data user's private key is associated with a set of attributes. Data are encrypted by an access policy expressed by an access tree, where the inner nodes are AND, OR gates, and leaf nodes are attributes.

CP-ABE Syntax

- **Setup** The setup algorithm takes no input other than the implicit security parameter. It outputs the public parameters PK and a master key MK.
- **Encryption** The encryption algorithm takes as input the public parameters PK, a message M, and an access structure $\mathbb{A}$ over the universe of attributes. The algorithm will encrypt M and produce a ciphertext CT, such that only a user that possesses a set of attributes that satisfies the access structure will be able to decrypt the message. We will assume that the ciphertext implicitly contains $\mathbb{A}$.
- **Key Generation** The key generation algorithm takes as input the master key MK and a set of attributes S that describe the key. It outputs a private key SK.
- **Decrypt** The decryption algorithm takes as input the public parameters PK, a ciphertext CT, which contains an access policy $\mathbb{A}$, and a private key SK, which is a private key for a set S of attributes. If the set S of attributes satisfies the access structure $\mathbb{A}$ then the algorithm will decrypt the ciphertext and return a message M.

Security Model

- **Setup** The challenger runs the setup algorithm and sends the public parameters to the adversary $\mathscr{A}$.
- **Phase 1** $\mathscr{A}$ makes repeated private keys corresponding to sets of attributes $S_1, \cdots, S_{q_1}$.
- **Challenge** The adversary submits two equal length messages M_0 and M_1. $\mathscr{A}$ chooses a challenge access structure $\mathscr{A}^*$ subject to the condition that none of queried sets $S_1, \cdots, S_{q_1}$ satisfy the access structure. The challenger flips a random coin b, and encrypts the chosen message M_b with the access structure $\mathscr{A}^*$. The generated ciphertext is given to $\mathscr{A}$.
- **Phase 2** Repeat Phase 1 with the restriction that the queried sets of attributes do not satisfy the access structure $\mathscr{A}^*$.
- **Guess** The adversary outputs a guess bit b' of b.

CP-ABE Construction

Similar to the KP-ABE scheme, it is also built upon a *bilinear* pairing group $\mathbb{G}_1 \times \mathbb{G}_1 \rightarrow \mathbb{G}_T$. The Lagrange coefficient $\Delta_{i,S}$ for $i \in \mathbb{Z}_p$ and a set S of elements in $\mathbb{Z}_p$ is denoted by $\Delta_{iS}(x) = \prod_{j \in S, j \neq i} \frac{x-j}{i-j}$. A hash function $H : \{0,1\}^* \rightarrow \mathbb{G}_1$, which maps attributes into elements of the group $\mathbb{G}_1$.

- **Setup** Choose a *bilinear* group $\mathbb{G}_0$ of prime order p and its generator g. Choose randomly two exponents $\alpha, \beta \in \mathbb{Z}_p$. Publish the generated public key below:
$$PK = \mathbb{G}_0, g, h = g^\beta, f = g^{1/\beta}, e(g,g)^\alpha$$

Keep secret the master key below:

$$MK = (\beta, g^{\alpha}).$$

- **Encryption** To encrypt a message M under an access tree T, first choose a polynomial q_x for each node x in the tree. The degree d_x of q_x is set to be $k_x - 1$, where if it is an AND gate, k_x is the number of children of the node x. Starting from the root node R of the tree, set $q_r(0) = s$, where $s \in \mathbb{Z}_p$ is a random number. Choose d_R other points of q_r randomly. For any other node x, set $q_x(0) = q_{parent(x)}(index(x))$, and then choose d_x other points randomly in the same way as for the root node R.

  Let Y denote the set of all leaf nodes in T. The generated ciphertext is as follows:

$$CT = (T, \widetilde{C} = Me(g,g)^{\alpha s}, C = h^s, \forall y \in Y : C_y = H(att(y))^{q_y(0)}).$$

- **Key Generation** Choose a random $r \in \mathbb{Z}_p$, $r_j \in$ for each attribute $j \in S$. The private key is:

$$SK = (D = g^{(\alpha+r)/\beta}), \forall j \in S : D_j = g^r \cdot H(j)^{r_j}, D'_j = g^{r_j}.$$

- **Decrypt** Define a cursive function $DecryptNode$ as follows. For a leaf node x, $i = att(x)$.

$$DecryptNode(CT, SK, x) = \begin{cases} \frac{e(D_i, C_x)}{e(D'_i, C'_x)} = e(g,g)^{r q_x(0)} & \text{if } i \in S \\ \bot & otherwise \end{cases}. \quad (1.3)$$

For non-leaf node x, call $DecryptNode(CT, SK, z)$ for all children nodes z and store the output as F_z. S_x denotes an arbitrary k_x-sized set of child nodes z so that $F_z \neq \bot$. If this set does not exist then the node is not satisfied, then the function returns $\bot$. Otherwise, compute:

$$F_x = \prod_{z \in S_x} F_z^{\Delta_{i,S'_x}(0)}, \text{ where } i = index(z), S'_x = \{index(z) : z \in S_x\}, \quad (1.4)$$
$$= e(g,g)^{s \cdot q_x(0)}$$

using polynomial interpolation.

If the tree T is satisfied by S, set $A = DecryptNode(CT, SK, r) = e(g,g)^{r q_R(0)} = e(g,g)^{rs}$. The ciphertext is decrypted in the following way:

$$\widetilde{C}/(e(C,D)/A) = \widetilde{C}/(e(h^s, g^{(\alpha+r)/\beta})/e(g,g)^{rs}) = M.$$

1.4 NOTATIONS

Table 1.1 provides the most commonly used symbols in different chapters of this book.

Table 1.1
Notations

Notation	Description
$\mathbb{A}$	access structure, which is a set of authorized attribute sets
ANC_i	the set $i's$ ancestor nodes on the path from root to i
A_i	i-th attribute in $\mathbb{U}$
A_x	the x-th row of A
b	a random element chosen by TA from G
CT	ciphertext
$e(\cdot,\cdot)$	*bilinear* mapping
$F : V \underset{g,h}{\rightarrow} V$	Multi-dimensional Range Derivation Function (MRDF) generator of a group
$G, \mathbb{G}_1, \mathbb{G}_2, \mathbb{G}_T$	a group
H	the number of layers in the identity structure tree
$\mathscr{H}_{\mathscr{P}_x}$	the cipher derived from encrypting the concatenation of $\mathscr{P}_x$ and $\mathscr{H}_{\mathscr{P}_{x-1}}$
$\mathscr{I}$	the identity set defined in the system, $\lvert \mathscr{I} \rvert = n$.
$\mathscr{I}_a$	the set of all the domain authority identities
$\mathscr{I}_{nr}$	the set of domain authority identities
$\mathscr{R}\mathscr{I}$	the set of root authorities
k	the number of attributes in $\mathbb{U}$
K	symmetric encryption key
$\mathscr{L}_u$	the data user u's attribute ranges
$\mathscr{L}_{x,u}$	the data user u's attribute ranges in $\mathbb{U}_x$
m	the number of attributes defined in the system
m_x	the number of attributes in $\mathbb{U}_x$
$\mathscr{M}$	message
M	an $l \times n$ matrix as part of an LSSS access structure
M_i	the xth row of matrix M
MK	master key
n	the number of identities in the system
N	the number of members in the Trust Coalition
n_i	the maximum number of attribute values in A_i
$n_{x,i}$	the maximum number of attribute values in $A_{x,i}$
p	the prime order of the multiplicative cyclic group G
$\mathscr{P}$	the data owner's access control policy
$\mathscr{P}_x$	the data owner's access control policy in $\mathbb{U}_x$
PK	public key
r	the number of identities involved in encryption
r_g	the number of revoked domain authorities
r_u	the number of revoked users

R, R', R^-, R^*	four different attribute range relationships
$S \in \mathbb{A}$	an authorized set in the access structure
SK	secret key
$t_{i,0}$	the dummy attribute value assigned to a user if he/she does not possess attribute A_i
t_{i,n_i}	the maximum attribute value in A_i
$[t_{i,a}, t_{i,b}]$	the attribute range on attribute A_i possessed by a data user
$[t_{i,j}, t_{i,k}]$	the range constraint on attribute A_i defined by $\mathscr{P}$
$[t_{x,i,j}, t_{x,i,k}]$	the range constraint on attribute $A_{x,i}$ defined by $\mathscr{P}_x$
$[t_{x,i,a}, t_{x,i,b}]$	the attribute range on attribute $A_{x,i}$ possessed by a data user
$\mathbb{U}, U$	the whole attribute set
$\mathbb{U}_{ID}$	the attribute set of a particular domain authority ID
$\mathbb{U}_x, A_{x,i}$	the x-th attribute domain and the i-th attribute in $\mathbb{U}_x$
X	the total number of attribute domains
α	a random element chosen by a Trusted Authority (TA) from G
$\mathbb{Z}_p^*$	$\{1, 2, \cdots, p-1\}$, where p is a prime
$\mathbb{Z}_p$	$\{0, 1, 2, \cdots, p-1\}$, where p is a prime
$\rho_i, \bar{\rho}_i$	the bound values associated with $[t_{i,j}, t_{i,k}]$; it depends on the range relation over A_i
$\rho_{x,i}, \bar{\rho}_{x,i}$	the bound values associated with $[t_{x,i,j}, t_{x,i,k}]$; it depends on the range relationship over $A_{x,i}$
$[1, n]$	$[1, n]$ denotes a set of integers i.e., $\{1, 2, \cdots, n\}$

1.5 SUMMARY

In this chapter, we first presented the concepts of ABE-based ABAC for cybersecurity applications. A set of design issues are presented to motivate the researchers to investigate into more flexible and feature-rich ABE solutions. We also introduced the mathematical foundations that are sufficient for readers to understand the presentations given in the rest of this book. It must be noted that we do not intend to provide an in-depth mathematical foundation to understand the basis of pairing-based cryptography and how to use it to construct the attribute-based solution. The goal of this chapter is to allow cybersecurity engineers to understand the presentation and basic format and logic of mathematical presentations to understand the presented ABE protocols in this book. For basic math foundations of ABE and pairing, interested readers should refer to alternate math and cryptography learning resources. Finally, a set of frequently used symbols in the rest of the book is presented as for an easy-access reference.

2 Comparable Attribute-Based Encryption

Data access control has been an increasing concern in the cloud environment where cloud users can compute, store and share their data. Cloud computing provides a scalable, location-independent, and high-performance solution by delegating computation tasks and storage into the resource-rich clouds [115, 130, 158, 134]. This overcomes the resource limitation of users with respect to data storage, data sharing and computation; especially when it comes to mobile devices considering their limitations of processing hardware, storage space, and battery life. However, in reality, the cloud is usually not fully trusted by data owners; moreover, the cloud service providers may be tempted to peek at users' sensitive data and produce trapdoors in computation for commercial interests. To enforce secure data access control on untrusted cloud servers, traditional methods (e.g., AES [65]) encrypt data before storing it in the cloud, but they incur high key-management overhead to provide dynamic group-based access control and significantly increases the system complexity.

Ciphertext-Policy Attribute-Based Encryption (CP-ABE) [25] can be used to provide a fine-grained access control for dynamic group formation in cloud-based data storage solutions. It enables the data owners to create access policies by designating attribute constraints and embedding the data access policies into the ciphertext, such that any data user has to satisfy the corresponding attributes to access the data. CP-ABE is designed to handle descriptive attributes, and it needs to convert comparative attributes into a bit-wise monotone access tree structure to enforce expressive access control of encrypted data. Green *et al.* [96] devised new methods for outsourcing decryption of ABE ciphertexts with significantly reduced decryption cost, but their encryption cost grows with the number of involved attributes, and bit-wise comparison has to be adopted for comparison. Generally speaking, most existing CP-ABE schemes suffer several drawbacks:

- They perform cryptographic comparison operations (such as $\preceq$ and $\succeq$) by following a series of bit-wise equal matching (e.g., 10*11*01) in a hierarchical tree structure, which involves a substantial amount of computational cost.
- They do not support effective range comparisons (e.g., $2 \preceq hours \preceq 4, 3 \preceq level \preceq 5$). In fact, an attribute could have a collection of possible values in a sequential partial order. In other words, certain attributes may take the form of range values.

To address the issues stated above, a new Constant-size Ciphertext Policy Comparative Attribute-Based Encryption (CCP-CABE) [219, 113, 114] is presented. A telemedicine example in Figure 2.1 is used to illustrate how it works in a real-world scenario. A patient periodically uploads his/her health records to the medical information service delivered by a cloud provider, and healthcare professionals in the designated clinic can monitor his/her health status based on his/her health records. This patient has a policy that only healthcare professionals with positions higher than *Nurse* can access his/her health info between time t_j and t_k. Thus, the data access can be specified by a policy $\mathscr{P} = [A_1 \wedge A_2]$, where $A_1 = rank$ and $A_2 = time$ are two attributes, and each attribute has a certain range, where $Rank = \{$Nurse, Attending Doctor, Senior Doctor, Clinic Director$\}$, and $Time = \{t_x | x \in Z\}$. Correspondingly, a *Senior Doctor* who has a higher rank can access the data if he/she has been authorized to the time interval that is contained in $[t_j, t_k]$.

Using CP-ABE scheme, the temporal comparison relies on bit-matching and incurs large sizes of data users' keys and overhead, resulting in high computational costs in encryption and decryption. Moreover, strict equal matching is not compliant with the ubiquitous partial-order relations. Zhu *et al.* [246] first proposed a flexible Comparison-Based Encryption (CBE) scheme to address this problem. It utilizes integer comparison to derive designated attribute range bound, and relies on a hierarchical attribute access tree without the support of negative attributes and wildcards. This results in linearly increasing computation and communication overhead with respect to the number of attributes on the side of data owners and data users, which is not suitable for resource-constrained mobile devices.

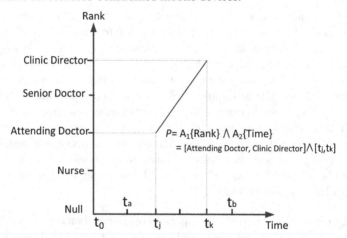

Figure 2.1: Two-dimensional attribute ranges in the telemedicine example.

CCP-CABE integrates all attribute ranges as a single encryption parameter and compares data users' attribute ranges against attribute constraints of the access policy designated by the data owner through Multi-dimensional Range Derivation Function (MRDF) . Consequently, the communication overhead is substantially reduced, as the packet size is constant regardless of the number of attributes. Furthermore, intensive encryption and decryption operations are delegated to the mobile cloud. As a result, the computation cost of resource-limited data owners and data users remains minimal. These features make the CCP-CABE approach suitable for data sensing and retrieval services running on lightweight mobile devices or sensors.

Here, an extended CCP-CABE is presented to satisfy the application requirement that the data owners need to share data with a policy written over attributes issued across various attribute domains. Both schemes are secure against various attacks, preventing honest-but-curious cloud service owners from decrypting ciphertext and countering key collusion attacks from multiple data owners and users. In summary, the salient features of CCP-CABE are presented as follows:

- CCP-CABE is a new comparative attribute-based encryption scheme to provide efficient and secure access control in a cloud environment. It leverages MRDF to compare data users' attribute ranges against attribute constraints designated by the data owner.
- CCP-CABE can predefine different range intersection relationships on different attributes. It also incorporates wildcards and negative attributes so it can handle more expressive types of access control.
- CCP-CABE minimizes the communication overhead to constant size regardless of the number of attributes and comparison ranges. It also minimizes the computation overhead on resource-constrained data owners and data users irrespective of the number of attributes due to secure computation delegation. The evaluation results show that the computation overhead of mobile devices remains small and constant irrespective of the associated attributes and comparison ranges.
- An extend CCP-CABE is presented to enforce the access control over multiple independent attribute domains. The encrypted access policy prioritizes the level of confidentiality of different attribute domains, and the data users can only start decryption from the least confidential domain to the most confidential one to help protect the privacy of the access policies. Towards the end of this chapter, the performance evaluations of communication and computation overhead are presented, which show the solution only grows with the number of trust authorities rather than the number of attributes.

2.1 CCP-CABE APPLICATION FRAMEWORK

The CCP-CABE framework consists of a central Trust Authority (TA), e.g., the government health agency, a trusted Encryption Service Provider, a cloud provider, data owners (e.g., patients), and data users (e.g., healthcare professionals). Its architecture is illustrated in Figure 2.2. In the telemedicine example, the patients have resource-limited bio-metric devices, and they need to distribute the sensitive Electronic Health Records (EHRs) to different storage servers hosted by cloud providers for healthcare professionals in remote places to review. The patients can specify different access policies with respect to healthcare professionals' attribute ranges (e.g., positions, length of service). To protect the patients' privacy, the government health agency issues keys to both patients and healthcare professionals for EHR encryption and decryption. Hence, the patients can embed their access policies into the health data with the keys, and only the eligible healthcare professionals can decrypt corresponding EHRs with their delegation/private keys based on their own attribute ranges.

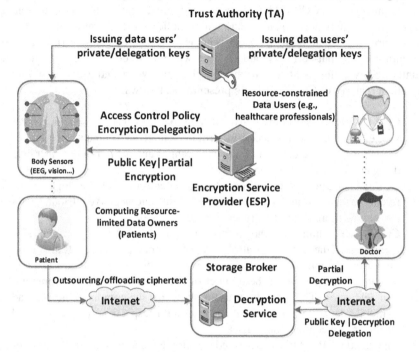

Figure 2.2: The CCP-CABE framework with central Trust Authority.

2.2 DEFINITION OF ATTRIBUTE RANGE AND PROBLEM FORMULATION

The comparison operations are shown as below:

- Let $\mathbb{U} = \{A_1, \cdots, A_m\}$ be a finite set of attributes, and each attribute $A_i \in \mathbb{U}$ contains a set of attribute values $T_i = \{t_{i,1}, t_{i,2}, \cdots, t_{i,n_i}\}$ consisting of discrete integer values, where n_i is the number of integer values for attribute A_i. Without loss of generality, assume that all elements in T_i are in ascending order such that $0 \leq t_{i,1} \leq t_{i,2} \leq \cdots \leq t_{i,n_i} \leq Z$, where Z is the maximum integer.
- Let $t_{A_i}(t_{i,j}, t_{i,k})$ represent the range constraint of attribute A_i on $[t_{i,j}, t_{i,k}]$, where $1 \leq j \leq k \leq n_i$, i.e., $t_{i,j} \leq t_{A_i} \leq t_{i,k}$.
- Let $\mathscr{P} = \{\bigwedge t_{A_i} | \forall A_i \in \mathbb{U}, t_{i,j} \leq t_{A_i} \leq t_{i,k}\}$, where $1 \leq j \leq k \leq n_i$ be the policy defined by the data owner over the set of attributes $\mathbb{U}$, and it is expressed as a series of AND operations.
- Let $\mathscr{L}_u = \{\bigwedge t_{A_i} | \forall A_i \in \mathbb{U}, t_{i,a} \leq t_{A_i} \leq t_{i,b}\}$, where $1 \leq a \leq b \leq n_i$ as the attribute ranges possessed by a data user u over the set of attributes $\mathbb{U}$.

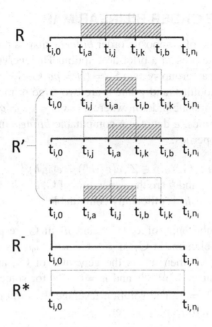

Figure 2.3: The range relations.

As illustrated in Figure 2.3, the data owner can apply any one of the following attribute range relations $\{R, R', R^-, R^*\}$ over each attribute A_i, such that the data user u's attribute ranges $\mathscr{L}_u$ has to satisfy the designated attribute range relations over all the attributes to access the resources. R implies that the attribute ranges $\mathscr{L}_u$ should completely satisfy $\mathscr{P}$ on A_i, and it holds if $([t_{i,j}, t_{i,k}] \setminus [t_{i,a}, t_{i,b}] = \emptyset) \wedge ([t_{i,j}, t_{i,k}] \cap [t_{i,a}, t_{i,b}] \neq \emptyset)$. On the contrary, R' implies that the attribute ranges $\mathscr{L}_u$ only needs to partially satisfy $\mathscr{P}$ on A_i, and it holds if $([t_{i,j}, t_{i,k}] \setminus [t_{i,a}, t_{i,b}] \neq \emptyset) \wedge ([t_{i,j}, t_{i,k}] \cap [t_{i,a}, t_{i,b}] \neq \emptyset)$.

In addition, R^- implies that the access control policy $\mathscr{P}$ designates the eligible data user must not own attribute A_i, which is classified as a negative attribute. Note that if the data user u does not own attribute A_i, he/she will be assigned a dummy integer value $t_{i,0}$, distinct from the other attribute integer values, such that $t_{i,a} = t_{i,b} = t_{i,0}$, and the system places $t_{i,0}$ ahead of $t_{i,1}$ to derive $\{t_{i,0}, t_{i,1}, \cdots, t_{i,n_i}\}$ in order to follow the ascending order. Accordingly, there exists $t_{i,j} = t_{i,k} = t_{i,0}$ in access control policy $\mathscr{P}$. Consequently, R^- is satisfied if and only if $[t_{i,j}, t_{i,k}] = [t_{i,a}, t_{i,b}] = \{t_{i,0}\}$ holds.

Furthermore, R^* implicates that the data owner does not care about attribute A_i, then there exist $t_{i,j} = t_{i,0}$ and $t_{i,k} = t_{t,n_i}$, and this attribute can be classified as a wildcard. If the data owner specifies A_i as a wildcard, then $[t_{i,j}, t_{i,k}]$ becomes $[t_{i,0}, t_{i,n_i}]$ and it always holds whatever the data user u's attribute range on A_i is. It implies $[t_{i,j}, t_{i,k}] \cap [t_{i,a}, t_{i,b}] \neq \emptyset$ always holds if $[t_{i,a}, t_{i,b}] \neq \emptyset$. In this manner, CCP-CABE is extended to be a comprehensive scheme to handle different range relations.

2.3   COMPOSITE ORDER BILINEAR MAP

The detailed description of composite order *bilinear* map is presented in Chapter 1, Section 1.2.3. Here, we present a brief description. The encryption system is based on the same *bilinear* map group system $\mathbb{S}_N = (N = pq, \mathbb{G}, \mathbb{G}_T, e)$ used in [246] where $N = pq$ is the RSA modulus, and p and q are two large primes. $\mathbb{G}$ and $\mathbb{G}_T$ are two cyclic groups with composite order n, where $n = sn' = s_1 s_2 p' q'$ and p, q, p', q', s_1, s_2 are all secret large primes. e denotes a computable *bilinear* map $e : \mathbb{G} \times \mathbb{G} \to \mathbb{G}_T$ with the following properties:

- Bilinearity: $\forall g, h \in \mathbb{G}, \forall a, b \in \mathbb{Z}, e(g^a, h^b) = e(g, h)^{ab}$;
- Nondegeneracy: g and h are the generators of $\mathbb{G}$, $e(g, h) \neq 1$;
- Computability: $e(g, h)$ is efficiently computable.

$\mathbb{G}_s$ and $\mathbb{G}_{n'}$ are subgroups of order s and n' in $\mathbb{G}$, respectively, and $e(g, h)$ becomes the identity element in $\mathbb{G}_T$ if $g \in \mathbb{G}_s$, $h \in \mathbb{G}_{n'}$. As an example, suppose w is the generator of $\mathbb{G}$, then $w^{n'}$ is the generator of $\mathbb{G}_s$, and w^s is the generator of $\mathbb{G}_{n'}$. assume that $g = (w^{n'})^{\rho_1}$ and $h = (w^s)^{\rho_2}$ for some ρ_1, ρ_2, it holds that $e(g, h) = e(w^{\rho_1}, w^{\rho_2})^{sn'} = 1$. The solution leverage the orthogonality between $\mathbb{G}_{n'}$ and $\mathbb{G}_s$ and keep $N, n, s, p, q, , p', q'$ secret.

2.4   MULTI-DIMENSIONAL RANGE DERIVATION FUNCTION

The presented Multi-dimensional Range Derivation Functions (MRDF) is based on Zhu's work [246]. The lower-bound and upper-bound integer values $t_{i,j}, t_{i,k}$ are selected out of the possible attribute range over each attribute $A_i \in \mathbb{U}$, and derive the integer set $U = \{t_{i,j}, t_{i,k}\}_{A_i \in \mathbb{U}}$. To construct a cryptographic algorithm for range comparison over multiple dimensions (or attributes), the order-preserving cryptographic map $\psi : U \to V$ is defined for MRDF where V takes the form of $v_{\{t_{i,j}, t_{i,k}\}_{A_i \in \mathbb{U}}}$. Note that $v_{\{t_{i,j}, t_{i,k}\}_{A_i \in \mathbb{U}}}$ is a cryptographic value reflecting the integer values of range bounds over each attribute $A_i \in \mathbb{U}$. The order-preserving cryptographic map ψ implies that there exists $v_{\{t_{i,j}, t_{i,k}\}_{A_i \in \mathbb{U}}} = \psi(\{t_{i,j}, t_{i,k}\}_{A_i \in \mathbb{U}}) \preceq v_{\{t'_{i,j}, t_{i,k}\}_{A_i \in \mathbb{U}}} = \psi(\{t'_{i,j}, t_{i,k}\}_{A_i \in \mathbb{U}})$ and $v_{\{t_{i,j}, t_{i,k}\}_{A_i \in \mathbb{U}}} = \psi(\{t_{i,j}, t_{i,k}\}_{A_i \in \mathbb{U}}) \preceq v_{\{t_{i,j}, t'_{i,k}\}_{A_i \in \mathbb{U}}} = \psi(\{t_{i,j}, t'_{i,k}\}_{A_i \in \mathbb{U}})$ if $t_{i,j} \leq t'_{i,j}$ and $t_{i,k} \geq t'_{i,k}$ hold for each $A_i \in \mathbb{U}$, where $\preceq$ denotes the partial-order relations.

To construct a cryptographic MRDF for integer comparisons over multiple attributes, we leverage the multiplicative group $\mathbb{G}_{n'}$ of RSA-type composite order $n' = p'q'$, where p' and q' are two large primes. A random generator φ is selected in the group $\mathbb{G}_{n'}$ where $\varphi^{n'} = 1$. Then two sets $\{\lambda_i, \mu_i\}_{A_i \in \mathbb{U}}$ are generated, where $\lambda_i, \mu_i \in \mathbb{Z}^*_{n'}$, and each λ_i, μ_i is relatively prime to all the other elements in $\{\lambda_i, \mu_i\}_{A_i \in \mathbb{U}}$ with sufficiently large order for all $A_i \in \mathbb{U}$. Consequently, the mapping function $\psi(\cdot)$ is defined to map the integer set U into V as shown below:

$$v_{\{t_{i,j}, t_{i,k}\}_{A_i \in \mathbb{U}}} \leftarrow \psi(\{t_{i,j}, t_{i,k}\}_{A_i \in \mathbb{U}})$$
$$= \varphi^{\Pi_{A_i \in \mathbb{U}} \lambda_i^{t_{i,j}} \mu_i^{Z - t_{i,k}}} \in \mathbb{G}_{n'}.$$

Accordingly, MRDF is defined as shown as follows:

Definition 2.1 (Multi-dimensional Range Derivation Function). A function $F : V \rightarrow V$ based on U is defined as a MRDF if it satisfies the following two conditions:

1. The function F can be computed in polynomial time, i.e., if $t_{i,j} \leq t'_{i,j}, t_{i,k} \geq t'_{i,k}, \forall A_i \in U$, then $v_{\{t'_{i,j}, t'_{i,k}\}_{A_i \in U}} \leftarrow F_{\{t_{i,j} \leq t_{i,k}, t_{i,k} \geq t'_{i,k}\}_{A_i \in U}}(v_{\{t_{i,j}, t_{i,k}\}_{A_i \in U}})$.

2. It is infeasible for any probabilistic polynomial time (PPT) algorithm to derive $v_{\{t'_{i,j}, t'_{i,k}\}_{A_i \in U}}$ from $v_{\{t_{i,j}, t_{i,k}\}_{A_i \in U}}$ if there exists $t_{i,j} > t'_{i,j}$ or $t_{i,k} < t'_{i,k}$ for some $A_i \in U$.

Specifically, $F(\cdot)$ takes the form as follows:

$$
\begin{aligned}
v_{\{t'_{i,j}, t'_{i,k}\}_{A_i \in U}} \quad &\leftarrow F_{\{t_{i,j} \leq t'_{i,j}, t_{i,k} \geq t'_{i,k}\}_{A_i \in U}}(v_{\{t_{i,j}, t_{i,k}\}_{A_i \in U}}) \\
&= (v_{\{t_{i,j}, t_{i,k}\}_{A_i \in U}})^{\Pi_{A_i \in U} \lambda_i^{t'_{i,j} - t_{i,j}} \mu_i^{t_{i,k} - t'_{i,k}}} \\
&= (\varphi^{\Pi_{A_i \in U} \lambda_i^{t_{i,j}} \mu_i^{Z - t_{i,k}}})^{\Pi_{A_i \in U} \lambda_i^{t'_{i,j} - t_{i,j}} \mu_i^{t_{i,k} - t'_{i,k}}} \\
&= \varphi^{\Pi_{A_i \in U} \lambda_i^{t'_{i,j}} \mu_i^{Z - t'_{i,k}}} \in \mathbf{G}_{n'}.
\end{aligned}
$$

Note that the ordering relationships among the integer values $t_{i,j}, t_{i,k}, t'_{i,j}, t'_{i,k}$ can be varied depending on the designated range relation R_i over each attribute A_i. Furthermore, it is infeasible to compute λ_i^{-1} and μ_i^{-1} in polynomial time due to the secrecy of n' under the RSA assumption. In addition, each λ_i is relatively prime to all the other elements in $\{\lambda_i\}_{A_i \in U}$, and each μ_i is also relatively prime to all the other elements in $\{\mu_i\}_{A_i \in U}$. Consequently, it is infeasible to compute $v_{\{t_{i,j}\}_{A_i \in U}}$ from $v_{\{t_{i,k}\}_{A_i \in U}}$, or derive $\bar{v}_{\{t_{i,k}\}_{A_i \in U}}$ from $\bar{v}_{\{t_{i,j}\}_{A_i \in U}}$ if there exist $t_{i,j} \leq t_{i,k}$ for some $A_i \in U$.

2.5 CCP-CABE OVERVIEW

The CCP-CABE scheme is comprised of six algorithms as shown below:

- *Setup*(κ, U): The *setup* algorithm takes input of the security parameter κ and the attribute set U. It outputs the global parameters GP for encryption and the master key MK;

- *KeyGen*$(GP, MK, u, \mathscr{L}_u)$: The *KeyGen* algorithm takes input of global parameters GP, master key MK, data user u's ID and corresponding attribute ranges $\mathscr{L}_u$ as the input. It outputs public keys PK_u and private keys SK_u for each data user;

- *EncDelegate*$(GP, MK, \mathscr{P})$: The *EncDelegate* algorithm takes GP, MK, and the the data owner's access control policy $\mathscr{P}$ as the input. It outputs the partially encrypted header $\widetilde{\mathscr{H}_{\mathscr{P}}}$ for the data owner to perform further encryption;

- *Encrypt*$(GP, \widetilde{\mathscr{H}_{\mathscr{P}}})$: The *encrypt* algorithm takes GP and $\widetilde{\mathscr{H}_{\mathscr{P}}}$ as the input. It creates a secret ε, outputs the session key K_ε and the ciphertext header $\mathscr{H}_{\mathscr{P}}$, such that only the data users with attribute ranges satisfying the access control policy can decrypt the message;

- *DecDelegate*($\mathcal{H}_{\mathcal{P}}, PK_u, \mathcal{L}_u, \mathcal{P}$): The *DecDelegate* algorithm takes input of the ciphertext header H, data user u's public key PK_u and the access control policy $\mathcal{P}$, and outputs the partially decrypted header $\mathcal{H}_{\mathcal{P}}$ to the data user for further decryption;
- *Decrypt*($SK_u, \mathcal{H}_{\mathcal{P}}$): The *decrypt* algorithm takes input of the partially decrypted ciphertext header $\mathcal{H}_{\mathcal{P}}$ and the data user's private key $PK_{\mathcal{L}_u}$. It performs further decryption over $\mathcal{H}_{\mathcal{P}}$ with $PK_{\mathcal{L}_u}$, and outputs the session key *ek* to decrypt the encrypted message.

2.6 SECURITY MODEL

In the presented security model, the Trust Authority (TA) and the Encryption Service Provider (ESP) are assumed that to be fully trustworthy, and they do not collude with other parties. However, data users attempt to obtain unauthorized access to data beyond their privileges, and the cloud provider is considered semi-honest [70]. Hence, the presented framework needs to be resistant against the following attacks:

Key Collusion Attack (KCA): In a normal case, each data user possesses a pre-assigned public key and private key from Trust Authority based on his/her attribute ranges. However, malicious data users may attempt to derive new private keys to reveal data protected by a multi-dimensional attribute range policy either individually or by collusion. Obviously, the latter is more threatening, so we only consider the collusion attack. The security under *KCA* is evaluated by the security game below:

Setup: The challenger runs the *setup* algorithm, gives the adversary the global parameters, and keeps private keys.

Learning: The adversary queries the challenger on behalf of a selected number of users $\{u_l\}_{1 \leq l \leq U}$ with attribute ranges $\{\mathcal{L}_{u_l}\}_{1 \leq l \leq U}$ by invoking *KeyGen* algorithm. The challenger responds with private keys $\{SK_{\mathcal{L}_{u_l}}, DK_{\mathcal{L}_{u_l}}\}_{1 \leq l \leq U}$ to the adversary in return.

Challenge: The challenger sends a challenge on behalf of user u' to the adversary.

Response: The adversary outputs $SK_{L_{u'}}$ with respect to user u'. If $SK_{L_{u'}}$ is valid and can bring more privileges for user u', then the adversary wins the game.

Chosen Delegation Key and Ciphertext Attack (CDKCA): The semi-honest cloud providers comply with protocols and output the correct results, but they are tempted to derive the information from the ciphertext header with the delegation keys from the data users without the permission of data owners. The security under Chosen Delegation Key and Ciphertext Attacks (CDKCA) is evaluated by the following security game:

Setup: The challenger runs the *setup* algorithm, gives the adversary the global parameters, and keeps private keys.

Learning: The adversary queries the challenger on behalf of a polynomial number of eligible users $\{u_l\}_{1 \leq l \leq U}$ with attribute ranges $\{\mathcal{L}_{u_l}\}_{1 \leq l \leq U}$ and $\mathcal{P}$ by invoking the *DecDelegate* algorithm. Note all the users are able to derive a session key from the ciphertext header. The challenger responds with delegation keys $\{DK_{\mathcal{L}_{u_l}}\}_{1 \leq l \leq U}$ to the adversary in return.

Challenge: The challenger sends a challenge ciphertext header to the adversary. The ciphertext header can be decrypted by the users mentioned above with their private keys.

Response: The adversary outputs the session key from the challenge ciphertext header. If the session key is valid, the adversary wins the game.

2.7 CONSTRUCTION

In this section, six algorithms of CCP-CABE scheme are presented in detail as follows:

2.7.1 SYSTEM SETUP (*SETUP*)

The central Trust Authority (TA) first chooses a *bilinear* map system $\mathbb{S}_N = (N = pq, \mathbb{G}, \mathbb{G}_T, e(\cdot, \cdot)$ of composite order $n = sn'$ and two subgroups $\mathbb{G}_s$ and $\mathbb{G}_{n'}$ of $\mathbb{G}$. Next, it selects random generators $w \in \mathbb{G}$, $g \in \mathbb{G}_s$ and $\varphi, \bar{\varphi} \in \mathbb{G}_{n'}$ such that there exist $e(g, \varphi) = e(g, \bar{\varphi}) = 1$, but $e(g, w) \neq 1$. The TA needs to choose $\lambda_i, \mu_i \in \mathbb{Z}_{n'}^*$ over each attribute $A_i \in \mathbb{U}$, and ensure that each λ_i, μ_i is relatively prime to all the other elements in $\{\lambda_i, \mu_i\}_{A_i \in \mathbb{U}}$. It also employs a cryptographic hash function $H : \{0, 1\}^* \to \mathbb{G}$ to convert a binary attribute string into an group element $\in \mathbb{G}$. In addition, the TA picks random exponents $\alpha, \beta \in \mathbb{Z}_n^*$ and generates $h = w^\beta$, $\eta = g^{\frac{1}{\beta}}$ and $e(g, w)^\alpha$. Consequently, the TA keeps its master key $MK = (p, q, n', \alpha, \beta)$ and publish the global parameters

$$GP = (\mathbb{S}, g, h, w, \eta, e(g, w)^\alpha, \varphi, \{\lambda_i, \mu_i\}_{A_i \in \mathbb{U}}, H(\cdot)).$$

2.7.2 KEY GENERATION (*KEYGEN*)

Each user u is labelled with a set of attribute ranges $\mathscr{L}_u = \{[t_{i,a}, t_{i,b}]\}_{A_i \in \mathbb{U}}$ with $t_{i,a} \leq t_{i,b}$ over all attributes. Specifically, if the user u does not possess the attribute A_i, then the TA sets $t_{i,a} = t_{i,b} = t_{i,0}$. The TA selects unique integers $\tau_u, r_u \in \mathbb{Z}$ to distinguish u from other users, and it concatenates the binary string forms of all the attributes and derive $\mathbf{A} = (A_1||A_2|| \cdots ||A_m)$. Consequently, for each user u with attribute ranges $\mathscr{L}_u$, his/her private key SK_u can be computed as

$$SK_u = (D_0^{(u)}, D_1^{(u)}, D_2^{(u)}) = (g^{\frac{\alpha + \tau_u}{\beta}}, g^{\tau_u}(H(\mathbf{A}))^{r_u}, w^{r_u}).$$

and his/her delegation key is computed as

$$DK_u = (v_{\mathscr{L}_u})^{r_u} = \varphi^{r_u \Pi_{A_i \in \mathbb{U}} \lambda_i^{t_{i,a}} \mu_i^{Z - t_{i,b}}},$$

where $v_{\mathscr{L}_u} = v_{\{t_{i,a}, t_{i,b}\}_{A_i \in \mathbb{U}}} = \varphi^{\Pi_{A_i \in \mathbb{U}} \lambda_i^{t_{i,a}} \mu_i^{Z - t_{i,b}}} \in \mathbb{G}_{n'}$. Afterwards, the keys are transmitted to the user u through secure channels.

2.7.3 ENCRYPTION DELEGATION (*ENCDELEGATE*)

The data owner first defines the access control policy of attribute constraints as $\mathscr{P} = \{\rho_i, \bar{\rho}_i\}_{A_i \in \mathbb{U}}$ over all attributes, and sends $\mathscr{P}$ to the trusted Encryption Service Provider to delegate the major part of encryption overhead if necessary. Note that $\{\rho_i, \bar{\rho}_i\}$ corresponds to the attribute constraint $[t_{i,j}, t_{i,k}]$ if the policy does not designate negative attributes or wildcards over A_i. Upon receiving $\mathscr{P}$, the Encryption Service Provider first sets ρ_i and $\bar{\rho}_i$ based on $\mathscr{P}$'s requirement of the range relationship $\mathscr{R}_i$ over the attribute A_i, respectively, as shown below:

- sets $\rho_i = t_{i,j}$ and $\bar{\rho}_i = t_{i,k}$ if there exists $\mathscr{R}_i := R$ over the attribute A_i;
- sets $\rho_i = t_{i,k}$ and $\bar{\rho}_i = t_{i,j}$ if there exists $\mathscr{R}_i := R'$ over the attribute A_i;
- sets $\rho_i = t_{i,0}$ and $\bar{\rho}_i = t_{i,0}$ if there exists $\mathscr{R}_i := R^-$ (negative attribute) over the attribute A_i;
- sets $\rho_i = t_{i,n_i}$ and $\bar{\rho}_i = t_{i,0}$ if there exists $\mathscr{R}_i := R^*$ (wildcard) over the attribute A_i.

Afterwards, the Encryption Service Provider computes $v_{\mathscr{P}} = v_{\{\rho_i, \bar{\rho}_i\}_{A_i \in \mathbb{U}}} = \varphi^{\Pi_{A_i \in \mathbb{U}} \lambda_i^{\rho_i} \mu_i^{Z - \bar{\rho}_i}}$. Accordingly, it generates the partially encrypted header $\widetilde{\mathscr{H}_{\mathscr{P}}}$ as shown below:

$$\widetilde{\mathscr{H}_{\mathscr{P}}} = (v_{\mathscr{P}}w, H(\mathbf{A}))$$

and sends it to the data owner for further encryption.

2.7.4 ENCRYPTION (*ENCRYPT*)

Upon receiving the partially encrypted header $\widetilde{\mathscr{H}_{\mathscr{P}}}$, the data owner generates a random secret $\varepsilon \in \mathbb{Z}_n$. Next, it computes $C = h^{\varepsilon}$ and the session key $ek = e(g^{\alpha}, w)^{\varepsilon}$. To improve efficiency, it first generates a random key ak to encrypt the target message and uses ek to encrypt the random key ak with symmetric key encryption $\mathbb{E}_{ak}(\cdot)$. Finally, it outputs the ciphertext header

$$\begin{aligned}
\mathscr{H}_{\mathscr{P}} &= (\mathbb{E}_{ek}(ak), C, E_{\varepsilon}, E_{\varepsilon}') \\
&= (\mathbb{E}_{ek}(ak), h^{\varepsilon}, (v_{\mathscr{P}}w)^{\varepsilon}, (H(\mathbf{A}))^{\varepsilon})
\end{aligned}$$

and transmits $\mathscr{H}_{\mathscr{P}}$ and the encrypted message along with $\mathscr{P}$ to the cloud for storage.

2.7.5 DECRYPTION DELEGATION (*DECDELEGATE*)

The data user u delegates his/her delegation key DK_u and claimed attribute ranges $\mathscr{L}_u$ to the cloud. Upon receiving DK_u and $\mathscr{L}_u$, the cloud checks if $\mathscr{L}_u$ satisfies $\mathscr{P}$ over all attributes. If so, it computes $(v_{\mathscr{P}})^{r_u}$ from $(v_{\mathscr{L}_u})^{r_u}$ as shown below:

$$\begin{aligned}
(v_{\mathscr{P}})^{r_u} &= (v_{\{\rho_i, \bar{\rho}_i\}_{A_i \in \mathbb{U}}})^{r_u} \\
&= F_{\{t_{i,a} \leq \rho_i, t_{i,b} \geq \bar{\rho}_i\}_{A_i \in \mathbb{U}}}((v_{\mathscr{L}_u})^{r_u}) \\
&= F_{\{t_{i,a} \leq \rho_i, t_{i,b} \geq \bar{\rho}_i\}_{A_i \in \mathbb{U}}}((v_{\{t_{i,a}, t_{i,b}\}_{A_i \in \mathbb{U}}})^{r_u}) \\
&= ((v_{\{t_{i,a}, t_{i,b}\}_{A_i \in \mathbb{U}}})^{r_u})^{\Pi_{A_i \in \mathbb{U}} \lambda_i^{\rho_i - t_{i,a}} \mu_i^{t_{i,b} - \bar{\rho}_i}} \\
&= (\varphi^{r_u \Pi_{A_i \in \mathbb{U}} \lambda_i^{t_{i,a}} \mu_i^{Z - t_{i,b}}})^{\Pi_{A_i \in \mathbb{U}} \lambda_i^{\rho_i - t_{i,a}} \mu_i^{t_{i,b} - \bar{\rho}_i}} \\
&= (\varphi^{\Pi_{A_i \in \mathbb{U}} \lambda_i^{\rho_i} \mu_i^{Z - \bar{\rho}_i}})^{r_u} \in \mathbb{G}_{n'},
\end{aligned}$$

where $v_{\mathscr{P}} = v_{\{\rho_i, \bar{\rho}_i\}_{A_i \in \mathbb{U}}}$ and $v_{\mathscr{L}_u} = v_{\{t_{i,a}, t_{i,b}\}_{A_i \in \mathbb{U}}}$. Afterwards, the cloud sends $\mathscr{H}_{\mathscr{P}} = ((v_{\mathscr{P}})^{r_u}, \mathscr{H}_{\mathscr{P}})$ along with the ciphertext to the data user for further decryption.

2.7.6 DECRYPTION (*DECRYPT*)

Upon receiving $\mathcal{H}_{\mathcal{P}}$ from the cloud, the data user first computes $(v_{\mathcal{P}})^{r_u}D_2^{(u)} = (v_{\mathcal{P}}w)^{r_u}$. Next, it computes

$$
\begin{aligned}
\Gamma(\varepsilon) \ &\leftarrow \ \frac{e(D_1^{(u)}, E_\varepsilon)}{e((v_{\mathcal{P}}w)^{r_u}, E'_\varepsilon)} \\
&= \frac{e(g^{\tau_u}(H(A))^{r_u}, (v_{\mathcal{P}}w)^\varepsilon)}{e((v_{\mathcal{P}}w)^{r_u}, (H(A))^\varepsilon)} \\
&= \frac{e(g^{\tau_u}, (v_{\mathcal{P}}w)^\varepsilon) \cdot e((H(A))^{r_u}, (v_{\mathcal{P}}w)^\varepsilon)}{e((v_{\mathcal{P}}w)^{r_u}, (H(A))^\varepsilon)} \\
&= e(g^{\tau_u}, (v_{\mathcal{P}})^\varepsilon) \cdot e(g^{\tau_u}, w^\varepsilon) \\
&= e(g^{\tau_u}, w^\varepsilon),
\end{aligned}
$$

where $e(g^{\tau_u}, (v_{\mathcal{P}})^\varepsilon) = 1$. Accordingly, the data user can derives the session key ek as shown below:

$$
ek = \frac{e(C, D_0^{(u)})}{\Gamma(\varepsilon)} = \frac{e((w^\beta)^\varepsilon, g^{\frac{\alpha + \tau_u}{\beta}})}{e(g, w)^{\tau_u \varepsilon}} = e(g^\alpha, w)^\varepsilon.
$$

With the session key ek, the data user can first retrieve the random key ak by decrypting $\mathbb{E}_{ek}(ak)$ and then derive the encrypted data with ak.

2.7.7 APPLICATION SCENARIOS

In this subsection, we use simple examples to illustrate how CCP-CABE can adapt for multiple different range relationships as shown in Figure 2.4.

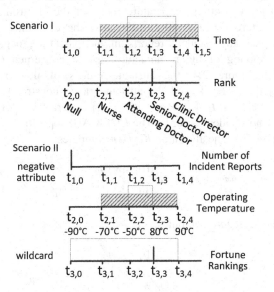

Figure 2.4: Different application scenarios.

Scenario I In the telemedicine example, the data owner applies the range relationship R, R' over attributes A_1, A_2 respectively in the access control policy $\mathscr{P}$. The "Time" attribute takes value out of the integer set $\{t_{1,0}, t_{1,1}, t_{1,2}, t_{1,3}, t_{1,4}, t_{1,5}\}$ representing different timestamps, and the "Rank" attribute takes value from the integer set $\{t_{2,0}, t_{2,1}, t_{2,2}, t_{2,3}, t_{2,4}\}$ representing different positions in a clinic. It can be learnt that the attribute ranges of the data user are $\mathscr{L}_u = \{[t_{1,1}, t_{1,5}], [t_{2,3}, t_{2,3}]\}$, and the attribute range constraints designated by the data owner are $\{[t_{1,2}, t_{1,4}], [t_{2,1}, t_{2,4}]\}$. The CCP-CABE operations associated with MRDF are listed below. The algorithm *KeyGen* computes:

$$v_{\mathscr{L}_u} = v_{\{t_{i,a}, t_{i,b}\}_{A_i \in \mathbb{U}}} = \varphi^{\prod_{A_i \in \mathbb{U}} \lambda_i^{t_{i,a}} \mu_i^{Z-t_{i,b}}}$$
$$= \varphi^{\lambda_1^{t_{1,1}} \lambda_2^{t_{2,3}} \mu_1^{Z-t_{1,5}} \mu_2^{Z-t_{2,3}}}.$$

The algorithm *EncDelegate* computes:

$$v_{\mathscr{P}} = v_{\{\rho_i, \bar{\rho}_i\}_{A_i \in \mathbb{U}}} = \varphi^{\prod_{A_i \in \mathbb{U}} \lambda_i^{\rho_i} \mu_i^{Z-\bar{\rho}_i}}$$
$$= \varphi^{\lambda_1^{t_{1,2}} \lambda_2^{t_{2,4}} \mu_1^{Z-t_{1,4}} \mu_2^{Z-t_{2,1}}}.$$

The algorithm *DecDelegate* computes:

$$(v_{\mathscr{P}})^{r_u} \leftarrow F_{\{t_{i,a} \leq \rho_i, t_{i,b} \geq \bar{\rho}_i\}_{A_i \in \mathbb{U}}}((v_{\mathscr{L}_u})^{r_u})$$
$$= (\varphi^{r_u \lambda_1^{t_{1,1}} \lambda_2^{t_{2,3}} \mu_1^{Z-t_{1,5}} \mu_2^{Z-t_{2,3}}})^{\Delta}$$
$$= (\varphi^{\lambda_1^{t_{1,2}} \lambda_2^{t_{2,4}} \mu_1^{Z-t_{1,4}} \mu_2^{Z-t_{2,1}}})^{r_u},$$

where $\Delta = \lambda_1^{t_{1,2}-t_{1,1}} \lambda_2^{t_{2,4}-t_{2,3}} \mu_1^{t_{1,5}-t_{1,4}} \mu_2^{t_{2,3}-t_{2,1}}$.

Scenario II In this scenario, an organization plans to select suppliers from electronic device manufacturers who produce electronic devices with the same intended use, and the products of the qualified manufacturers should meet three requirements: i) the operating temperature range of the electronic devices must cover the temperature range $[-50\,°C, 80\,°C]$; ii) the electronic devices should have never received any incident reports in the past (i.e., negative attribute); iii) the fortune ranking of the manufacturer is not concerned (i.e., wildcard). As illustrated in Figure 2.4, the attribute ranges of the manufacturer are $\{[t_{1,0}, t_{1,0}], [t_{2,1}, t_{2,4}], [t_{3,3}, t_{3,3}]\}$, and the attribute range constraints designated by the organization are $\{[t_{1,0}, t_{1,0}], [t_{2,2}, t_{2,3}], [t_{3,0}, t_{3,4}]\}$, where $t_{1,0}, t_{2,0}, t_{3,0} > 0$ and $t_{1,0}$ implies there are no incident records. The CCP-CABE operations associated with MRDF are listed below. The algorithm *KeyGen* computes:

$$v_{\mathscr{L}_u} = v_{\{t_{i,a}, t_{i,b}\}_{A_i \in \mathbb{U}}} = \varphi^{\prod_{A_i \in \mathbb{U}} \lambda_i^{t_{i,a}} \mu_i^{Z-t_{i,b}}}$$
$$= \varphi^{\lambda_1^{t_{1,0}} \lambda_2^{t_{2,1}} \lambda_3^{t_{3,3}} \mu_1^{Z-t_{1,0}} \mu_2^{Z-t_{2,4}} \mu_3^{Z-t_{3,3}}}.$$

The algorithm *EncDelegate* computes:

$$v_{\mathscr{P}} = v_{\{\rho_i, \bar{\rho}_i\}_{A_i \in \mathbb{U}}} = \varphi^{\prod_{A_i \in \mathbb{U}} \lambda_i^{\rho_i} \mu_i^{Z-\bar{\rho}_i}}$$
$$= \varphi^{\lambda_1^{t_{1,0}} \lambda_2^{t_{2,2}} \lambda_3^{t_{3,4}} \mu_1^{Z-t_{1,0}} \mu_2^{Z-t_{2,3}} \mu_3^{Z-t_{3,0}}}.$$

The algorithm *DecDelegate* computes:

$$(v_{\mathscr{D}})^{r_u} \leftarrow F_{\{t_{i,a} \leq \rho_i, t_{i,b} \geq \bar{\rho}_i\}_{A_i \in \mathbb{U}}}((v_{\mathscr{L}_u})^{r_u})$$

$$= (\varphi^{r_u \lambda_1^{t_{1,0}} \lambda_2^{t_{2,1}} \lambda_3^{t_{3,3}} \mu_1^{Z-t_{1,0}} \mu_2^{Z-t_{2,4}} \mu_3^{Z-t_{3,3}}})^{\Delta}$$

$$= (\varphi^{\lambda_1^{t_{1,0}} \lambda_2^{t_{2,2}} \lambda_3^{t_{3,4}} \mu_1^{Z-t_{1,0}} \mu_2^{Z-t_{2,3}} \mu_3^{Z-t_{3,0}}})^{r_u},$$

where $\Delta = \lambda_2^{t_{2,2}-t_{2,1}} \lambda_3^{t_{3,4}-t_{3,3}} \mu_2^{t_{2,4}-t_{2,3}} \mu_3^{t_{3,3}-t_{3,0}}$.

2.8   EXTENDED CONSTRUCTION

In this section, we discuss how to extend the use of CCP-CABE over multiple attribute domains. In some cases, multiple attribute domains are required by independent organizations, such that each organization can run an Attribute Authority (AA) to host its own attribute domain. Correspondingly, each AA hands out secret keys for a distinct set of attributes to reflect the users' attribute values within an attribute domain, and the failure of some attribute authorities does not impact the operation of other AAs. Accordingly, only the users with attribute ranges that satisfy the attribute constraints across multiple attribute domains can access that data. In addition, different attribute domains are at different levels of confidentiality from the perspectives of different data owners, and the data owners should be able to embed the levels of confidentiality associated with attribute domains into the access control policy dynamically. As an example, a military student's attributes associated with the army are more confidential than his/her attributes associated with the enrolled university. Therefore, CCP-CABE can be used as the building block and propose Extended CCP-CABE (ECCP-CABE) to prioritize different attribute domains to reflect different levels of confidentiality across domains. In ECCP-CABE, if one attribute range of the data user cannot satisfy the access policy in the corresponding attribute domain, then the decryption process stops and the access policy over the remaining attribute domains is still hidden.

In ECCP-CABE, each AA generates the master key and global parameters along with users' keys associate in the AA's own attribute domain using the same *setup* and *KeyGen* in CCP-CABE. Next, the data owners can delegate the encryption overhead to the trusted Encryption Service provider as with *EncDelegate* in CCP-CABE. The differences between CCP-CABE and ECCP-CABE lie in the algorithms of Encryption and Decryption as follows:

2.8.1   ECCP-CABE ENCRYPTION

From the perspective of the data owner, different attribute domains are at different levels of confidentiality. Accordingly, the data owner sorts AAs in descending order from the most confidential attribute domain to the least confidential attribute domain and derives $(\mathbb{U}_1, \cdots, \mathbb{U}_X)$. Upon receiving the partially encrypted header $\mathscr{H}_{\mathscr{D}}$, the data owner generates a random secret $\varepsilon_x \in \mathbb{Z}_n$ for each $\mathbb{U}_x$. Next, it computes $C_x = h_x^{\varepsilon_x}$ and $ek_x = H_1(e(g_x^{\alpha_x}, w_x)^{\varepsilon_x})$ for each $\mathbb{U}_x$ with $H_1 : \mathbb{G}_T \to \{0,1\}^*$, and generates a random key ak to encrypt the target message.

To embed the levels of confidentiality into the policy, the data owner first starts from the most confidential $\mathbb{U}_1$ and uses ek_1 to encrypt ak to get $\mathscr{H}_{\mathscr{P}_1} = \mathscr{P}_1 || \mathbb{E}_{ek_1}(ak)$ where $\mathscr{P}_1$ denotes the policy over $\mathbb{U}_1$ and $\mathbb{E}_{ek_1}(\cdot)$ denotes the symmetric encryption using ek_1. Next, the data owner moves on to the second most confidential $\mathbb{U}_2$ and computes $\mathscr{H}_{\mathscr{P}_2} = \mathscr{P}_2 || \mathbb{E}_{ek_2}(\mathscr{H}_{\mathscr{P}_1})$. The process goes on until the data owner moves on to the least confidential $\mathbb{U}_X$ and computes $\mathscr{H}_{\mathscr{P}_X} = \mathscr{P}_X || \mathbb{E}_{ek_X}(\mathscr{H}_{\mathscr{P}_{X-1}})$.

Finally, it outputs the ciphertext header

$$\mathscr{H}_{\mathscr{P}} = (\mathscr{H}_{\mathscr{P}_X}, \{C_x, E_{\varepsilon_x}, E'_{\varepsilon_x}\}_{1 \leq x \leq X}),$$

where

$$(E_{\varepsilon_x}, E'_{\varepsilon_x}) = ((v_{\mathscr{P}_x} w_x)^{\varepsilon_x}, (H(A_x))^{\varepsilon_x}),$$

and transmits $\mathscr{H}_{\mathscr{P}}$ and the encrypted message to the cloud for storage.

2.8.2 ECCP-CABE DECRYPTION

The cloud first transmits $\mathscr{H}_{\mathscr{P}_X}$ to the data user u, such that the data user u knows the corresponding policy $\mathscr{P}_X$ over the least confidential attribute domain A_X. Upon receiving $\mathscr{H}_{\mathscr{P}_X}$, the data user u checks if $\mathscr{L}_{X,u}$ satisfies $\mathscr{P}_X$. If so, the data user u delegate his/her delegation key $DK_{X,u}$ and claimed attribute ranges $\mathscr{L}_{X,u}$ to the cloud.

- *DecDelegate*: Upon receiving $DK_{X,u}$ and $\mathscr{L}_{X,u}$, the cloud derives $(v_{\mathscr{P}_X})^{r_{X,u}}$ from $(v_{\mathscr{L}_{X,u}})^{r_{X,u}}$ in the same manner as CCP-CABE, and then sends $(v_{\mathscr{P}_X})^{r_{X,u}}$ to the data user for further decryption.
- *Decrypt*: As with CCP-CABE, the data user u computes:

$$\Gamma(\varepsilon_X) = \frac{e(D_X^{(u)}, E_{\varepsilon_X})}{e((v_{\mathscr{P}_X} w_X)^{r_{X,u}}, E'_{\varepsilon_X})}.$$

Next, the data user u computes:

$$e(g_X^{\alpha_X}, w_X)^{\varepsilon_X} = \frac{e(C_X, D_{X,0}^{(u)})}{\Gamma(\varepsilon_X)},$$

where

$$D_{X,0}^{(u)} = g_X^{\frac{\alpha_X + \tau_{X,u}}{\beta_X}}.$$

Then, the data user u computes:

$$ak_X = H_1(e(g_X^{\alpha_X}, w_X)^{\varepsilon_X}),$$

and finally derives: $\mathscr{H}_{\mathscr{P}_{X-1}}$.

Afterwards, the data owner u and the cloud move on to A_{X-1} and invoke the algorithms *DecDelegate* and *Decrypt* again. This process proceeds recursively until they reach A_1 and retrieve the session key ek. After retrieving the session key, the data user can derive the encrypted data. It can be seen that this onion-like decryption enables a gradual exposure of the access control policy from the least confidential attribute domain to the most confidential attribute domain. It significantly preserves the privacy of access control policy, as the data user is unable to decrypt one more level to discover the policy over the next more confidential attribute domain if his/her attribute ranges cannot satisfy the policy over the current attribute domain.

2.9 PERFORMANCE EVALUATION

In this section, we analyze the complexity of CCP-CABE and ECCP-CABE in detail with evaluation results.

2.9.1 COMPLEXITY ANALYSIS

We compare the presented CCP-CABE scheme with CBE [246], ABE-AL [248], and CP-ABE [25] for complexity analysis. CBE and ABE-AL utilize different forward/backward derivation functions for comparison-based encryption and decryption, while CP-ABE and its variants use bit-wise matching method to implement integer comparison for comparison-based access control. Similarly to CBE, CCP-CABE only focus on the pairing and exponentiation operations while neglecting the hash and multiplication cost in both G and G_T as well as symmetric encryption/decryption cost, since they are much faster compared to the paring and exponentiation operations. We use similar notations as with CBE for these operations in both G and G_T. B indicates the bit form of the upper and lower bound values of the attribute range for comparison in CP-ABE. P denotes *bilinear* pairing cost. $E(G)$ and $E(G_T)$ refer to the exponential computation overhead in G and G_T, respectively. $E(\mathbb{Z}_n^*)$ refers to the exponential computation overhead in $\mathbb{Z}_n^*$. $\mathcal{T}$ represents the number of leaves in the access tree, and S represents the attributes involved in encryption and decryption. L is the ciphertext size resulting from symmetric encryption with the session key ek.

We demonstrate the differences of key size and ciphertext size between these schemes in Table 2.1. It is clear that the key size in CP-ABE, CBE, and ABE-AL grow linearly with the number of associate attributes S, and the ciphertext size in these three schemes also increase proportionally with the number of attributes $\mathcal{T}$ in the access tree. In contrast, CCP-CABE keeps both the key size and ciphertext size constant irrespective of the number of involved attributes. Table 2.2 gives the comparison between these schemes regarding the total communication cost on mobile devices including key generation, delegation, encryption and decryption, and we can learn that the communication cost of the first three schemes also grow with the number of related attributes, while the communication cost of CCP-CABE remains constant regardless of the number of attributes. Note that in CCP-CABE, it does not consider the communication overhead caused by the transmission of $\mathscr{P}$ and $\mathscr{L}_u$, as they are cleartext, which could be pre-distributed and compressed into a very small size.

The comparison of the computation overhead of encryption and decryption on mobile devices is presented in Table 2.3 and Table 2.4, respectively, as assume that

Table 2.1
Comparison of key size and ciphertext size

Scheme	Key Size	Ciphertext Size
CP-ABE	$(1+2|S||B|)l_G$	$l_{G_T} + (2|\mathscr{T}||B|+1)l_G$
CBE	$(1+4|S|)l_G$	$(4|\mathscr{T}|+1)l_G$
ABE-AL	$(1+|S|)l_G + |S|l_{\mathbb{Z}_n^*}$	$l_{G_T} + (2|\mathscr{T}|+1)l_G$
CCP-CABE	$4l_G$	$L+3l_G$

Table 2.2
Comparison of communication overhead

Scheme	Communication Cost
CP-ABE	$2l_{G_T} + (2|S||B|+4|\mathscr{T}||B|+3)l_G$
CBE	$3l_{G_T} + (3+10|S|+8|\mathscr{T}|)l_G$
ABE-AL	$2l_{G_T} + (2+|S|+4|\mathscr{T}|)l_G + |S|l_{\mathbb{Z}_n^*}$
CCP-CABE	$2L + l_{G_T} + 15l_G$

both the cloud providers and the Encryption Service Provider are resource-rich in computation capability, so the computation overhead on mobile devices are the only concern. It can be learned that the encryption and decryption overhead in CCP-CABE stay the same irrespective of the number of attributes involved in that all the computation-intensive operations are offloaded to the resource-rich Encryption Service Provider and cloud providers, while the computation cost of the other three schemes increase with the number of associated attributes.

ECCP-CABE uses CCP-CABE as the building blocks, and it reveals the policy domain by domain unless it reaches the most sensitive attribute domain. Correspondingly, Gradual Identity Exposure (GIE) proposed in [117, 244], also presented in Chapter 9, the variant of CP-ABE, enables the exposure of the access policy, attribute by attribute. Similar to the previous assumption, we assume that B indicates the bit form of the upper and lower bound values of the attribute range for comparison, $\mathscr{T}$ represents the number of leaves in the tree, and S represents the attributes involved in encryption and decryption in GIE. In addition, we assume that there exist X attribute domains in ECCP-CABE and the size of $\mathscr{H}_{\mathscr{D}_X}$ is L. Therefore, we compare ECCP-CABE with GIE in terms of key size, ciphertext size, and communication cost associated with encryption, delegation, and decryption as shown in Table 2.5.

Table 2.3

Comparison of encryption overhead

Scheme	Encryption
CP-ABE	$P + (1 + 2|\mathcal{T}|\,|B|)E(\mathbb{G})$
CBE	$(1 + 4|\mathcal{T}|)E(\mathbb{G}) + E(\mathbb{G}_T)$
ABE-AL	$(2|\mathcal{T}| + 1)E(\mathbb{G}) + E(\mathbb{G}_T)$
CCP-CABE	$3E(\mathbb{G}) + E(\mathbb{G}_T)$

Table 2.4

Comparison of decryption overhead

Scheme	Decryption
CP-ABE	$(2 + 3|S|\,|B|)E(\mathbb{G}_T) + 2|S|\,|B|P$
CBE	$P + (5|S| + 1)E(\mathbb{G})$
ABE-AL	$2|S|P + (|S| + 2)E(\mathbb{G}_T) + 2|S|E(\mathbb{G})$
CCP-CABE	$3P$

Table 2.5

Comparison of key size, ciphertext size, and communication cost between GIE and ECCP-CABE

Metric	GIE	ECCP-CABE
Key Size	$(1 + 2|S|\,|B|)l_{\mathbb{G}}$	$4Xl_{\mathbb{G}}$
Ciphertext Size	$l_{\mathbb{G}_T} + (2|\mathcal{T}|\,|B| + 1)l_{\mathbb{G}}$	$L + 3Xl_{\mathbb{G}}$
Comm. Cost	$2l_{\mathbb{G}_T} + (2|S|\,|B| + 4|\mathcal{T}|\,|B| + 3)l_{\mathbb{G}}$	$l_{\mathbb{G}_T} + (1 + X)L + 17Xl_{\mathbb{G}}$

The comparison regarding the computation cost between GIE and ECCP-CABE is shown in Table 2.6.

It can be seen that the key size, ciphertext size, and communication cost of GIE grow linearly with the number of associated attributes, while those of ECCP-CABE increase with the number of attribute domains. This also applies to GIE and ECCP-CABE in terms of encryption and decryption cost. In a real-world scenario, the number of attribute domains is usually smaller than the number of attributes, thus ECCP-CABE is generally more efficient than GIE in terms of communication and computation cost. On the other hand, GIE can provide more fine-grained privacy preservation of access control policy as it reveals the policy attribute by attribute.

Table 2.6

Comparison of computation cost between GIE and ECCP-CABE

Operation	GIE	ECCP-CABE
Encryption	$P+(1+2|\mathscr{T}||B|)E(\mathrm{G})$	$3XE(\mathrm{G})+XE(\mathrm{G}_T)$
Decryption	$(2\ +\ 3|S||B|)E(\mathrm{G}_T)\ +$ $2|S||B|P$	$3Xl_{\mathrm{G}}$

2.9.2 EXPERIMENT

CCP-CABE scheme is implemented on Linux virtual machines and android smartphones. The Trust Authority, Encryption Service Provider, and cloud provider are simulated by virtual machines with Intel Core i3-2100 CPU at 3.10GHz and 4GB memory running 64-bit Ubuntu Precise Pangolin (Ubuntu 12.04). The mobile device is a Samsung Galaxy Note with Quad core ARM Cortex-A9 at 1.6GHz and 2GB memory running Android 4.2 (Jelly Bean). The experiment utilizes the Java Pairing-Based Cryptography (jPBC) library [67]. As with [246], the same *bilinear* map system $ of composite order n is used, where $n = s_1 s_2 p' q'$ and $|p'| = |q'| = 256$ bits. The tests demonstrate that the Ubuntu virtual machines run more than 20 times faster than the smartphone on average regarding the algorithms in CCP-CABE and ECCP-CABE.

Figure 2.5 illustrates the impact of the range of integer comparison on the computational cost of the algorithms in CCP-CABE where the total number of attributes is set as 10. We assume that the value range of each attribute is $[1, Z]$, and Z takes the form of 2^x. The data owner adopts the range relationship R and designates $[\frac{3}{8}Z, \frac{5}{8}Z]$ as the attribute constraint over each attribute, and the attribute range of the data user is $[\frac{1}{8}Z, \frac{7}{8}Z]$. Therefore, the comparison range is $\frac{Z}{4}$, and it grows from 2 to 2^{12} as x increases from 3 to 14. It demonstrates that the comparison range has negligible impact over the computational cost of the algorithms in CCP-CABE. This implies that each attribute can have many integer values for comparison without increasing the computational overhead in real-world settings. In Figure 2.6, the comparison range is fixed as 2^4. It shows that the computational cost of *KeyGen*, *EncDelegate*, and *DecDelegate* running on the server grows almost linearly as the number of attributes increases from 1 to 12. Meanwhile, the computational cost of *encrypt* and *decrypt* remain the same irrespective of the number of attributes, which is suitable for resource-constrained mobile devices.

In Figure 2.7, we assume that each attribute domain has 6 attributes and the comparison range of each attribute is 2^4. In addition, the experiment uses AES-128 for the recursive encryption and decryption over the attribute domains. As each attribute authority is only responsible for *setup*, *KeyGen*, and *EncDelegate* in its own domain, its performance is approximately the same as that in CCP-CAB. Figure 2.7 shows that the computational cost of *encrypt*, *DecDelegate*, and *decrypt* grows with the

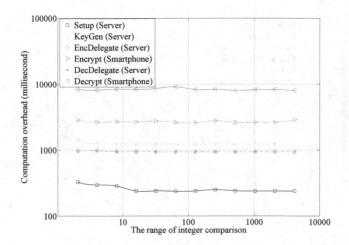

Figure 2.5: Computational cost of CCP-CABE algorithms with different comparison range.

number of attribute domains. As the number of attribute domains is usually much smaller than the number of attributes in real world, the computational overhead is still acceptable. It can be learned that the data owner should associate the policy only with the concerned attribute domains to reduce overhead.

2.10 SECURITY ANALYSIS

In this section, the security analyses of CCP-CABE and ECCP-CAB are presented. The hardness assumptions is presented below for the theorem proofs in the following subsections.

Definition 2.2 (RSA Assumption). For the RSA delegation key (N, e) and the ciphertext $C = M^e \in G_{n'}$, it is intractable to compute the plaintext M.   □

Definition 2.3 (Co-CDH Assumption). Given a quadruple $(g_1, g_1^y, g_2, g_2^z) \in G^4$ where $y, z \in \mathbb{Z}_n^*$, it is intractable to compute g_2^{yz}.   □

Definition 2.4 (Bilinear Co-CDH Assumption). Given a quintuple $(g_1, g_1^y, g_1^z, g_2, g_2^z) \in G^5$ where $y, z \in \mathbb{Z}_n^*$, it is intractable to compute $e(g_1^z, g_2^{yz})$.   □

Definition 2.5 (Discrete Logarithm Problem(DLP) Assumption). Given $(g_1, g_1^y) \in G$ where $y \in \mathbb{Z}_n^*$, it is intractable to compute y.   □

In ECCP-CABE, each attribute authority generates parameters and operate independently in its own attribute domain as with CCP-CABE. Accordingly, the security of ECCP-CABE fully depends on CCP-CABE, and the security proof of CCP-CABE also applies to ECCP-CABE.

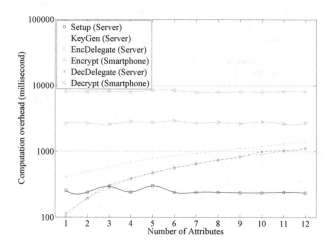

Figure 2.6: Computational cost of CCP-CABE algorithms with different number of attributes.

2.10.1 SECURITY FOR MRDF

MRDF $F(\cdot)$ is based on the correctness of forward and backward derivation function $(f, \bar{f})$ defined in [246], and it can be expressed as shown below:

$$F_{\{t_{i,j} \leq t'_{i,j}, t_{i,k} \geq t'_{i,k}\}_{A_i \in \mathbb{U}}} \left(v_{\{t_{i,j}, t_{i,k}\}_{A_i \in \mathbb{U}}}\right)$$
$$= \left(\varphi^{\prod_{A_i \in \mathbb{U}} \lambda_i^{t_{i,j}} \mu_i^{Z - t_{i,k}}}\right) \prod_{A_i \in \mathbb{U}} \lambda_i^{t'_{i,j} - t_{i,j}} \mu_i^{t_{i,k} - t'_{i,k}}$$
$$= \varphi^{\prod_{A_i \in \mathbb{U}} \lambda_i^{t'_{i,j}} \mu_i^{Z - t'_{i,k}}}$$
$$= v_{\{t'_{i,j}, t'_{i,k}\}_{A_i \in \mathbb{U}}}.$$

It is easy to compute $(\varphi^{\lambda_i^{t_{i,j}}})^{\lambda_i^{t'_{i,j} - t_{i,j}}}$ and $(\varphi^{\mu_i^{Z - t_{i,k}}})^{\mu_i^{t_{i,k} - t'_{i,k}}}$ if $t_{i,j} \leq t'_{i,j}$ and $t_{i,k} \geq t'_{i,k}$ hold. On the contrary, λ_i^{-1} and μ_i^{-1} cannot be efficiently derived by any PPT algorithms under the RSA assumption due to the secrecy of n', Hence, if $t_{i,j} \leq t'_{i,j}$, it is intractable to derive $\varphi^{\lambda_i^{t_{i,j}}}$ from $\varphi^{\lambda_i^{t'_{i,j}}}$ for any $A_i \in \mathbb{U}$. It is also intractable to derive $\varphi^{\lambda_i^{Z - t_{i,k}}}$ from $\varphi^{\lambda_i^{Z - t'_{i,k}}}$ if $t'_{i,k} > t'_{i,k}$ holds for any $A_i \in \mathbb{U}$. In addition, the value of the product $\prod_{A_i \in \mathbb{U}} \lambda_i^{t_{i,k}}$ is unique, which implies that $\prod_{A_i \in \mathbb{U}} \lambda_i^{t_{i,j}} \mu_i^{Z - t_{i,k}} \neq \prod_{A_i \in \mathbb{U}} \lambda_i^{t'_{i,j}} \mu_i^{Z - t'_{i,k}}$ if there exists $t_{i,j} \neq t'_{i,j}$ or $t_{i,k} \neq t'_{i,k}$ over some $A_i \in \mathbb{U}$ due to the relative primality between any two numbers in $\{\lambda_i, \mu_i\}_{A_i \in \mathbb{U}}$. Accordingly, given $|n'| = 512$, the collision probability of $\{v_{\{t_{i,j}, t_{i,k}\}_{A_i \in \mathbb{U}}} = v_{\{t'_{i,j}, t'_{i,k}\}_{A_i \in \mathbb{U}}}\}$ is less than $2^{-2n'} = 2^{-1024}$, which is negligible. As a result, it is infeasible to deduce $v_{\{t_{i,j}, t_{i,k}\}_{A_i \in \mathbb{U}}}$ from $v_{\{t'_{i,j}, t'_{i,k}\}_{A_i \in \mathbb{U}}}$ if $t_{i,j} > t'_{i,j}$ or $t_{i,k} < t'_{i,k}$ holds. Therefore, MRDF is hard to invert, and its one-way property can be guaranteed.

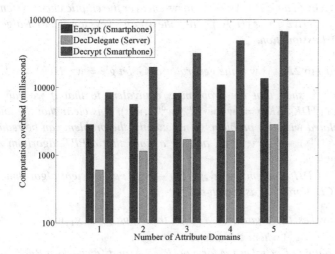

Figure 2.7: Computational cost of ECCP-CABE algorithms with different number of attribute domains.

2.10.2 SECURITY FOR KEY COLLISION ATTACKS

The security of CCP-CABE and ECCP-CABE schemes against the Key Collision Attack (KCA) relies on the confidentiality of r_u associated with user u's identity. A user could leverage key collision attacks to extend his/her attribute range and increase privileges. For the sake of brevity, we do not consider negative attributes and wildcards, as they are actually special cases of the range relations R and R'. In addition, we only consider KCA committed by two adversaries in CCP-CABE, which can be extended to all cases. For example, a user u' with attribute ranges $\mathcal{L}_{u'} = \{[t'_{i,a}, t'_{i,b}]\}_{A_i \in \mathbb{U}}$ attempts to transfer another user u's attribute ranges $\mathcal{L}_u = \{[t_{i,a}, t_{i,b}]\}_{A_i \in \mathbb{U}}$ into his/her own key, such that he/she can obtain more privilege over some attribute A_i as there exists $t_{i,a} < t'_{i,a} < t'_{i,b} < t_{i,b}$. In other words, user u' depends on the prior knowledge of

$$
\begin{aligned}
(SK_u, DK_u) &= (D_0^{(u)}, D_1^{(u)}, D_2^{(u)}, DK_u) \\
&= (g^{\frac{\alpha + \tau_u}{\beta}}, g^{\tau_u}(H(\mathbf{A}))^{r_u}, w^{r_u}, (v_{\mathcal{L}_u})^{r_u}), \\
(SK_{u'}, DK_{u'}) &= (D_0^{(u')}, D_1^{(u')}, D_2^{(u')}, DK_{u'}) \\
&= (g^{\frac{\alpha + \tau_{u'}}{\beta}}, g^{\tau_{u'}}(H(\mathbf{A}))^{r_{u'}}, w^{r_{u'}}, (v_{\mathcal{L}_{u'}})^{r_{u'}}),
\end{aligned}
$$

and he/she launches KCA-I attacks to derive new keys $(g^{\frac{\alpha + \tau_{u'}}{\beta}}, g^{\tau_{u'}}(H(\mathbf{A}))^{r_u}, w^{r_u}, (v_{\mathcal{L}_u})^{r_u}$ by exchanging $g^{\tau_{u'}}$ or $(H(\mathbf{A}))^{r_{u'}}$ with some known keys. We can prove the presented schemes are resistant against KCA-I attacks with the following theorem:

Theorem 2.1. *Given a CCP-CABE cryptosystem over the elliptic curve system $\mathbb{S}_N$, it is intractable to derive g^{τ_u} or $(H(A))^{r_u}$ from the user u's keys for any PPT algorithms under Co-CDH assumption.* □

Proof of Theorem 2.1. *As w is the generator of $\mathbb{G}$, let $g^{\tau_u} = w^{\xi}, H(A) = w^{\delta}, v_{\mathscr{L}_u} = w^{\delta_1}, g^{\frac{\alpha+\tau_{u'}}{\beta}} = w^{\gamma}$, such that the problem is equivalent to that if we can derive $(w^{\xi}, w^{\delta r_u})$ from $(DK_u, SK_u) = (w^{\gamma}, w^{\xi+\delta r_u}, w^{\delta_1 r_u}, w^{r_u})$. It is clear that the unknown δ_1 is uncorrelated with this problem. Consequently, the problem can be reduced to $(w, w^{r_u}, w^{\delta}, w^{\xi+\delta r_u}) \longrightarrow (w^{\xi}, w^{\delta r_u})$, and we assume that a PPT algorithm $\mathscr{A}$ can break this problem.*

Given a Co-CDH problem $(g_1, g_1^y, g_2, g_2^z) \longrightarrow g_2^{yz}$, an efficient algorithm $\mathscr{B}$ can solve the Co-CDH problem as shown below:

1$\mathscr{B}$ invokes the algorithm $\mathscr{A}$ with the input $(w = g_1, w^{r_u} = g_1^y, w^{\delta} = g_2^z, w^{\xi+\delta r_u} = g_2^{\tau})$ where τ is a random integer;

2The output of $\mathscr{A}$ takes the form of (R_1, R_2), and $\mathscr{B}$ checks if $R_1 \cdot R_2 = g_2^{\tau}$ and $e(g_1, R_2) = e(g_1^y, g_2^z)$ are valid. If not, $\mathscr{B}$ repeats step (1).

3$\mathscr{B}$ returns R_2 as the output.

The equations above hold as there exist $r_u = y, R_2 = w^{\delta r_u}, R_1 \cdot R_2 = g_2^{\xi+\delta r_u} = g_2^{\tau}$, and $e(g_1, R_2) = e(g_1, g_2^{\delta r_u}) = e(g_1^y, g_2^z)$. It implicates $\mathscr{B}$ can solve Co-CDH problem if $\mathscr{A}$ is a PPT algorithm, which contradicts the hardness assumption of Co-CDH. This implies it is intractable to derive $(w^{\xi}, w^{\delta r_u})$. □

In addition, the colluders could also commit KCA-II attacks to forge new keys $\{g^{\frac{\alpha+\tau_{u'}}{\beta}}, g^{\tau_{u'}} (H(A))^{r_{u'}}, w^{r_{u'}}, (v_{\mathscr{L}_{u'}})^{r_{u'}}\}$ by replacing $v_{\mathscr{L}_{u'}}$ with some new $v_{\mathscr{L}_u}$ to get some advantage in their privileges, where there exists $t_{i,a} < t'_{i,a} < t'_{i,b} < t_{i,b}$ for some attribute A_i in $\mathscr{L}_u$. We can prove the presented schemes are secure against KCA-II attacks with the theorem below:

Theorem 2.2. *Given a multi-tuple $(N, \varphi, \{\lambda_i, \mu_i, t_{i,j}\}_{A_i \in \mathbb{U}}, \varphi^{r_u} \Pi_{A_i \in \mathbb{U}} \lambda_i^{t_{i,j}} \mu_i^{Z-t_{i,k}})$ over the elliptic curve system $\mathbb{S}_N$, it is intractable to compute $(\{t'_{i,j}, t'_{i,k}\}_{A_i \in \mathbb{U}}, \varphi^{r_u} \Pi_{A_i \in \mathbb{U}} \lambda_i^{t'_{i,j}} \mu_i^{Z-t'_{i,k}})$ if there exists $t_{i,j} > t'_{i,j}$ or $t_{i,k} < t'_{i,k}$ for some $A_i \in \mathbb{U}$ under the RSA assumption.* □

Proof of Theorem 2.2. *We first assume that there exists $t_{i',j} > t'_{i',j}$ over one attribute $A_{i'} \in \mathbb{U}$. Assume that a PPT algorithm $\mathscr{A}$ takes $(N, \varphi, \{\lambda_i, \mu_i, t_{i,j}\}_{A_i \in \mathbb{U}}, \varphi^{r_u} \Pi_{A_i \in \mathbb{U}} \lambda_i^{t_{i,j}} \mu_i^{Z-t_{i,k}})$ as the input and outputs $(\{t'_{i,j}, t'_{i,k}\}_{A_i \in \mathbb{U}}, \varphi^{r_u} \Pi_{A_i \in \mathbb{U}} \lambda_i^{t'_{i,j}} \mu_i^{Z-t'_{i,k}})$. Without loss of generality, we assume that a PPT algorithm $\mathscr{B}$ can derive the plaintext $M = C^{e^{-1}}$ with the public key $(\mathbb{G}, N, e)$ and the ciphertext C over some attribute A_i based on $\mathscr{A}$, and breaks the RSA problem as follows:*

1$\mathcal{B}$ invokes the algorithm $\mathcal{A}$ with the input $(N, \varphi, e = \lambda_{i'}, t_{i',j}, C = \varphi^{r_u} \Pi_{i \neq i', A_i \in \mathbb{U}} \lambda_i^{t_{i,j}} \Pi_{A_i \in \mathbb{U}} \mu_i^{Z-t_{i,k}}$ where $\varphi = C^{r'_u}$ is a random element in $\mathbb{G}$ and $t_{i',j}$ is a random integer;

2$\mathcal{B}$ checks $\mathcal{A}$'s outputs $(t'_{i',j}, R)$ if $t'_{i',j} - t_{i',j} - 1 \geq 0$ and $R^{\lambda_{i'}^{t'_{i',j} - t_{i',j}}} = C$ hold. If not, $\mathcal{B}$ repeats step (1);

3$\mathcal{B}$ computes $M = R^{e^{t'_{i',j} - t_{i',j} - 1}}$ and outputs the plaintext M.

The output of $\mathcal{B}$ is valid as $M^e = C = R^{e^{t'_{i',j} - t_{i',j} - 1}}$ holds. It is intractable to computes $r_u = (r'_u \lambda_{i'}^{t'_{i',j}})^{-1}$ mod n' due to the hardness of factoring large number N. This implicates $\mathcal{B}$ can solve RSA problem in polynomial time if $\mathcal{A}$ is a PPT algorithm, which contradicts the hardness of RSA. By the same token, it is intractable to deduce $(\{t'_{i,j}, t'_{i,k}\}_{A_i \in \mathbb{U}}, \varphi^{r_u} \Pi_{A_i \in \mathbb{U}} \lambda_i^{t'_{i,j}} \mu_i^{Z-t'_{i,k}})$ if there exists $t_{i',k} < t'_{i',k}$ over one attribute $A_{i'} \in \mathbb{U}$, which can be further applied to multiple attributes in the same manner. Hence, it is intractable to derive $(\{t'_{i,j}, t'_{i,k}\}_{A_i \in \mathbb{U}}, \varphi^{r_u} \Pi_{A_i \in \mathbb{U}} \lambda_i^{t'_{i,j}} \mu_i^{Z-t'_{i,k}})$ if there exists $t_{i,j} > t'_{i,j}$ or $t_{i,k} < t'_{i,k}$ for some $A_i \in \mathbb{U}$. $\square$

Consequently, it is infeasible for the users to forge new keys with more privileges by key collusion.

2.10.3 SECURITY FOR CHOSEN DELEGATION KEY AND CIPHERTEXT ATTACKS

The DLP assumption makes it is hard for the cloud provider to derive ε from the ciphertext header $(C = h^\varepsilon, E_\varepsilon = (v_{\mathscr{P}} w)^\varepsilon, E'_\varepsilon = (H(\mathsf{A}))^\varepsilon)$. We can also prove that the cloud provider cannot obtain any advantage in CDKCA with a polynomial number of delegation keys and ciphertext headers. The delegation keys $DK_{\mathscr{L}_u}$ contains only part of the information, and r_u prevents applying one user's delegation key to another user's decryption process. Additionally, the secret keys are not disclosed to the cloud providers, so it is infeasible to cancel out r_u, τ_u and derive $ek = e(g^\alpha, w)^\varepsilon$ without the secret keys. It can be formally proved under bilinear Co-CDH assumption with the theorem below:

Theorem 2.3. Given the RSA-type elliptic curve system $\mathbb{S}_N$, CCP-CABE is semantically secure against CDKCA under the bilinear Co-CDH assumption.

Proof of Theorem 2.3. Assume that a PPT algorithm $\mathcal{A}$ can break this algorithm. Given a bilinear Co-CDH problem $(g_1, g_1^y, g_1^z, g_2, g_2^z) \longrightarrow e(g_1^z, g_2^{yz})$, an efficient algorithm $\mathcal{B}$ can be constructed to solve the bilinear Co-CDH problem based on $\mathcal{A}$ as follows:

Setup: $\mathcal{B}$ randomly chooses an integer θ and defines $\alpha = yz$, $\beta = \frac{\theta}{\iota}$, $g = g_1$, $w = g_2$, $h = w^z = g_1^\theta$, $\zeta = e(g_1^y, g_2^z)$, $\varphi = g_2^\rho$ where $\iota = \log_{g_1} g_2$ and $s|\rho$. Accordingly, $\mathcal{B}$ sends $GP = (\mathbb{S}, g, h, \eta, e(g, w)^\alpha, \varphi, \{\lambda_i, \mu_i\}_{A_i \in \mathbb{U}}, H(\cdot))$ to $\mathcal{A}$.

Learning: $\mathscr{A}$ sends a polynomial number of *DecDelegate* queries with $\{\mathscr{L}_{u_l}\}_{1\le l\le U}$ and $\mathscr{P}$. For each query, $\mathscr{B}$ computes $(v_{\mathscr{P}})^{r_u} = (\varphi^{\Pi_{A_i\in U}\lambda_i^{\rho_i}\mu_i^{Z-\rho_i}})^{r_u}$, and sends it to $\mathscr{A}$.

Challenge: $\mathscr{B}$ sets $\varepsilon = z$ and computes $h^\varepsilon = (g_1^z)^\theta$, $E_\varepsilon = (g_2^{\varepsilon_1})^{\rho_1\Pi_{A_i\in U}\lambda_i^{\rho_i}\mu_i^{Z-\rho_i}+1}$, $E'_\varepsilon = g_2^{\varepsilon\varsigma}$, where $H(\mathsf{A}) = g_2^\varsigma$, and sends them as the challenge ciphertext to $\mathscr{A}$.

Response: $\mathscr{A}$ outputs a session key ek' to $\mathscr{B}$, and $\mathscr{B}$ outputs it as the result.

The output of $\mathscr{B}$ is valid if $\mathscr{A}$ outputs correct result, as there exists $ek' = e(g_1^y, g_2^{yz}) = e(g^\alpha, w^\varepsilon)$. This implies $\mathscr{B}$ can solve Bilinear *Co-CDH* problem if $\mathscr{A}$ is a *PPT* algorithm, which contradict the hardness assumption of Bilinear *Co-CDH*. □

Consequently, it is infeasible for the honest-but-curious cloud provider to reveal the encrypted content by taking advantage of the ciphertext and the delegation keys.

2.11 SUMMARY

Attributes are describable terms and usually they have deterministic meanings such as *Teacher, Student, Street Name*, etc. However, sometimes, attributes can be used to describe a range such as *Rank: General, Captain, Lieutenant, Sergeant, Time Range: 8:0 AM to 5 PM, Geo-location region*, etc. In practice, using ABE, these attributes require a comparable feature embedded in the ABE scheme to check if a subject's or object's attributes satisfy the given comparable range. In this chapter, we introduce the problem of ABE with comparable attributes, and a solution to address this problem. The solution includes proposed algorithms, proof, and efficiency evaluation.

3 Privacy-Preserving Attribute-Based Encryption

Ciphertext Policy Attribute-Based Encryption (CP-ABE) enforces expressive data access policies, and each policy consists of a number of attributes. Most existing CP-ABE schemes incur a very large ciphertext size, which increases linearly with respect to the number of attributes in the access policy. Herranz et al. [101] proposed a construction of CP-ABE with constant ciphertext. However, [101] does not consider the recipients' anonymity and the access policies are exposed to potential malicious attackers. On the other hand, existing privacy preserving schemes [127, 165] protect the anonymity but require bulky, linearly-increasing ciphertext size.

In this chapter, a new construction of CP-ABE, named Privacy-Preserving Constant CP-ABE (PP-CP-ABE) is presented, in which it can reduce the ciphertext to a constant size with any given number of attributes. PP-CP-ABE leverages a hidden policy construction such that the recipients' privacy is preserved efficiently. In addition, a Privacy-Preserving Attribute-Based Broadcast Encryption (PP-AB-BE) scheme is presented, in which compared to existing Broadcast Encryption (BE) schemes, PP-AB-BE is more flexible because a broadcasted message can be encrypted by an expressive hidden access policy, either with or without explicitly specifying the receivers. Moreover, PP-AB-BE significantly reduces the storage and communication overhead to the order of $O(\log N)$, where N is the system size. Furthermore, it can be approved, using information theoretical approaches, PP-AB-BE thus attains minimal bound on storage overhead for each user to cover all possible subgroups in the communication system.

3.1 INTRODUCTION

Ciphtertext Policy Attribute-Based Encryption (CP-ABE) has been a very active research area in recent years [25, 56, 36, 127]. In the construction of CP-ABE, each attribute is a descriptive string and each entity may be tagged with multiple attributes. Many entities may share common attributes, which allow message encryptors to specify a secure data access policy over the shared attributes to reach a group of receivers. A decryptor's attributes need to satisfy the access policy in order to recover the message. These unique features make CP-ABE solutions appealing in many systems, where expressive data access control is required for a large number of users.

One major problem of existing CP-ABE schemes is bulky, linearly-increasing ciphertext. In the CP-ABE schemes reported in [25, 36, 127], the size of a ciphertext proliferates linearly with respect to the number of included attributes. For example, the message size in BSW CP-ABE [25] starts at about 630 bytes, and each additional attribute adds about 250-300 bytes.

Herranz et al. [101] proposed a CP-ABE that requires constant ciphertext size. However, it does not consider the anonymity of data recipients, and the data access policies are attached to the ciphertext in plaintext form. Thus, passive attackers can track a user or infer the sensitivity of ciphertext by eavesdropping on the access policies. In many environments, it is also critical to protect the access policies, as well as the data content. For example, the access policy "General" **AND** "Pentagon" disclose the recipient's roles or positions and implies the sensitivities of the message. On the other hand, existing privacy-preserving schemes [127, 165] protect the access policies but require large, linearly-increasing ciphertext size. To the best of the knowledge, there is no work that can achieve privacy-preservation and constant ciphertext size at the same time.

In this chapter, the Privacy-Preserving Constant-Size Ciphertext Policy Attribute Based Encryption (PP-CP-ABE) is presented, in which it enforces hidden access policies with wildcards and incurs constant-size conjunctive headers, regardless of the number of attributes. Each conjunctive ciphertext header only requires two *bilinear* group elements, which are bounded by 100 bytes in total. The actual size of *bilinear* group depends on the chosen parameters for the cryptosystem. In the implementation, we use Type-D MNT curves with element compression [152]. To support disjunctive or more flexible access policies, multiple constant-size conjunctive headers can be attached to the same ciphertext message. It should be noted that we restricted each ciphertext header to be conjunctive in order to avoid ambiguity while preserving receivers' anonymity. Moreover, PP-CP-ABE supports non-monotonic data access control policy.

In existing BE schemes, e.g., [33], a sender encrypts a message for a specified set of receivers who are listening on a broadcast channel. Each receiver in the specified set can decrypt the message while all other listeners cannot decrypt even though they collude together. However, in large-scale systems, identifying every receiver, and acquiring and storing their public keys, are not easy tasks. For example, to broadcast a message to all CS students in a university, the encryptor needs to query the CS department roster and acquire the public key of every student in the roster; this process could be very expensive and time-consuming.

Using the presented Privacy-Preserving Attribute-Based Broadcast Encryption (PP-AB-BE) approach, an encryptor has the flexibility to encrypt the broadcasted data either with or without the exact information of intended receivers. For example, Alice can specify a hidden access policy: "CS" **AND** "Student" to restrict the broadcast message to all CS students without specifying the receivers explicitly. Accordingly, Bob, who has attributes {"EE", "Student"}, cannot decrypt the data while Carol, who has attributes {"CS" , "Student"} can access the data. Moreover, Alice can also encrypt the broadcasted message to any arbitrary set of receivers such as {"Bob", "Carol"}.

PP-AB-BE also significantly reduces the storage overhead compared to many existing BE schemes, where cryptographic key materials required by encryption or decryption increase linearly or sublinearly on the number of receivers. For example, in a BGW scheme [33], the public key size is $O(N)$ or $O(N^{1/2})$, where N is the number of users in the system. PP-AB-BE reduces the key storage overhead problem by optimizing the organization of attribute hierarchy. In a system with N users, the storage overhead is $O(\log N + m)$, where m is a constant number and $m \ll N$. We also proved from the information theoretical perspective that PP-AB-BE achieves storage lower bound to satisfy all possible subgroup formations, and thus it can be applied to storage constrained systems.

The presented scheme is a unified privacy-preserving attribute-based solution considering constraints on both communication and storage, and the solution is provably secure. It is also worth noting that PP-CP-ABE can be used to implement an identity-based encryption with wildcards (WIBE) [13] to achieve the first constant ciphertext size WIBE construction with privacy preserving features. In a summary, the salient features of the presented solutions are:

- *PP-CP-ABE*: an efficient Privacy-Preserving Constant Ciphertext Policy Attribute-Based Encryption (PP-CP-ABE) scheme is constructed that enforces hidden conjunctive access policies with wildcards in constant ciphertext size.

- *PP-AB-BE*: Based on PP-CP-ABE, an Privacy-Preserving Attribute-Based Broadcast Encryption (PP-AB-BE) scheme is presented. Compared with existing BE schemes, PP-AB-BE is flexible as it uses both descriptive and non-descriptive attributes, which enables a user to specify the decryptors based on different abstraction levels, with or without exact information of intended receivers. Moreover, PP-AB-BE demands less storage overhead compared to existing BE schemes. We proved that the construction requires minimal storage to support all the possible user group formations for BE applications.

The rest of this chapter is organized as follows. The related work of privacy-preserving solutions for ABE is presented in Section 3.2. Then, in Section 3.3, system models used in this chapter are presented. The detailed PP-CP-ABE construction is presented in Section 3.4. Section 3.5 presents the construction of PP-AB-BE and the storage analysis using an information theoretical approach. In Section 3.6, the performance of PP-AB-BE is presented through both theoretical analysis and experimental studies. Finally, we summarize the presented solutions in Section 3.7.

3.2 RELATED WORKS

Attribute-Based Encryption (ABE) was first proposed as a fuzzy version of IBE in [186], where an identity is viewed as a set of descriptive attributes. The private key for an identity w can decrypt the message encrypted by the identity w' if and only if w and w' are closer to each other than a pre-defined threshold in terms of set overlap distance metric. In the paper [173], the authors further generalize the threshold-based set overlap distance metric to expressive access policies with *AND* and *OR* gates. There are two main variants of ABE proposed so far, namely Key Policy Attribute-Based Encryption (KP-ABE [95]) and Ciphertext Policy Attribute-Based Encryption

(CP-ABE [25]). In KP-ABE, each ciphertext is associated with a set of attributes and each user's private key is embedded with an access policy. Decryption is enabled only if the attributes on the ciphertext satisfy the access policy of the user's private key. In CP-ABE [25, 56, 127, 168, 94, 184, 240], each user's private key is associated with a set of attributes, and each ciphertext is encrypted by an access policy. To decrypt the message, the attributes in the user private key need to satisfy the access policy. The key difference between identity and attribute is identities are many-to-one mapped to users while attributes are many-to-many mapped to users. Thus, to simulate a constant size conjunctive header, one needs to encrypt the message using each receiver's identity and the size of ciphertext is linearly increasing.

In [79], the authors proposed a CP-ABE scheme with constant-size conjunctive headers and a constant number of pairing operations. It must be noted that they did not seek to address the issues of recipient anonymity. One drawback of their scheme does not support wildcards (or do-not-care) in the conjunctive access policies. To decrypt a ciphertext, the decryptor's attributes need to be identical to the access policy. In other words, the model is still one-to-one, i.e., an access policy is satisfied by one attribute list or ID, which makes the number of access policies increase exponentially. Thus, their scheme can be simply implemented using IBE schemes with same efficiency by using each user's attribute list as his/her ID. We should note that in a system with n attributes, the number of attribute combinations is 2^n. As the result, without using wildcards, there needs 2^n access policies to express all combinations. With wildcards, one can use a single access policy to express many combinations of attributes. Herranz et al. [101] proposed a construction of CP-ABE scheme with constant ciphertext. Their proposed scheme achieves constant ciphertext with any monotonic threshold data access policy, e.g., n-of-n (AND), 1-of-n (OR), and m-of-n. However, compared with the presented PP-CP-ABE, their scheme does not consider the recipient anonymity as one of the design goals.

To protect the privacy of access policy, a KSW scheme [127], NYO scheme [165], RC scheme [184], and YRL1 scheme [229] were proposed, where the encryptor specified access policy is hidden. specifically, the attribute names in both [184] and [229] are explicitly disclosed in the access policy, while only the eligible attribute values are hidden. Also, a YRL2 scheme was proposed in [232] based on a BSW scheme [25] as a group key management scheme providing group membership anonymity. In [109], we presented a novel alternative to the hidden policy to preserve privacy efficiently. The main difference between the scheme and existing hidden policy attribute-based encryption schemes is PP-CP-ABE significantly reduced the size of ciphertext to constant, while all existing hidden policy solutions require ciphertext that is linearly increasing on the number of attributes in the hidden policy.

It must be noted that the construction in this chapter is developed based on the research work presented in [240, 242, 245]. Through the development of this scheme, some major improvements presented in this chapter include: 1) the privacy-preserving requirements for ABE and incorporate the privacy-preserving solutions into the previous approaches; 2) a PP-AB-BE with an information theoretical analysis to address its complexity; and 3) a comprehensive performance evaluation.

ABE can be used as a perfect cryptographic building block to realize Broadcast Encryption (BE), which was introduced by Fiat and Naor in [81]. The encryptor in the existing BE schemes need to specify the receiver list for a particular message. In many scenarios, it is very hard to know the complete receiver list, and it is desirable to be able to encrypt without exact knowledge of possible receivers. Also, existing BE schemes [33, 68] can only support a simple receiver list. It is hard to support flexible, expressive access control policies. A broadcast encryption with an attribute-based mechanism was proposed in [150], where an expressive attribute-based access policy replaces the flat receiver list. Also, in [55, 56], the authors proposed to use a CP-ABE [25, 56] and flat-table [49] mechanism to minimize the number of messages and support expressive access policies. Compared with these works, the presented scheme significantly reduces the size of ciphertext from linear to constant.

3.3 MODELS

This section first describes how to use attributes to form a data access policy, followed by the concept of the broadcast encryption based on an attribute-based mechanism; the *bilinear* map is presented, in which it is the building block of ABE schemes. Finally, the complexity assumption that will be used for the security proof is presented.

3.3.1 ATTRIBUTES, POLICY, AND ANONYMITY

Let $U = \{A_i\}_{i \in [1,k]}$ be the *universe* of attributes in the system. Each A_i has three values: $\{A_i^+, A_i^-, A_i^*\}$. When a user u joins the system, u is tagged with an attribute list defined as follows:

Definition 3.1. *A user's attribute list is defined as $L = \{L[i]_{i \in [1,k]}\}$, where $L[i] \in \{A_i^+, A_i^-\}$ and k is the number of attributes in the universe.*   □

Intuitively, A_i^+ denotes the user has A_i; A_i^- denotes the user does not have A_i or A_i is not a proper attribute of this user. For example, suppose $U = \{A_1 = \text{CS}, A_2 = \text{EE}, A_3 = \text{Faculty}, A_4 = \text{Student}\}$. Alice is a student in the CS department; Bob is a faculty member in the EE department; Carol is a faculty member holding a joint position in both the EE and CS departments. Their attribute lists are illustrated in Table 3.1.

As the actual data access policy is hidden in the ciphertext header, effective measures are required to avoid ambiguity. In other words, when a decryptor receives a ciphertext header without knowing the access policy, he/she should NOT try a large number of access policies when performing decryption. To this end, the solution adopts an AND-gate policy construction so that each decryptor only needs to try once on each ciphertext header.

The hidden AND-gate access policy is defined as below:

Definition 3.2. *Let $W = \{W[i]\}_{i \in [1,k]}$ be an AND-gate access policy, where $W[i] \in \{A_i^+, A_i^-, A_i^*\}$. It uses the notation $L \models W$ to denote that the attribute list L of a user*

Table 3.1

Attribute examples

Attributes	$L[1]$	$L[2]$	$L[3]$	$L[4]$
Description	CS	EE	Faculty	Student
Alice	A_1^+	A_2^-	A_3^-	A_4^+
Bob	A_1^-	A_2^+	A_3^+	A_4^-
Carol	A_1^+	A_2^+	A_3^+	A_4^-

satisfies W, as:

$$L \models W \iff W \subset L \bigcup \{A_i^*\}_{i \in [1,k]}.$$

□

A_i^+ or A_i^- requires the exact same attribute in the user's attribute list. As for A_i^*, it denotes a wildcard value, which means the policy does not care about the value of attribute A_i. Effectively, each user with either A_i^+ or A_i^- fulfills A_i^* automatically.

Accordingly, we also define an anonymized AND-gate policy that removes all identifying attribute values, i.e., $\{A_i^+, A_i^-\}$, except do-not-care values, i.e., A_i^*. Formally, we define an anonymized AND-gate policy as follows:

Definition 3.3. *Let $\overline{W} = W \cap \{A_i^*\}_{i \in [1,k]}$ be an anonymized AND-gate access policy.*

It must be noted that the do-not-care attribute values are included in the anonymized access policy. If we hide the wildcard attributes, the decryptor will need to guess 2^k possible access policies if there are k attributes in the policy, i.e., for each attribute, its value can be either A_i^* or the specific value (A_i^+ or A_i^-) assigned to the decryptor. This would make the scheme infeasible in terms of performance. As an example shown in Table 3.2, to specify an access policy W_1 for all CS students and an access policy W_2 for all CS people:

Table 3.2

An example of the access policies and anonymized policies

Attributes	$W[1]$	$W[2]$	$W[3]$	$W[4]$
Description	CS	EE	Faculty	Student
W_1	A_1^+	A_2^-	A_3^-	A_4^+
$\overline{W}_1$	✠	✠	✠	✠
W_2	A_1^+	A_2^-	A_3^*	A_4^*
$\overline{W}_2$	✠	✠	A_3^*	A_4^*

* where ✠ represents *"do not care"*.

The anonymity policy is defined as *the state of being not identifiable within a set of subjects, i.e., the anonymity set.* As the access policy is one-to-many mapped to users, we can define "anonymity set" as:

Definition 3.4. *The anonymity set of a blinded policy $\overline{W}$ is the set of access policies that are identically blinded to $\overline{W}$.*

Here, we briefly analyze the anonymity level of the blinded access policy. First, if there are no wildcards in the original access policy (hidden), the blinded policy $\overline{W}$ will be empty. In this case, the size of anonymity set is 2^k, as there are 2^k possible access policies blinded to $\overline{W}$. If there are j wildcards in the original access policy (hidden), the size of anonymity set is 2^{k-j}.

3.3.2 BROADCAST WITH ATTRIBUTE-BASED ENCRYPTION

A broadcast encryption is usually applied in the scenario wherein a broadcaster sends messages to multiple receivers through an insecure channel. The broadcaster should be able to select a subset of users with certain policies from all receivers, and consequently only the eligible users are able to decrypt the ciphertexts and read the messages. It is possible that the number of all possible receivers is infinite, and the subset of privileged receivers changes dramatically in each broadcast based on the content of the message and the will of the broadcaster.

The notion of Attribute-based Encryption [186] can be utilized to address this problem. In ABE, all the possible receivers are ascribed by an attribute set. As such, the broadcaster can specify an expressive policy and select a group of privileged receivers defined by their attributes. Consequently, only the receivers whose attributes satisfy the policy embedded into the access structure are able to decrypt the ciphertexts transmitted through the insecure broadcast channel.

3.3.3 BILINEAR MAPS

A pairing is a bilinear map $e : \mathbb{G}_0 \times \mathbb{G}_0 \to \mathbb{G}_1$, where $\mathbb{G}_0$ and $\mathbb{G}_1$ are two multiplicative cyclic groups with large prime order p. The discrete logarithm problem on both $\mathbb{G}_0$ and $\mathbb{G}_1$ is hard. Pairing has the following properties:

- *Bilinearity*:

$$e(P^a, Q^b) = e(P, Q)^{ab}, \quad \forall P, Q \in \mathbb{G}_0, \forall a, b \in \mathbb{Z}_p^*.$$

- *Nondegeneracy*:
  $e(g, g) \neq 1$, where g is the generator of $\mathbb{G}_0$.
- *Computability*:
  There exist an efficient algorithm to compute the pairing.

3.3.4 COMPLEXITY ASSUMPTION

The security of the presented constructions is based on a complexity assumption called the Bilinear Diffie-Hellman Exponent assumption (BDHE) [30].

Let $\mathbb{G}_0$ be a *bilinear* group of prime order p. The K-BDHE problem in $\mathbb{G}_0$ is stated as follows: given the following vector of $2K + 1$ elements (note that the $g^{\alpha^{K+1}}$ is not in the list):

$$(h, g, g^{\alpha}, g^{(\alpha^2)}, \cdots, g^{\alpha^K}, g^{\alpha^{K+2}}, \cdots, g^{\alpha^{2K}}) \in \mathbb{G}_0^{2K+1},$$

as the input and the goal of the computational K-BDHE problem is to output $e(g,h)^{\alpha^{(K+1)}}$. We can denote the the the set as:

$$Y_{g,\alpha,K} = \{g^{\alpha}, g^{(\alpha^2)}, \cdots, g^{\alpha^K}, g^{\alpha^{K+2}}, \cdots, g^{\alpha^{2K}}\}.$$

Definition 3.5. *(Decisional K-BDHE) The decisional K-BDHE assumption is said to be held in $\mathbb{G}_0$ if there is no probabilistic polynomial time adversary who is able to distinguish*

$$< h, g, Y_{g,\alpha,K}, e(g,h)^{\alpha^{(K+1)}} >$$

and

$$< h, g, Y_{g,\alpha,K}, e(g,h)^R >$$

with non-negligible advantage, where $\alpha, R \in \mathbb{Z}_p$ and $g, h \in \mathbb{G}_0$ are chosen independently and uniformly at random. □

3.4 PP-CP-ABE CONSTRUCTION

In this section, the construction of the PP-CP-ABE scheme is presented.

3.4.1 PP-CP-ABE CONSTRUCTION OVERVIEW

The PP-CP-ABE scheme consists of four fundamental algorithms:

- **Setup$(1^{\lambda}, k)$**
  The **setup** algorithm takes input of the security parameter 1^{λ} and the number of attributes in the system k. It returns a public key PK and a master key MK. The public key is used for encryption while the master key is used for private key generation.
- **KeyGen(PK, MK, L)**
  The **KeyGen** algorithm takes the public key PK, the master key MK, and the user's attribute list L as input. It outputs the private key of the user.
- **Encrypt(PK, W, M)**
  The **encrypt** algorithm takes the public key PK, the specified access policy W, and the message M as input. The algorithm outputs ciphertext CT such that only a user with an attribute list satisfying the access policy can decrypt the message. The ciphertext also associates the anonymized access policy $\overline{W}$.

- **Decrypt(PK, SK, CT)**
  The **decrypt** algorithm decrypts the ciphertext when the user's attribute list satisfies the access policy. It takes the public key PK, the private key SK of the user, and the ciphertext CT, which only includes the anonymized access policy $\overline{W}$ as input. It returns a valid plaintext M if $L \models W$, where L is the user's attribute list and W is the access policy hidden from the ciphertext.

Boneh et al. proposed a broadcast encryption construction with constant ciphertext size in [33], where the broadcast encryptor uses the public key list corresponding to intended receivers to perform encryption. To make the ciphertext constant, each receiver's public key is multiplied together, assuming a multiplicative group structure. Thus, the resulting ciphertext is still an element on the group, i.e., the size of the ciphertext is constant. We use a similar strategy to achieve constant ciphertext in the presented scheme.

In the construction, each public key is mapped to an attribute value, including A_i. To encrypt a message, the encryptor specifies an access policy W by assigning an attribute value ($A_i \in \{1, 0, *\}$) for each of the n attributes in the Universe and encrypts the message using public keys of the attribute values in the W. Each decryptor is generated as a set of private key components corresponding to his/her attribute list L. All the private key components of the same user are tied together by a common random factor to prevent collusion attacks.

3.4.2 SETUP

Assuming there are k attributes $\{A_1, A_2, \cdots, A_k\}$ in the system, it has $K = 3k$ attributes values since each attribute A_i has three values: $\{A_i^+, A_i^-, A_i^*\}$. For ease of presentation, the attribute values can be mapped to integer numbers as depicted in the Table 3.3.

Table 3.3

Mapping attribute values to numbers

Attributes	A_1	A_2	A_3	$\cdots$	A_k
A_i^+	1	2	3	$\cdots$	k
A_i^-	$k+1$	$k+2$	$k+3$	$\cdots$	$2k$
A_i^*	$2k+1$	$2k+2$	$2k+3$	$\cdots$	$3k$

Trusted Authority (TA) first chooses two *bilinear* groups $\mathbb{G}_0$ and $\mathbb{G}_1$ of prime order p (such that p is λ bits long) and a *Bilinear* map $e : \mathbb{G}_0 \times \mathbb{G}_0 \to \mathbb{G}_1$. TA then picks a random generator $g \in \mathbb{G}_0$ and a random $\alpha \in \mathbb{Z}_p$. It computes $g_i = g^{(\alpha^i)}$ for $i = 1, 2, \cdots, K, K+2, \cdots, 2K$, where $K = 3k$. Next, TA picks a random $\gamma \in \mathbb{Z}_p$ and sets $v = g^\gamma \in \mathbb{G}_0$. The public key is:

$$PK = (g, g_1, \ldots, g_K, g_{K+2}, \ldots, g_{2K}, v) \in \mathbb{G}_0^{2K+1}.$$

The master key $MK = \{\gamma, \alpha\}$ is guarded by the TA.

3.4.3 KEY GENERATION

Each user u is tagged with the attribute list $L_u = \{L_u[i]_{i \in [1,k]}\}$ when joining the system, where $1 \leq L_u[i] \leq 2k$. The TA first selects k random numbers $\{r_i\}_{i \in [1,k]}$ from $\mathbb{Z}_p$ and calculates $r = \sum_{i=1}^{k} r_i$.

The TA computes $D = g^{\gamma r} = v^r$. For $\forall i \in [1,k]$, TA calculates $D_i = g^{\gamma(\alpha^{L_u[i]} + r_i)} = g_{L_u[i]}^{\gamma} \cdot g^{\gamma r_i}$ and $F_i = g^{\gamma(\alpha^{2k+i} + r_i)} = g_{2k+i}^{\gamma} \cdot g^{\gamma r_i}$.

The private key for user u is computed as:

$$SK_u = (D, \{D_i\}_{i \in [1,k]}, \{F_i\}_{i \in [1,k]}).$$

3.4.4 ENCRYPTION

The encryptor picks a random t in $\mathbb{Z}_p$ and sets the one-time symmetric encryption key, $Key = e(g_K, g_1)^{kt}$. Suppose AND-gate policy is W with k attributes. Each attribute is either positive/negative or wildcards.

The encryptor first encrypts the message using the symmetric key Key as $\{M\}_{Key}$. The encryptor also sets $C_0 = g^t$. Then, it calculates $C_1 = (v \prod_{j \in W} g_{K+1-j})^t$. Also, the encryptor anonymizes the access policy W by removing all attribute values except do-not-care values, i.e., A_i^*, and outputs $\overline{W} = W \bigcap \{A_i^*\}_{i \in [1,k]}$.

Finally, the ciphertext is:

$$CT = (\overline{W}, \{M\}_{Key}, g^t, (v \prod_{j \in W} g_{K+1-j})^t)$$

$$= (\overline{W}, \{M\}_{Key}, Hdr),$$

where the ciphertext header $Hdr = \{C_0, C_1\}$.

3.4.5 DECRYPTION

Before performing decryption, the decryptor u has has little information about the access policy that enforced the ciphertext. Only if $L_u \models W$ can u successfully recover the valid plaintext and access policy. Otherwise, u can only get a random string, which can be easily detected. Moreover, the access policy remains unknown to the unsuccessful decryptors.

First of all, u constructs a local guess of access policy, denoted as $\widetilde{W}$, as specified in Algorithm 3.1. Essentially, this algorithm constructs only one guess by replacing hidden attributes in the anonymized access policy $\overline{W}$ with the corresponding attribute values of the receiver. If the receiver satisfies the access policy, Algorithm 3.1 will always produce the correct guess and the decryption will succeed. On the other hand, if a guess is not identical to the actual access policy, the decryption will fail and the decryptor does not need to try other guesses.

Algorithm 3.1 Construct local guess $\widetilde{W}$

Initialize $\widetilde{W} = \overline{W}$
for $i = 1$ to k **do**
   if $\overline{W}[i] == $ ✠ **then**
      $\widetilde{W}[i] = L_u[i]$;
   end if
end for
return $\widetilde{W}$;

For $\forall i \in [1,k]$, u calculates the T_0 and T_1 as follows.

$$T_0 = e(g_{\widetilde{W}[i]}, C_1)$$
$$= e(g^{\alpha^{\widetilde{W}[i]}}, g^{t(\gamma + \sum_{j \in \widetilde{W}} \alpha^{K+1-j})})$$
$$= e(g,g)^{t\gamma\alpha^{\widetilde{W}[i]} + t\sum_{j \in \widetilde{W}} \alpha^{K+1-j+\widetilde{W}[i]}};$$

and if $\widetilde{W}[i] \in L_u$, u computes:

$$T_1 = e(D[i] \cdot \prod_{j \in \widetilde{W}, j \neq \widetilde{W}[i]} g_{K+1-j+\widetilde{W}[i]}, C_0)$$
$$= e(g^t, g^{\gamma(\alpha^{\widetilde{W}[i]} + r_i) + \sum_{j \in \widetilde{W}, j \neq \widetilde{W}[i]} \alpha^{K+1-j+\widetilde{W}[i]}})$$
$$= e(g,g)^{t\gamma(\alpha^{\widetilde{W}[i]} + r_i) + t\sum_{j \in \widetilde{W}, j \neq \widetilde{W}[i]} \alpha^{K+1-j+\widetilde{W}[i]}};$$

else, if $\widetilde{W}[i] \in \{A_i^*\}_{i \in [1,k]}$, u computes:

$$T_1 = e(F[i] \cdot \prod_{j \in \widetilde{W}, j \neq \widetilde{W}[i]} g_{K+1-j+\widetilde{W}[i]}, C_0)$$
$$= e(g^t, g^{\gamma(\alpha^{\widetilde{W}[i]} + r_i) + \sum_{j \in \widetilde{W}, j \neq \widetilde{W}[i]} \alpha^{K+1-j+\widetilde{W}[i]}})$$
$$= e(g,g)^{t\gamma(\alpha^{\widetilde{W}[i]} + r_i) + t\sum_{j \in \widetilde{W}, j \neq \widetilde{W}[i]} \alpha^{K+1-j+\widetilde{W}[i]}}.$$

Then, calculate

$$T_0/T_1 = e(g,g)^{-t\gamma r_i + t\alpha^{K+1}}.$$

After u calculate all k terms, the approach makes a production of all the quotient terms and gets:

$$e(g,g)^{-t\gamma(r_1 + r_2 + \cdots + r_k) + kt\alpha^{k+1}} = e(g,g)^{-t\gamma r + kt\alpha^{K+1}}.$$

u calculates:

$$e(D, C_0) = e(g,g)^{t\gamma r}.$$

Then, u produces these two terms and obtains $Key = e(g,g)^{kt\alpha^{K+1}} = e(g_K, g_1)^{kt}$ and decrypts the message. If the decrypted message is valid, $\overline{W} = W$ and u decrypt the ciphertext successfully. Otherwise, u has no information on the W, and the anonymity set of $\overline{W}$ does not change.

3.4.6 SECURITY ANALYSIS

In the security analysis, it reduces Chosen Plaintext Attack (CPA) security of the presented scheme to decisional K-BDHE assumption. Let's first define the decryption proxy to model collusion attackers.

Security Game for PP-CP-ABE

A CP-ABE scheme is considered to be secure against chosen CPA if no probabilistic polynomial-time adversaries have non-negligible advantages in this game.

Init: The adversary choose the challenge access policy W and give it to challenger.

Setup: The challenger runs the setup algorithm and gives the adversary the PK.

Phase 1: The adversary submits L for a KeyGen query, where $L \not\models W$. The challenger answers with a secret key SK for L. This can be repeated adaptively.

Challenge: The challenger runs the encrypt algorithm to obtain $\{< C_0, C_1 >, Key\}$. Next, the challenger picks a random $b \in \{0,1\}$. It sets $Key_0 = Key$ and picks a random Key_1 with same length to Key_0 in $\mathbb{G}_1$. It then gives $\{< C_0, C_1 >, Key_b\}$ to the adversary.

Phase 2: Same as Phase 1.

Guess: The adversary outputs its guess $b' \in \{0,1\}$, and it wins the game if $b' = b$.

Note that the adversary may make multiple secret key queries both before and after the challenge, which results in the collusion resistance in the presented scheme. It should be noted that this CPA security game is called as selective ID security, because the adversary must submit a challenge access structure before the setup phase.

Theorem 3.1. *If a probabilistic polynomial-time adversary wins the CPA game with non-negligible advantage, then it can construct a simulator that distinguish a K-DBHE tuple with non-negligible advantage.* □

Proof of Theorem 3.1. *The proof reduces CPA security of the presented scheme to decisional K-BDHE assumptions. It first defines the decryption proxy to model collusion attackers.*

Definition 3.6. *(Decryption Proxy) In order to model the collusion attacks, a 2k decrypting proxy is defined in the security game. Each decrypting proxy $p_i(r) = g^{\gamma(\alpha^i + r)}$, where $r \in \mathbb{Z}_p$ and $i \in \{1, \cdots, 2k\}$, i.e., a private key component corresponding to a particular attribute value.*

In collusion attacks against access policy W, a user with attribute list $L \not\models W$ colludes with $x \leq k$ decryption proxies to attack the ciphertext. The colluding with x decryption proxy is defined as x-collusion. Intuitively, x-collusion means the attacker needs x attributes values, say $\{i_1, i_2, \cdots, i_x\}$ to add to his attribute list L such that $L \cup \{i_1, i_2, \cdots, i_x\} \models W$. Note that 0-collusion means no decryption proxy is used and the user does not collude.

Suppose that an adversary $\mathscr{A}$ wins the selective game for PP-CP-ABE with the advantage ε. Then, a simulator $\mathscr{B}$ is constructed to break decisional K-BDHE assumption with the advantage $\max\{\varepsilon/2, (1 - q/p)^l \varepsilon/2, (1 - (1 - (1 - q/p)^l)^m)\varepsilon/2\}$. The simulator $\mathscr{B}$ takes an input a random decisional K-BDHE challenge

$$< h, g, Y_{g,\alpha,K}, Z >,$$

where Z is either $e(g,h)^{\alpha^{(K+1)}}$ or a random element on $\mathbf{G}_0$. $\mathscr{B}$ now plays the role of challenger in the pre-defined CPA game:

Init: $\mathscr{A}$ sends to $\mathscr{B}$ the access policy W that $\mathscr{A}$ wants to be challenged.
Setup: $\mathscr{B}$ runs the **setup** algorithm to generate PK. $\mathscr{B}$ chooses random $d \in \mathbb{Z}_p$ and generates:

$$v = g^d \Big(\prod_{j \in W} g_{K+1-j} \Big)^{-1} = g^{d - \Sigma_{j \in W} \alpha^{K+1-j}} = g^\gamma.$$

The $\mathscr{B}$ outputs the PK as:

$$PK = (g, Y_{g,\alpha,K}, v) \in \mathbf{G}_0^{2K+1}.$$

Phase 1: The adversary $\mathscr{A}$ submits an attribute list L for a private key query, where $L \not\models W$. Otherwise, the simulator quits.

The simulator $\mathscr{B}$ first selects k random numbers $r_i \in \mathbb{Z}_p$ for $i = 1 \ldots k$ and set $r = r_1 + \cdots + r_k$. Then, $\mathscr{B}$ generates

$$D = \Big(g^d \prod_{j \in W} (g_{K+1-j})^{-1} \Big)^r$$
$$= g^{(d - \Sigma_{j \in W} \alpha^{K+1-j})r}$$
$$= g^{\gamma r}.$$

Then, for $\forall i \in [1, k]$ and $W[i]! = L[i]$, $\mathscr{B}$ generates:

$$D_i = g_{L[i]}^d \prod_{j \in W} (g_{K+1-j+L[i]})^{-1} g^{u r_i} \prod_{j \in W} (g_{K+1-j})^{-r_i},$$

Then, for $\forall i \in [1, k]$ and $W[i]! = A_i^*$, $\mathscr{B}$ generates:

$$F_i = g_{2k+i}^d \prod_{j \in W} (g_{K+1-j+2k+i})^{-1} g^{u r_i} \prod_{j \in W} (g_{K+1-j})^{-r_i},$$

Note that each for each D_i or F_i is valid since:

$$D_i = \Big(g^d \Big(\prod_{j \in W} g_{K+1-j} \Big)^{-1} \Big)^{(\alpha^{L[i]} + r_i)} = g^{\gamma(\alpha^{L[i]} + r_i)},$$

and

$$F_i = (g^d(\prod_{j \in W} g_{K+1-j})^{-1})^{(\alpha^{2k+i}+r_i)} = g^{\gamma(\alpha^{2k+i}+r_i)}.$$

Challenge: *The simulator $\mathcal{B}$ sets $< C_0, C_1 >$ as $< h, h^d >$. It then gives the challenge $\{< C_0, C_1 >, Z^k\}$ to $\mathcal{A}$.*

To see the validity of challenge, $C_0 = h = g^t$ for some unknown t. Then:

$$\begin{aligned} h^d &= (g^d)^t \\ &= (g^d \prod_{j \in W} (g_{K+1-j})^{-1} \prod_{j \in W} (g_{K+1-j}))^t \\ &= (v \prod_{j \in W} (g_{K+1-j}))^t, \end{aligned}$$

and if $Z = e(g,h)^{\alpha^{(K+1)}}$, then $Z^k = Key$.

Phase 2: *Repeat as* **Phase 1**.

Guess: *The adversary $\mathcal{A}$ output a guess b' of b. When $b' = 0$, $\mathcal{A}$ guesses that $Z = e(g,h)^{\alpha^{(K+1)}}$. When $b' = 1$, $\mathcal{A}$ guesses Z is a random element.*

If Z is a random element, then the $\mathbf{Pr}[\mathcal{B}(h,g,Y_{g,\alpha,K},Z) = 0] = \frac{1}{2}$.

Before considering the case when $Z == e(g,h)^{\alpha^{(K+1)}}$, how to use decryption proxy in the proof must be explained. Each decryption proxy, $p_i(r)$, simulates a legal private key component embedded with random number r. When calling $p_i(r)$, $\mathcal{A}$ passes a random r as a guess of the $r_{i'}$, which is the random number embedded in the D_i or F_i, where $i \in W$. As a matter of fact, the procedure of calling the decryption proxy mimics the collusion of multiple users, who combine their private key components.

Lemma 3.1. *Suppose the $\mathcal{A}$ has issued q private queries and there is only one attribute $i \notin W$, $\mathcal{A}$ queries $p_i(r)$ l times. The possibility that the none of the queries returns a legal private key component of any q is $(1 - q/p)^l$.*   □

Proof of Lemma 3.1. *The possibility that the one query does not return a legal private key component of any q is $1 - q/p$. Thus, if none of the l query succeed, the probability $\mathbf{Pr}[r \neq r_{i'}] = (1 - q/p)^l$, where r is the random number in decryption proxy, $r_{i'}$ is the random number embedded in the private key, q is the number of private key queries in phase 1 and phase 2, l is the number of calling decryption proxy with different r, and p is the order of $\mathbb{Z}_p$.*   □

Lemma 3.2. *Suppose the $\mathcal{A}$ has issued q private queries and there is m attributes violate the W, $\mathcal{A}$ queries each of the m decryption proxy $p_{i_1}(r_1), p_{i_2}(r_2), \cdots, p_{i_m}(r_m)$ l times. The possibility that the none of the queries returns a legal private key component of any q is $(1 - (1 - q/p)^l)^m$.*   □

Proof of Lemma 3.2. *The probability that 1 decryption proxy fails is $\mathbf{Pr}[r \neq r_{i'}] = (1 - q/p)^l$. The probability that all the m decryption proxy successfully return legal components is $(1 - (1 - (q/p)^l))^m$. In the case when not all m succeed, the probability is $\mathbf{Pr}[r_{i_j} \neq r_{i'_j}, \exists j \leq m] = 1 - (1 - (1 - q/p)^l)^m$.*   □

If $Z == e(g,h)^{\alpha^{(K+1)}}$, the following cases are to be considered:

- 0-collusion: If no decryption proxy is used, $\mathscr{A}$ has at least $\varepsilon/2$ advantage in breaking the scheme, then $\mathscr{B}$ has at least ε advantage in breaking K-BDHE, i.e.,

$$|\mathbf{Pr}[\mathscr{B}(h,g,Y_{g,\alpha,K},Z) = 0] - \frac{1}{2}| \geq \varepsilon/2.$$

- 1-collusion: If 1 decryption proxy, say $p_i(r)$ is used, $\mathbf{Pr}[r \neq r_{i'}] = (1-q/p)^l$, where r is the random number in decryption proxy, $r_{i'}$ is the random number embedded in the private key, q is the number of private key queries in phase 1 and phase 2, l is the number of calling decryption proxy with different r, and p is the order of $\mathbb{Z}_p$. Note that if $r = r_{i'}$, $\mathscr{A}$ can use $p_i(r)$ as a valid private key component to compromise the ciphertext.
  If the $\mathscr{A}$ has at least ε advantage in breaking the scheme, then $\mathscr{B}$ has at least $(1-q/p)^l\varepsilon/2$ advantage in breaking K-BDHE.
- m-collusion: If m decryption proxies, say

$$p_{i_1}(r_1), p_{i_2}(r_2), \cdots, p_{i_m}(r_m)$$

are used. The possibility that $\mathbf{Pr}[r_{i_j} \neq r_{i'_j}, \exists j \leq m] = (1-(1-(q/p)^l))^m$, where r_m is the random number in m decryption proxy $p_{i_m}(r_{i_m})$ for the private key component i_m, $r_{i'_m}$ is the random number generated for the $\mathscr{A}$, q is the number of private key queries in phase 1 or phase 2, l is the number of calling m decryption proxies with different r's, and p is the order of $\mathbb{Z}_p$.
  If the $\mathscr{A}$ has at least ε advantage in breaking the scheme, then $\mathscr{B}$ has at least $(1-(1-(1-q/p)^l)^m)\varepsilon/2$ advantage in breaking K-BDHE.

This concludes the proof.   □

3.5 PRIVACY-PRESERVING ATTRIBUTE-BASED BROADCAST ENCRYPTION

Based on the construction of PP-CP-ABE, we construct an efficient and flexible Broadcast Encryption (BE) scheme–Privacy-Preserving Attribute-Based Broadcast Encryption (PP-AB-BE), where the size of any single ciphertext is still constant.

Compared to existing BE schemes, using PP-AB-BE, encryptor does not need to store a large number of key materials, i.e., public key and private key. By carefully organizing the attributes in the system, we will show that the storage overhead of each user can be reduced from $O(N)$ to $O(\log N + m)$, where N is the number of users in the system, and $m \ll N$ is the number of descriptive attributes in the system.

Also, in PP-AB-BE, an encryptor enjoys the flexibility of encrypting broadcast data using either a specific list of decryptors or an access policy without giving an exact list of decryptors.

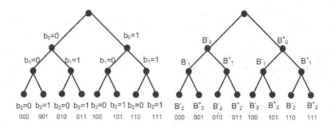

Figure 3.1: An illustration of ID and bit-assignment attributes distribution.

3.5.1 PP-AB-BE SETUP

In PP-AB-BE with N users, each user is issued an n-bit binary ID $b_0b_1\cdots b_n$, where b_i represents the i'th bit in the user's binary ID, where $n = \log N$. Accordingly, we can define n bit-assignment attributes $\{B_1, B_2, \cdots, B_n\}$. Each user is assigned n bit-assignment attribute values according to his/her ID. If the $b_i = 1$, he/she is assigned the B_i^+, if the $b_i = 0$, he/she is assigned the B_i^-. For example, in a system with eight possible users, each user is assigned three bit-assignment attributes to represent the bit values in their ID, as illustrated in Figure 3.1.

Given the $n = \log N$ the bit-assignment attributes, the TA generates $3n$ attributes values, i.e., bit-assignment attribute B_i has $\{B_i^+, B_i^-, B_i^*\}$ values.

In addition to the bit-assignment attributes, the TA also chooses m descriptive attributes for the system. These descriptive attributes present the real properties or features of an entity, which can be used to describe the decryptors' social or role features, e.g., "CS", "EE", "Student", "Faculty", etc. Each of the m descriptive attributes has $\{1, 0, *\}$ values.

With the $3n + 3m$ attribute values, the authority runs the **Setup(n + m)** algorithm and generates public keys and private keys.

3.5.2 BROADCAST ENCRYPTION

In order to control the access to the broadcasted message, the sender needs to specify an access policy using either the descriptive attributes or bit-assignment attributes. For example, in Table 3.4, if Alice wants send message to all CS students, she can specify the descriptive policy W_1 in the table. Or she wants to send message to Bob and Carol, whose IDs are 100 and 101, respectively, she can use the bit-assignment policy W_2, which is equivalent to enumerate every receivers.

Here, we focus on how an encryptor can specify the list of receivers explicitly using n bit-assignment attributes. We first define some of the terms used in the following presentations:

- *Literal*: A variable or its complement, e.g., $b_1, \overline{b_1}$, etc.
- *Product Term*: Literals connected by AND, e.g., $\overline{b_2}b_1\overline{b_0}$.
- *Sum-of-Product Expression (SOPE)*: Product terms connected by OR, e.g., $\overline{b_2}b_1b_0 + b_2$.

Table 3.4

Sample policies

	CS	EE	Student	Faculty	B_0	B_1	B_2
W_1	A_1^+	A_2^-	A_3^+	A_4^-	B_0^*	B_1^*	B_2^*
W_2	A_1^*	A_2^*	A_3^*	A_4^*	B_0^+	B_1^-	B_2^*

Given the set of receivers S, the membership function $f_S()$, which is in the form of SOPE, specifies the list of receivers:

$$f_S(b_1^u, b_2^u, \ldots, b_n^u) = \begin{cases} 1 & \text{iff } u \in S, \\ 0 & \text{iff } u \notin S. \end{cases}$$

For example, if the subgroup $S = \{000, 001, 011, 111\}$, then $f_S = \bar{b}_0\bar{b}_1\bar{b}_2 + \bar{b}_0\bar{b}_1 b_2 + \bar{b}_0 b_1 b_2 + b_0 b_1 b_2$.

Then, the broadcast encryptor runs the Quine-McCluskey algorithm [157] to reduce f_S to minimal SOPE f_S^{min}. The reduction can consider *do not care* values $*$ on those IDs that are not currently assigned to any receiver to further reduce the number of product terms in the membership function. For example, if $S = \{000, 001, 011, 111\}$, $f_S^{min} = \bar{b}_0\bar{b}_1 + b_1 b_2$.

Since f_S^{min} is in the form of SOPE, encryption is performed on each product term. That is, for each product term E in f_S^{min}, the encryptor specifies an AND-gate access policy W using the following rules:

1. For positive literal $b_i \in f_S^{min}$, set B_i^+ in the access policy W.
2. For negative literal $\bar{b}_i \in f_S^{min}$, set B_i^- in the access policy W.
3. Set B_i^* for the rest of bit-assignment attributes.

For each W, the encryptor uses **Encrypt(PK, W, M)** algorithm to encrypt the message. The total number of encrypted message equals to the number of product terms in f_S^{min}.

For example, if $S = \{000, 001, 011, 111\}$, $f_S^{min} = \bar{b}_0\bar{b}_1 + b_1 b_2$. The access policies W_1 and W_2 are shown in Table 3.5:

Table 3.5

Access policy example

	CS	EE	Student	Faculty	B_0	B_1	B_2
W_1	A_1^*	A_2^*	A_3^*	A_4^*	B_0^-	B_1^-	B_2^*
W_2	A_1^*	A_2^*	A_3^*	A_4^*	B_0^*	B_1^+	B_2^+

We can find that f_S^{min} contains two product terms: the message M for S can be encrypted into two ciphertexts with W_1 and W_2, respectively.

3.5.3   INFORMATION THEORETICAL OPTIMALITY

In this section, we present the optimality of PP-AB-BE through an information theoretical approach similar to the models in [175]. In Section 3.5.3, we proved that PP-AB-BE attains information theoretical lower bound of storage requirements with $O(\log N)$ bit-assignment attributes. In Section 3.5.3, we also compare the BGW [33] BE scheme [33] and PP-AB-BE from a theoretical perspective.

Optimal Storage

To be uniquely identified, each user's ID should not be a prefix of any other user's, i.e., *prefix-free*. For example, suppose a user u' is issued an ID 00, which is a prefix of u_1 with ID 000 and u_2 with ID 001. When an encryptor tries to reach u_1 and u_2, the minimized membership function is $M = \overline{B}_0\overline{B}_1$, which is also satisfied by u'. Similarly, it is also imperative that a user's bit-assignment attributes should not be a subset of any other users'. The prefix-free condition is a necessary and sufficient condition for addressing any user with their bit-assignment attributes.

Theorem 3.2. *[62] For any instantaneous code (prefix code) over an alphabet of size D, the codeword lengths $\ell_1, \ell_2, \cdots, \ell_m$ must satisfy the inequality*

$$\sum_{i=1}^{N} D^{-\ell_i} \leq 1.$$

Proof of Theorem 3.2. *[62] Consider a D-ary tree in which each node has D children. Let the branches of the tree represent the symbols of the codeword. For example, the D branches arising from the root node represent the D possible values of the first symbol of the codeword. Then each codeword is represented by a leaf on the tree. The path from the root traces out the symbols of the codeword.*

The prefix condition on the codewords implies that no codeword is an ancestor of any other codeword on the tree. Hence, each codeword eliminates its descendants as possible codewords.

Let ℓ_{max} be the length of the longest codeword of the set of codewords. Consider all nodes of the tree at level ℓ_{max}. Some of them are codewords, some are descendants of codewords, and some are neither. A codeword at level ℓ_i has $D^{\ell_{max}-\ell_i}$ descendants at level ℓ_{max}. Each of these descendant sets must be disjoint. Also, the total number of nodes in these sets must be less than or equal to $D^{\ell_{max}}$. Hence, summing over all the codewords, we have $\sum D^{\ell_{max}-\ell_i} \leq D^{\ell_{max}}$ or $\sum D^{-\ell_i} \leq 1$.

Theorem 3.3. *If we denote the number of bit-assignment attributes for a user u_i by l_i. For a broadcast encryption system with N users and satisfy the prefix-free condition, the set $\{l_1, l_2, \ldots, l_N\}$ satisfies the Kraft inequality:*

$$\sum_{i=1}^{N} 2^{-l_i} \leq 1.$$

□

Proof of Theorem 3.3. *Refer to proof of Theorem 3.2 and let* $D = 2$.

Assuming l_i bit-assignments are required to identify u_i and the probability to send a message to u_i is p_i, we can model the storage overhead as:

$$\sum_{i=1}^{N} p_i l_i. \tag{3.1}$$

Intuitively, this formation argues that the storage overhead from a sender's perspective is the average number of bit-assignments required to address to any particular receiver. Thus, an optimization problem is formulated to minimize the storage overhead for a broadcast encryption system:

$$\min_{l_i} \sum_{i=1}^{N} p_i l_i$$

s.t.

$$\sum_{i=1}^{N} 2^{-l_i} \leq 1.$$

This optimization problem is identical to the optimal codeword-length selection problem [62] in information theory. Before giving the solution to this optimization problem, we define the entropy of targeting one user in the system:

Definition 3.7. *The entropy H of targeting a user is*

$$H = -\sum_{i=1}^{N} p_i \log p_i.$$

□

Theorem 3.4. *For a system of N users with prefix-free distribution of bit-assignments, the optimal (i.e., minimal) storage overhead required for a sender to address a receiver, written as $\sum_{i=1}^{N} p_i l_i$, can be given by the binary entropy:*

$$H = -\sum_{i=1}^{N} p_i \log p_i.$$

□

Proof of Theorem 3.4. *The theorem is equivalent to an optimal codeword-length selection problem and proof is available in [62].* □

Since the average number of bit-assignment attributes required for addressing one particular receiver is given by the entropy of targeting a user, we now try to derive the upper and lower bounds of the entropy:

$$\max_{p_i} -\sum_{i=i}^{N} p_i \log p_i$$

and

$$\min_{p_i} -\sum_{i=i}^{N} p_i \log p_i$$

s.t.

$$\sum_{i=1}^{N} p_i = 1.$$

The upper bound $H_{max} = -\sum_{i=1}^{N} \frac{1}{N} \log N = \log N$ is yielded when $p_i = 1/N$, $\forall i \in \{1, 2, \dots, N\}$, when each user has equal possibility to be addressed as the receiver. When there is no apriori information about the probability distribution of targeting one of the users, $l = H_{max} = \log_d N$ correspond to the optimal strategy to minimize the average storage overhead required for each user. On the other hand, the lower bound $H_{min} = 0$ is achieved when $p_i = 1$ for $\exists i \in \{1, 2, \dots, N\}$, which is an extreme case where there is no randomness and only one user is reachable.

Compare with BGW BE scheme

We denote the "optimal bit-assignment attributes assignment" as using the least number of bit-assignment for attributes to identify each user. We can refer to the BGW scheme in [33] as maximalist. In a BGW scheme, for a system with N users, each user is mapped to a unique public key. Given all N public keys, the number of combinations is $2^N - 1$, which equals the number of receiver subsets in the system. Thus, each encryptor needs a maximal number of public keys to perform broadcast encryption.

To compare the minimalist and maximalist storage strategy, we can define entropy of an attribute or a public key defined as:

$$H(p) = p \log p^{-1} + (1 - p) \log(1 - p)^{-1}.$$

where p as the percentage of totals users who have this attribute or public key. We see the entropy of each attribute in minimalist strategy as $H(1/2) = 1$, since for each particular attribute, exact half of the users have it while the other half do not have it. On the other hand, the entropy of a public key in a maximalist strategy is $H(1/N) = (1/N)\log(N) + ((N-1)/N)\log(N/(N-1)) < 1$. Hence, we can conclude that minimalist strategy attains maximal binary entropy while the maximalist strategy attains minimal binary entropy.

3.6 SYSTEM PERFORMANCE ASSESSMENT

In this section, we analyze the performance of PP-AB-BE and compare it with several related solutions: a subset-difference broadcast encryption scheme (Subset-Diff) [81], BGW [33], and FT implemented using CP-ABE (FT-ABE) [55]. We also compare some works in tree-based multicast group key distribution domain where a group controller removes some group members by selectively multicasting key update messages to all remaining members. Those solutions can be broadly divided into two categories: flat-table (FT) schemes [49] and non-flat-table schemes, including OFT [197], LKH [224], and ELK [171].

Table 3.6

Comparison of communication overhead and storage overhead in different broadcast encryption schemes and group key management schemes

Scheme	Communication Overhead		Storage Overhead	
	single receiver	multiple receivers	Center	User
PP-AB-BE	$O(1)$	$\approx O(\log N)$	N/A	$O(\log N + m)$
Subset-Diff	$O(t^2 \cdot log^2 t \cdot \log N)$	$O(t^2 \cdot log^2 t \cdot \log N)$	$O(N)$	$O(t \log t \log N)$
BGW$_1$	$O(1)$	$O(1)$	N/A	$O(N)$
BGW$_2$	$O(N^{\frac{1}{2}})$	$O(N^{\frac{1}{2}})$	N/A	$O(N^{\frac{1}{2}})$
Flat-Table	$O(\log N)$	$\approx O(\log N)$	$O(\log N)/O(N)$	$O(\log N)$
Flat-Table-ABE	$O(\log N)$	$\approx O(\log^2 N)$	$O(\log N)/O(N)$	$O(\log N)$
Non-Flat-Table-Tree	$O(\log N)$	$O(l \cdot \log N)$	$O(N)$	$O(\log N)$

N: the number of group members; l: the number of leaving members; t: maximum number of colluding users to compromise the ciphertext.

3.6.1 COMMUNICATION OVERHEAD

The complexity analysis of communication overhead for various schemes is summarized in Table 3.6. In a Subset-Diff scheme, the communication overhead is $O(t^2 \cdot log^2 t \cdot \log N)$, with t as maximum number of colluding users to compromise the ciphertext. For a BGW scheme, the message size is $O(N^{\frac{1}{2}})$ as reported in [33]. In an ACP scheme, the size of the message depends on the degree of access control polynomial, which equals the number of current receivers. Thus, the message size is $O(N)$.

For non-flat-table tree-based multicast key distribution schemes such as OFT [197], LKH [224], ELK [171], etc., the communication overhead for removing members depends on the number of keys in the tree that need to be updated [202, 171]. In the case of removing a single member, $O(\log N)$ messages are required since the center needs to update $\log N$ auxiliary keys distributed to the removed member. Some tree-based schemes tried to optimize the number of messages to update all the affected keys in the case of multiple *leaves*. In ELK [171], which is known to be one of the most efficient tree-based schemes, the communication overhead for multiple *leaves* is $O(a - l)$, where $a \approx l \log N$ is the number of affected keys and l is the number of leaving members. Thus, the complexity can be written as $O(l \log N)$.

For flat-table tree-based scheme [49], the complexity of removing a single member is also $O(\log N)$. The main benefit of flat-table, however, is the minimal number of messages for batch-removing multiple members. In fact, the scheme requires the same number of messages compared to flat-table scheme, thus they both achieved information theoretical optimality. However, flat-table is vulnerable to collusion attacks. In [55], the authors proposed to implement flat-table using CP-ABE [25] to counter collusion attacks.

To control a set of receivers S using PP-AB-BE, the number of messages depends on the number of product terms in the f_S^{min}. In [189], the authors derived an upper bound and lower bound on the average number of product terms in a minimized SOPE. Experimentally, the average number of messages required is $\approx \log N$ [55].

Number of Messages: Worst Cases

We examine some cases when maximal number of messages is required to reach multiple receivers.

Lemma 3.3 (multiple receivers worst case). *The worst case of reaching multiple receivers happens when both of following conditions hold: 1) the number of distinct receivers is $N/2$; 2) the Hamming distance between IDs of any two receivers is at least 2. In the worst case, the number of key updating messages is $N/2$.*   □

Proof of Lemma 3.3. *[49] To show this, first consider the case when the number of departing members is greater or equal to $N/2$ and thus the number of remaining members is $N/2$ or less. Clearly, the number of messages required is no more than the number of remaining members, since at worst one has to send out one mesage to update a remaining member or one message per minterm of the membership function. Hence, in this case the number of messages required will be at most $N/2$.*

Now consider the case when the number of departing members is less than $N/2$ and thus, the number of remaining memebers is greater than $N/2$. Looking at Figure 3.2, for every additional member that remains in the group or, equivalently, for each additional minterm of the membership function, there exists (at least) one previously existing minterm of the function with which it can be grouped. Grouping can be done in a systematic way so that every additional member is grouped with one and only one of the existing ones. For example, by pairing with the member with the largest UID smaller than that of itself. This argument holds even if the existing $N/2$ minterms are not placed as shown in Figure 3.2, since moving any of the minterms shown to any other position in the table will reduce the number of messages by 1.

In this case, the number of messages is $N - N/2 = N/2$ using PP-AB-BE. However, we can see that the worst cases happens in extremely low probability:

Lemma 3.4 (worst-case possibility). *When communicating all subgroups with uniform opportunity, the worst-case scenario happens with probability $\frac{1}{2^{N-1}}$.*   □

Proof of Lemma 3.4. *In the worst case, the Hamming distance of IDs of $N/2$ receivers should be at least 2. As shown in the Karnaugh table in Figure 3.2, each cell represents an ID. For any cell marked 0 and any cell marked 1, the Hamming distance is at least 2. Thus, the worst cases happen in two cases: (1) the encryptor wants to reach $N/2$ receivers marked 1 in Figure 3.2; (2) the encryptor wants to reach $N/2$ receivers marked 0 in Figure 3.2.*

                                                                                        □

Lemma 3.5. *[49] Excluding C_1 and C_2, every member of the secure multicast group can decrypt at least one of the n messages.*

b_0b_1 \ b_2b_3	00	01	11	10
00	1	0	1	0
01	0	1	0	1
11	1	0	1	0
10	0	1	0	1

Figure 3.2: Worst cases of broadcast encryption to $N/2$ receivers.

Proof of Lemma 3.5. *Consider an arbitrary remaining member of the group C, with UID $y_{n-1}y_{n-2}\cdots y_1$, which is obviously different from the IDs of C_1 and C_2. Let m be the highest-order bit in which the IDs of C and C_1 differ, i.e.,*

$$y_i = x_i, i = n-1, n-2, \cdots, m+1, y_m = \bar{x}_m$$

If $m < n-1$, C possesses both k_{m+1} and $\bar{k}_m$ and, therefore, can decrypt the $(n-(m+1))$th message.

If $m = n-1$, that is $y_{n-1} = \bar{x}_{n-1}$, let ℓ be the lowest-order bit in which the IDs of C and C_1 match, that is,

$$y_\ell = x_\ell, 0 \le \ell < n-1, y_i = \bar{x}_i, i = \ell-1, \ell-2, \cdots, 0.$$

If $\ell = 0$, C possesses both k_0 and $\bar{k}_{n-1}$, so it can decrypt the nth message. Otherwise, C possesses both k_ℓ and $\bar{k}_{\ell-1}$, so it can decrypt the $(n-\ell)$th message.

Theorem 3.5. *[49] Re-keying a secure multicast group of size 2^n when two group members are to be removed requires at most n messages.*

Proof of Theorem 3.5. *[49] If the IDs of the two users, denoted by C_1 and C_2, differ in all bits (i.e., have maximum Hamming distance), Lemma 3.5 applies.*

Otherwise, the IDs have at least one bit in common; let this be X_i. Observe that a message encrypted with the key that corresponds to the complement of X_i, i.e., k_i if $X_i = 0$ or $\bar{k}_i$ if $X_i = 1$, is sufficient to distribute the new keying information to half (2^{n-1}) of the group members, while excluding C_1 and C_2. The remaining $2^{n-1} - 2$ members, which also belong to the group, together with C_1 and C_2, all have X_i as the i-th bit in their IDs. Hence, this bit can be effectively ignored, and thus the problem reduces itself to that of removing two users from a group of 2^{n-1} members, whose IDs have $(n-1)$ bits.

This procedure can be applied recursively, yielding one message for every common bit in C_1 and C_2. After the i-th message, 2^{n-i} users are left, including C_1 and C_2, the solution is comprised of two steps.

- *Generate k messages which convey the new keying information to $2^n - 2^{n-k}$ clients.*

- *Re-key a group of 2^{n-k} clients whose IDs are of size $(n-k)$ bits and where the Hamming distance between the IDs of the two users removed is maximum. As shown in Lemma 3.5, this problem is solved with $(n-k)$ messages.*

Summing over the two steps, this solution requires n messages. If we assume, without loss of generality that the common bits are the k first, as follows,

$$C_1 = x_{n-1}x_{n-2}\cdots x_{n-k}x_{n-(k+1)}\cdots x_0$$

$$C_2 = x_{n-1}x_{n-2}\cdots x_{n-k}\bar{x}_{n-(k+1)}\cdots \bar{x}_0,$$

then these messages can be expressed as follows.

$\{SK(r+1)\}_{f(\bar{k}_{n-1})}, \{SK(r+1)\}_{f(\bar{k}_{n-2})}, \{SK(r+1)\}_{f(\bar{k}_{n-k})},$
$\{SK(r+1)\}_{f(\bar{k}_{n-(k+1)},\bar{k}_{n-(k+2)})}, \cdots, \{SK(r+1)\}_{f(k_0,\bar{k}_{n-(k+1)})}$

Note that if $k = n-1$, the k first messages are sufficient, as can be easily verified.

We also have the worst case for communicating the majority of users.

Lemma 3.6 (Worst case of reaching $N-2$ receivers). *When reaching $N-2$ receivers, the maximal number of messages required is $n = \log N$, when the Hamming distance between 2 non-receivers is n.*   □

Proof of Lemma 3.6. *Please refer to [49].*   □

Number of Messages: Average Case

To investigate the average case, we simulated PP-AB-BE in a system with 512 users and 1024 users, and the number of messages required are shown in Figure 3.3 and Figure 3.4, respectively. In the simulation, we consider the cases of 0%, 5%, 25%, and 50% IDs that are not assigned (i.e., *do not care* value). For each case, different percentages of receivers are randomly selected from the group. We repeat 100 times to average the results. As shown in Figure 3.3 and Figure 3.4, PP-AB-BE achieves roughly $O(\log N)$ complexity, where the number of messages is bounded by $9\log N$ for the 512-member group and $18\log N$ for the 1024-member group.

Total Message Size

Finally, we look into the message size of PP-AB-BE, with comparison to FT-CP-ABE [55], which are presented in Figures 3.5 and 3.6. As mentioned in [55], in FT-CP-ABE, the size of ciphertext grows linearly based on the increase of the number of attributes in the access policy [55, 25]. Experimentally, the message size in FT-CP-ABE starts at about 630 bytes, and each additional attribute adds about 300 bytes. In a system with 10 bit ID or 1024 users, the number of attributes using FT-CP-ABE ciphertext is at most 10 and the message size may be as large as $630 + 9 \cdot 300 = 3330$ bytes. Since the number of attributes in the access policy is bounded by $\log N$, we can conclude that the communication overhead of FT-CP-ABE is in the order

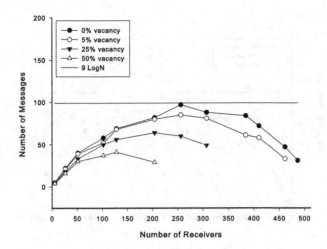

Figure 3.3: Number of messages in a system with 512 users.

of $O(\log^2 N)$. In PP-AB-BE, every ciphertext contains exactly two group members on $\mathbb{G}_0$. Empirically, the size of one element on $\mathbb{G}_0$ is about 128 bytes. Thus, the ciphertext header in PP-AB-BE is bounded within 300 bytes, which is significantly smaller than the ciphertext size reported in FT-CP-ABE [55]. Moreover, since the component C_0 in the ciphertext can be shared by multiple messages, we can further reduce the message size of PP-AB-BE with efficient communication protocol design.

3.6.2 STORAGE OVERHEAD

In PP-AB-BE, there are $6\log N + 1$ elements on $\mathbb{G}_0$ in the *PK*. Also, a user needs to store $m \ll N$ descriptive attributes. Thus, the storage overhead is $O(\log N + m)$, assuming a user does not store any IDs of other users. Although the broadcast encryptor may need the list of receivers' IDs along with the list of *do not care* IDs to perform Boolean function minimization, we can argue that this does not incur extra storage overhead.

- The encryptors do not need to store the receiver's IDs after the broadcast; thus, the storage space can be released.
- The TA can periodically publish the minimized SOPE of all *do not care* IDs, which can be used by encryptors to further reduce number of messages.
- If IDs are assigned to users sequentially, i.e., from low to high, TA can simply publish the lowest unassigned IDs to all users, who can use all the higher IDs as *do not care* values.
- Even if a user needs to store N IDs, the space is merely $N\log N$ bits. If $N = 2^{20}$.

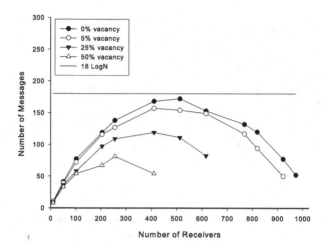

Figure 3.4: Number of messages in a system with 1024 users.

- If a broadcast encryptor cannot utilize *do not care* values to further reduce the membership function in SOPE form, the communication overhead might be a little higher. As shown in Figure 3.3 and Figure 3.4, the curve of 0% vacancy can also be used as the number of messages required if a broadcast encryptor does not know the *do not care* IDs.

3.6.3 COMPUTATION OVERHEAD

In this section, we compare the computation overhead of those asymmetric key-based schemes and the summarized results are presented in Table 3.7. In ACP scheme, the author reports that the encryption needs $O(N^2)$ finite field operations when the subgroup size is N; in the BGW scheme, the encryption and decryption require $O(N)$ operations on the *bilinear* group, which are heavier than finite field operations [99]. In PP-AB-BE, each encryption requires $\log N$ operations on the $\mathbb{G}_0$, and the decryption requires $2\log N + 1$ pairings and $\log N(\log N - 1) + \log N$ operations on $\mathbb{G}_0$ and $\log N$ operations on $\mathbb{G}_1$. Thus, the complexities of encryption and decryption are bounded by $O(\log N)$. Although the problem of minimizing SOPE is NP-hard, efficient approximations are widely known. Thus, PP-AB-BE is much more efficient than ACP and BGW when group size is large.

In Table 3.8, we summarize the computation overhead based on the benchmark evaluations for PP-CP-ABE operations. The benchmark was performed on a modern workstation, which has a 3.0GHz Pentium 4 CPU with 2MB cache and 1.5 GB memory and runs Linux 2.6.32 kernel. In the performance evaluation, the Type-D curve [152] is used in the testing. We run each of the algorithm 100 times, and the result is the average value. Since the encryption algorithm only requires $\log N$ operations

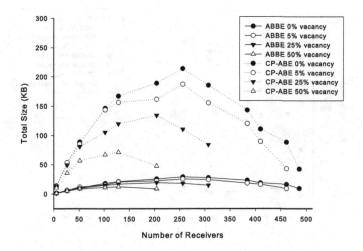

Figure 3.5: Total size of messages in a system with 512 users.

Table 3.7

Comparison of computation complexity in different broadcast encryption schemes

Scheme	Computation Overhead	
	Encryption	Decryption
PP-AB-BE	$O(\log N)$	$O(\log N)$
BGW	$O(M)$	$O(M)$
ACP	$O(M^2)$	$O(1)$

N: the number of group members; M: the number of receivers.

on the $\mathbb{G}_0$ group, the encryption time differences between the 1024 group and 4096 group is very small. On the other hand, the decryption algorithm requires $2\log N + 1$ expensive pairings operations, and $\log N(\log N - 1) + \log N$ operations on the $\mathbb{G}_0$ group. Thus, the decryption on 4096 groups requires 4 more pairing operations than the decryption on 1024 group and each pairing requires around 20 ms in the experiment. Overall, the experiment results are consistent without complexity analysis.

3.7 SUMMARY

In this chapter, a Constant Ciphertext Policy Attribute-Based Encryption (PP-CP-ABE) was presented. Compared with existing CP-ABE constructions, PP-CP-ABE significantly reduces the ciphertext size from linear to constant and supports

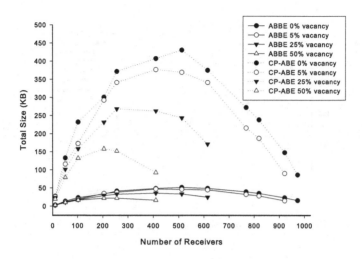

Figure 3.6: Total size of messages in a system with 1024 users.

Table 3.8

Computation overhead for 1024 **and** 4096 **group**

	1024 group	4096 group
Encrypt (ms)	12	13
Decrypt (ms)	360	455

expressive access policies. Thus, PP-CP-ABE can be used in many communication-constrained environments.

Based on PP-CP-ABE, we further presented an Attribute-Based Broadcast Encryption (PP-AB-BE) scheme that attains information with theoretical minimal storage overhead. Thus, a storage-restricted user can easily pre-install all required key materials to perform encryption and decryption. Through theoretical analysis and simulation, we compared PP-AB-BE with many existing BE solutions, and we showed that PP-AB-BE achieve better trade-offs between storage and communication overhead.

The security of PP-CP-ABE is based on selective-ID attackers. One open problem is constructing constant CP-ABE that is secure against adaptive adversaries. Another limitation is the PP-CP-ABE is constructed and proved following the BGW [33] model. We are looking for new constructions with equal or a stronger security level. Also, in this chapter, we only proved PP-AB-BE is minimalist in terms of storage overhead. We are working on more information-theoretical analysis that considers both storage and communication overheads in BE schemes.

The future research of this work will have two directions: First, the presented solution only supports conjunctive access policy. An important improvement is to extend access policies to support more flexible forms, e.g., including disjunctive normal form and non-monotonic form. Second, the wildcard attribute is not hidden in the access policy to avoid ambiguity for the decryptor; an interesting enhancement is to support a complete hidden access policy without needing to identify the involved wildcard attributes.

4 Identity Revocable CP-ABE

Cipher-Text Policy Attribute-Based Encryption (CP-ABE) provides a flexible data access control approach, where access policies are enforced based on users' assigned attributes. Attributes can be used to compose security policies in an attribute policy tree structure to define a privilege that a user needs to have in order to decrypt cipher-text. Using CP-ABE, attributes' keys are considered as long-term keys, and they are predistributed. During a communication, a secure group does not need to be pre-defined, and data access policies are defined in the attribute policy tree sent along with the ciphertext. Thus, attributes are used to define users' access groups that provide group-based access control rather than a fine-grained access control approach requiring a group setup phase by existing ID-based or role-based solutions.

CP-ABE provides an efficient group-based access control by removing the group setup phase; however, it suffers three major attribute management issues. First, it is not always able to derive one or multiple attribute policy trees to construct an access policy to include desired group members; second, multiple attribute policy trees may be required to define an access group, in which the number of involved attributes can be very large; and third, there is no existing CP-ABE-based solutions allowing to revoke one or multiple users who have been assigned attributes and corresponding private keys. To address these issues, in this chapter, we present a new CP-ABE scheme called Identify Revocable CP-ABE (IR-CP-ABE) that incorporates an ID-Based Revocation capability. IR-CP-ABE can be used to revoke subgroup members defined by one or multiple attribute policy trees in order to construct access policies for all possible subgroups; moreover, IR-CP-ABE can be used to reduce attributes used by attribute policy trees; finally, IR-CP-ABE can revoke one or multiple group members. By the end of this chapter, we provide a security proof for IR-CP-ABE and evaluate the computation, storage, and communication overhead incurred in IR-CP-ABE scheme.

4.1 INTRODUCTION

Some roots of Ciphertext-Policy Attribute-Based Encryption (CP-ABE) can be traced back to the introduction of Identity-Based Encryption (IBE) in [32], which could be considered a special case of CP-ABE. In IBE, an identity or ID is a string one-to-one mapped to each user as his/her public key, and a user can acquire a private key corresponding this ID from Trusted Authority (TA). The encryption-decryption scheme in IBE is one-to-one, where the message encrypted by a particular ID can only decrypted by the user with corresponding private key. Accordingly, IBE enables secure fine-grained peer-to-peer communication for a single recipient while preventing other ineligible users from decrypting the ciphertext.

To securely communicate with multiple recipients, the message has to be separately encrypted for each recipient, thereby bringing significant system overhead in multi-user communication scenario.

CP-ABE is a milestone in the cryptographic research field in the past ten years, and it is an extension from IBE by enabling expressive access policy to control the decryption process [25]. Its popularity over other ABE schemes due to its flexible key management and data access control can be easily incorporated into many existing data access mechanisms. It constructs an attribute policy tree structure to defined a group of recipients who can access to the root of the tree through any given tree leaves (i.e., incorporated attributes) rather than using individual identities. Therefore, CP-ABE allows secure coarse-grained group communication with less key management overhead. However, one major issue of using CP-ABE is due to the fact that it is not always able to derive one or multiple attribute policy trees to construct an access policy to include all required group members. CP-ABE cannot pinpoint individual recipients when attributes are shared by multiple users. For example, there is an attribute set $\{student, male, female\}$ assigned to a group of students $\{Alice, Bob, Carol\}$. It is obvious that there is no way to construct an attribute policy tree for student subgroups such as $\{Alice, Bob\}$, $\{Bob, Carol\}$ by using the given attributes. The second issue of CP-ABE is that sometimes, over-complicated attribute literal expressions in conjunctive normal form or disjunctive normal form are needed to construct a specific group. For example, there may exist multiple attribute policy trees to define one subgroup, in which the worst case of involved attributes is linearly proportional to the subgroup size; i.e., each individual user has a unique attribute. Finally, there is no existing CP-ABE-based solutions allowing revocation of one or multiple users who have been assigned attributes and corresponding private keys.

To address these described issues, in this chapter, we present a new Identity-Revocable CP-ABE scheme (short for IR-CP-ABE) that incorporates an ID-based revocation capability to enforce secure group communications on both coarse-grained and fine-grained levels. Revoking users' IDs has been presented by Lewko *et. al* in a broadcast encryption [136] system, which designed ID revocation methods in public key broadcast encryption systems. Yu *et al.* [52] and Zhou *et al.* [245] proposed an attribute-based data-sharing scheme, where each attribute gets three distinct values for its positive form, negative form, as well as "don't care" form. This approach is inefficient since a user needs to have an attribute form for all the attributes that have been involved in the system. IR-CP-ABE relieves this restriction by incorporating identity revocation into CP-ABE scheme, where attributes are allocated normally according to users' privilege, and a unique ID is assigned to each user. It works by first specifying attribute literals in conjunctive/disjunctive normal forms as an attribute policy tree to cover the recipients of the target group with the minimum redundancy, and then removing the ineligible recipients by incorporating ID revocation. As such, it simplifies the attribute literals in some cases and makes it possible to build all possible subgroups that may not be constructed solely by attributes.

4.1.1 RESEARCH CONTRIBUTION

Most existing research work relies on either key regeneration or complicated tree structure based on presumed relationships between users and attributes, thereby resulting in overwhelming overhead in a dynamic system where users frequently join or leave. We propose an effective identity-revocable CP-ABE scheme to enable efficient group-policy management by incorporating user revocation mechanism into CP-ABE cryptosystem. In the presented scheme, key regeneration for non-revoked users is not needed, and identities and attributes are uncorrelated, thereby ensuring the greatest flexibility. The presented scheme does not change the storage, computation, and communication complexity of CP-ABE scheme. Moreover, key delegation feature is realized, such that the subordinates enjoy same-access privileges as their delegation authority, and these subordinates' access privileges can be revoked when their delegation authority's ID is revoked. Thus, the presented approach provides great flexibility for delegation revocation that involves a large number of group members. The key features of the presented solutions are summarized in follows:

- IR-CP-ABE is an integrated scheme to provide a large group access control through CP-ABE, and in the meantime, it also provides revocations of individual and multiple users to prune the group defined by the CP-ABE access policies;
- IR-CP-ABE simplifies the group policy construction by reducing the number of attributes and/or attribute policy trees;
- IR-CP-ABE provides delegation services, and thus a large group of users can be evicted by revoking delegator's ID;
- The presented solution is proved to be secure under selective security model;
- Performance analysis and evaluation shows that IR-CP-ABE does not increase the storage, computation, and communication complexity of existing CP-ABE.

4.2   FLEXIBLE GROUP CONSTRUCTION

We herein illustrate how IR-CP-ABE simplifies the attribute literal expression of group construction and even makes it possible where the group construction cannot solely rely on attribute literals.

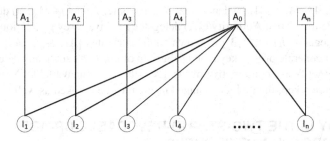

Figure 4.1: An example of multiple users sharing one attribute.

In Figure 4.1, $A_i(i \in [1,n])$ denotes the i-th attribute and $I_j(j \in [1,n])$ denotes the j-th user identity, where n is the total number of attributes or users. We can construct the following user group $\{I_1, \cdots, I_n\} \setminus I_k (k \in [1,n])$ by using only the attributes in the form of *Sum-of-Product Expression* as follows:

$$\{I_1, I_2, \cdots, I_n\} \setminus I_k = A_1 + \cdots + A_{k-1} + A_{k+1} + \cdots + A_n.$$

In IR-CP-ABE, it can be simplified as follows:

$$\{I_1, I_2, \cdots, I_n\} \setminus I_k = A_0 \bar{I}_k,$$

where $\bar{I}_k$ represents that user I_k is revoked, or it represents a negative ID attribute and in this specific example, the complexity of the total number of involved attribute literals is reduced from $O(n)$ to $O(1)$.

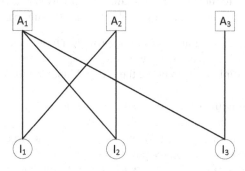

Figure 4.2: An example of subgroups cannot be established based on users' assigned attributes.

In Figure 4.2, $A_i(i \in [1,3])$ denotes the i-th attribute and $I_j(j \in [1,3])$ denotes the j-th user identity. This example shows that it is impossible to construct the user groups $\{I_1, I_3\}$, $\{I_2, I_3\}$ by using only attributes. Nonetheless, with the help of identity revocation, they can be expressed in the form of *Sum-of-Product Expression* as follows:

$$\{I_1, I_3\} = A_1 \bar{I}_2,$$
$$\{I_2, I_3\} = A_1 \bar{I}_1.$$

In Figure 4.3, $A_i(i \in [1,4])$ denotes the i-th attribute, and $I_j(j \in [1,12])$ denotes the j-th user identity. Trust Authority (TA) delegates its private key generation privilege to a few delegators $I_k(k \in [1,3])$ who own all the attributes $\{A_1, A_2, A_3, A_4\}$, and each of them can generate private keys for a subgroup of users. As such, any delegator can temporarily work as an alternative of the TA if the TA is unavailable. Note that if a delegator is revoked, then all its managed users must be revoked as well.

4.3 WHY IS THE TWO-STEP ID-REVOCABLE CP-ABE APPROACH NOT SECURE?

Here, we present one naïve solution by simply constructing a two-step encryption solution that combines a CP-ABE scheme and an identity revocation scheme, and then show why this approach is not secure.

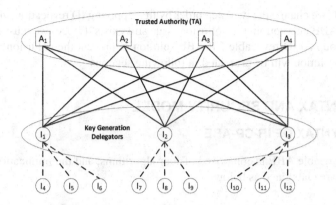

Figure 4.3: An example of attribute delegation.

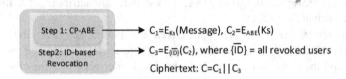

Figure 4.4: A naïve two-step revocation scheme.

The two-step revocation framework is presented in Figure 4.4 and is described as follows:

- *CP-ABE*: Enforcing CP-ABE-based access control by applying CP-ABE scheme [25][220] and encryption and decryption algorithms.
- *Identity-Based Revocation*: $\forall U_{ID} \in \{ID\}$, where $\{ID\}$ represents the set including all users' IDs, and $\{\overline{ID}\}$ is the collection of users' IDs to be revoked. This enforces ID-based access control by imposing the constraint that specific users $\forall U_{ID} \in \{\overline{ID}\}$ cannot access the encrypted data, and K_s is a short-term data encrypting key.
- *Attribute-ID-Based Revocation Scheme*: This applies two-step encryption, i.e., the inner layer provides attribute-based encryption to enforce the security policies applied on ciphertext; the outer layer provides the membership access control on the encrypted data.

The loosely-coupled two-step ID-revocable ABE solution is actually subject to collusion attacks. Assume a CP-ABE system has two attributes $\{A_1 = teacher, A_2 = student\}$ and two users $\{ID_1 = Alice, ID_2 = Bob\}$, where *Alice* is a teacher and *Bob* is a student. The access control policy $A_1 \bar{I}_1$ states that all the teachers except *Alice* can decrypt the ciphertext. Nonetheless, the two ineligible users *Alice* and *Bob* can collude to decrypt the ciphertext as follows: i) *Bob* first derives C_2 by decrypting C_3 and passes C_2 to *Alice*; ii) *Alice* then can derive K_s by decrypting C_2 and get

Message. If we change the two-step order by first applying ID revocation and second using CP-ABE encryption, the collusion issue still exists. Hence, we must construct a tightly-coupled ID-revocable CP-ABE solution to prevent the collusion attack, in which the solution will be presented in following sections.

4.4   SYNTAX AND SECURITY MODEL

4.4.1   SYNTAX OF IR-CP-ABE

The ID-revocable ABE is comprised of four algorithms, and the implications of the algorithm are elaborated as below:

- *Setup*($\mathscr{U}$, $\mathscr{I}$): Given the attribute universe description $\mathscr{U}$ and the identity set $\mathscr{I}$, the TA publishes its public key *PK* but keeps its master key *MSK*;
- *KeyGen*(*MSK*, *S*, *ID*): Given *MSK*, the user's ID and attribute set *S*, the TA issues private keys *SK*;
- *Encrypt*(*PK*, (*M*, ρ), $\mathscr{M}$, {*ID*$_j$}): Given the public key *PK*, the LSSS matrix *M* and its corresponding mapping ρ to each attribute, the message $\mathscr{M}$, and the revoked ID set {*ID*$_j$}, the data owner generates the ciphertext and sends it to a public place for storage;
- *Decrypt*(*CT*, *SK*): Given the ciphertext *CT*, the data user derives the message $\mathscr{M}$ by decrypting with its private key *SK*.

4.4.2   SECURITY MODEL

We now present the full security definition for IR-CP-ABE systems which derive from the security definitions for identity-based revocation framework [136] and general CP-ABE systems [220]. In the IR-CP-ABE security definition, we need to consider stronger adversaries whose attributes satisfy the attribute access policy of the challenge ciphertext but whose identity is in the revocation set.

Setup.   The challenger runs the setup algorithm and gives the public parameters, PK to the adversary.

Phase 1.   The adversary makes repeated private keys query corresponding sets of attributes $S_1, ..., S_{q1}$, where each attribute set is owned by an entity with the revoked identity *ID*.

Challenge.   The adversary submits two equal length messages $\mathscr{M}_0$ and $\mathscr{M}_1$. In addition, the adversary gives a challenge access structure *M* and a set $\mathbb{I}$ of revoked identities, such that $\mathbb{I}$ must include all identities that were queried. The challenger picks up a random coin *b*, and encrypts $\mathscr{M}_b$ under the access structure *M* and the revoked identity set $\mathbb{I}$. The ciphertext *CT* is given to the adversary.

Phase 2.   Phase 1 is repeated with the restriction that queried sets of attributes $S_{q1+1}, ..., S_q$ owned by the entity whose identity is in the revocation set $\mathbb{I}$.

Guess.   The adversary outputs a guess b' of b.

The advantage of an adversary $\mathscr{A}$ in this game is defined as $Pr[b' = b] - 1$. Note that this model can easily be extended to handle chosen-ciphertext attacks by allowing for decryption queries in Phase 1 and Phase 2.

Definition 4.1. An identity revocable CP-ABE scheme is secure if all polynomial time adversaries have at most a negligible advantage in the above game.

We say that a system is selectively secure if we add an Init stage before setup where the adversary commits to the challenge access structure M and the revocation ID set S. All of the constructions will be proved secure in the selective security model.

4.4.3 ASSUMPTIONS

We now present the q-type complexity assumptions that we will depend on to prove the security of the presented systems. This assumption is formulated on prime order *bilinear* groups, denoted by modified decisional q-parallel BDHE, which is similar to the Decisional Parallel *Bilinear* Diffie-Hellman Exponent (q-parallel BDHE) Assumption [220].

The modified decisional q-parallel *Bilinear* Diffie-Hellman problem is defined as follows. We select a group $\mathbb{G}$ of prime order p and a random generator g of $\mathbb{G}$ and random exponents $a, s, b_1, b_2, b_q \in \mathbb{Z}_p$. Given

$$y = \{g, g^s, g^a, g^{(a^q)}, , g^{(a^{q+2})}, \cdots, g^{(a^{2q})}\}$$
$$\forall_{1 \leq j \leq q} g^{a/b_j}, \cdots, g^{a^q/b_j}, , g^{a^{q+2}/b_j}, ..., g^{a^{2q}/b_j}$$
$$\forall_{1 \leq j \leq q} g^{a \cdot s/b_j}, \cdots, g^{(a^q \cdot s/b_j)},$$

it is hard to distinguish $e(g, g)^{a^{q+1}s} \in \mathbb{G}_T$ from any random element in $\mathbb{G}_T$.

An algorithm $\mathbb{B}$ that outputs $z \in \{0, 1\}$ has advantage ε in solving the modified decisional q-parallel BDHE in $\mathbb{G}$ if

$$|Pr[\mathbb{B}(y, T = e(g, g)^{a^{q+1}s}) = 0] - Pr[\mathbb{B}(y, T = R) = 0]| \geq \varepsilon.$$

The modified decisional q-parallel BDHE assumption holds if only negligible advantage exists for any algorithm to solve the modified decisional q-parallel BDHE problem in polynomial time.

Obviously, the generic security of the modified decisional q-parallel BDHE assumption can be easily reduced to the decisional q-parallel BDHE assumption.

4.5 SCHEME CONSTRUCTION

In this section, we present the IR-CP-ABE scheme. We first present the construction of one ID revocation, which is represented as OIDR-CP-ABE. A multiple-ID revocation scheme, denoted as MIDR-CP-ABE, is also presented. The main challenge to achieve multiple-ID revocation is due to the collusion problem.

4.5.1   ONE-ID REVOCATION FOR CP-ABE SCHEME (OIDR-CP-ABE)

In the following, four algorithms are presented for OIDR-CP-ABE scheme.

a. Setup$(\mathscr{U}, \mathscr{I})$

The *Setup* algorithm takes an attribute set $\mathscr{U}$ and an identity set $\mathscr{I}$ as inputs, where $|\mathscr{U}| = m$ and $|\mathscr{I}| = 1$. It chooses a group $\mathbb{G}$ of prime order p, a generator g, and m random group elements $h_1, h_2, \cdots, h_m \in \mathbb{G}$ that are associated with the m attributes in the system. It also chooses random exponents $\alpha, b \in \mathbb{Z}_p$.

Therefore, the public key is in the form:

$$PK = \{g, g^b, g^{b^2}, e(g,g)^\alpha, h_1^b, \cdots, h_m^b\}.$$

The master secret key is in the form:

$$MSK = \{\alpha, b\}.$$

b. KeyGen$(\mathbf{MSK}, \mathbf{S}, \mathbf{ID})$

S is the attribute set of user $ID \in \mathscr{I}$. *KeyGen* algorithm chooses a random $t \in \mathbb{Z}_p$ and generates secret keys for user ID as follows:

$$SK = (K = g^\alpha g^{b^2 t}, \{K_x = (g^{b \cdot ID} h_x)^t\}_{\forall x \in S}, L = g^{-t}).$$

c. Encrypt$(\mathbf{PK}, (\mathbf{M}, \rho), \mathscr{M}, \mathbf{ID_j})$

Encrypt algorithm takes inputs as an LSSS access structure (M, ρ) and the function ρ associates each row of M to corresponding attributes. ID_j is the identity to be revoked. Let M be an $l \times n'$ matrix. The *encrypt* algorithm first chooses a random vector $v = (s, y_2, \cdots, y_{n'}) \in \mathbb{Z}_p^{n'}$. These values will be used to share an encryption exponent s. For $x \in [1, l]$, it calculates $\lambda_x = v \cdot M_x$, where M_x is the vector corresponding to the x-th row of M. The *encrypt* algorithm chooses random $r_1, \cdots, r_l \in \mathbb{Z}_p$. Then, for message $\mathscr{M}$, the ciphertext is presented as follows:

$$C = \mathscr{M} e(g,g)^{\alpha s},$$

$$C_0 = g^s,$$

$$\hat{C} = \{C_k^* = g^{b \cdot \lambda_k}, C_k' = (g^{b^2 \cdot ID_j} h_{\rho(k)}^b)^{\lambda_k}\}_{\forall k = 1, \dots, l}.$$

d. Decrypt$(\mathbf{CT}, \mathbf{SK})$

CT is the input ciphertext with access structure (M, ρ), and SK is a private key for a set S:

$$CT = (C, C_0, \hat{C}, (M, \rho)).$$

Suppose that S satisfies the access structure and let $I \subset \{1, 2, \dots, l\}$ be defined as $I = \{i : \rho(i) \in S\}$. Let $\{\omega_i \in \mathbb{Z}_p\}_{i \in I}$ be a set of constants, such that if $\{\lambda_i\}$ are valid shares of any secret s according to M, then $\Sigma_{i \in I} \omega_i \lambda_i = s$. If the identity ID combined in the SK is not equal to the revocation identity ID_j in the ciphertext, we can perform

$$\frac{e(C_0, K)}{(\prod_{i \in I}[e(K_{\rho(i)}, C_i^*) \cdot e(L, C_i')]^{\omega_i})^{1/(ID - ID_j)}}$$

$$= e(g^s, g^\alpha g^{b^2 t}) / (\prod_{i \in I}[e((g^{b \cdot ID} h_{\rho(i)})^t, g^{b \cdot \lambda_i})$$
$$\cdot e(g^{-t}, (g^{b^2 \cdot ID_j} h_{\rho(i)}^b)^{\lambda_i})]^{\omega_i})^{1/(ID - ID_j)}$$

$$= e(g^s, g^\alpha) \cdot e(g^s, g^{b^2 t}) / (\prod_{i \in I}[e(g^{b \cdot ID \cdot t}, g^{b \cdot \lambda_i}) \cdot e(h_{\rho(i)}^t, g^{b \cdot \lambda_i})$$
$$\cdot e(g^{-t}, g^{b^2 \cdot ID_j \cdot \lambda_i}) \cdot e(g^{-t}, h_{\rho(i)}^{b \cdot \lambda_i})]^{\omega_i})^{1/(ID - ID_j)}$$

$$= e(g, g)^{\alpha s} \cdot e(g, g)^{b^2 st} / (\prod_{i \in I}[e(g^{b \cdot ID \cdot t}, g^{b \cdot \lambda_i}) \cdot$$
$$e(g^{-t}, g^{b^2 \cdot ID_j \cdot \lambda_i})]^{\omega_i})^{1/(ID - ID_j)}$$

$$= e(g, g)^{\alpha s} \cdot e(g, g)^{b^2 st}$$
$$\cdot 1 / (\prod_{i \in I}[e(g, g)^{b^2 t \lambda_i (ID - ID_j)}]^{\omega_i})^{1/(ID - ID_j)}$$

$$= e(g, g)^{\alpha s} \cdot e(g, g)^{b^2 st} / (\prod_{i \in I} e(g, g)^{b^2 t \lambda_i \omega_i})$$

$$= e(g, g)^{\alpha s} \cdot e(g, g)^{b^2 st} / e(g, g)^{b^2 t \Sigma_{i \in I} \lambda_i \omega_i}$$

$$= e(g, g)^{\alpha s}.$$

4.5.2 MULTIPLE-ID REVOCATION FOR CP-ABE SCHEME (MIDR-CP-ABE)

a. Setup$(\mathcal{U}, \mathcal{I})$

The algorithm takes an attribute set $\mathcal{U}$ and an identity set $\mathcal{I}$ as input where $|\mathcal{U}| = m$ and $|\mathcal{I}| = n$. It chooses a group $\mathbb{G}$ of prime order p, a generator p, and m random group elements $h_1, h_2, \cdots, h_m \in \mathbb{G}$ that are associated with the m attributes in the system. It also chooses random exponents $\alpha, b \in \mathbb{Z}_p$.

Therefore, the public keys are output as:

$$PK = \{g, g^b, g^{b^2}, e(g, g)^\alpha, h_1^b, \cdots, h_m^b\}.$$

The master secret key is: $MSK = \{\alpha, b\}$.

b. KeyGen$(\mathbf{MSK}, \mathbf{S}, \mathbf{ID})$

S is the attribute set of user $ID \in \mathcal{I}$. The algorithm chooses a random $t \in \mathbb{Z}_p$ and derives the secret keys as follows:

$$SK = (K = g^\alpha g^{b^2 t}, \{K_x = (g^{b \cdot ID} h_x)^t\}_{\forall x \in S}, L = g^{-t}).$$

c. Encrypt$(\mathbf{PK}, (\mathbf{M}, \rho), \mathcal{M}, \mathbf{S})$

It takes the input as an LSSS access structure (M, ρ) and the function ρ associates rows of M to attributes. ID_j is assumed to be the identity which will be revoked. Let M be an $l \times n'$ matrix. The algorithm first chooses a random vector $v = (s, y_2, \cdots, y_{n'}) \in \mathbb{Z}_p^{n'}$. These values will be used to share the encryption exponent

s. For $x \in [1, l]$, it calculates $\lambda_x = v \cdot M_x$, where M_x is the vector corresponding to the x-th row of M. Let $r = |S|$ and ID_j denote the j-th identity in S. The algorithm chooses random $\mu_1, ..., \mu_r \in \mathbb{Z}_p$ such that $\mu = \mu_1 + ... + \mu_r$. It generates the first part of ciphertext:

$$C = \mathcal{M} e(g,g)^{\alpha s \mu}, C_0 = g^{s \mu},$$

$$
\begin{aligned}
C_{1,1}^* &= g^{b \cdot \lambda_1 \mu_1}, \quad C_{1,1}' = (g^{b^2 \cdot ID_1} h_{\rho(1)}^b)^{\lambda_1 \mu_1} \\
\cdots \\
&\qquad C_{l,1}^* = g^{b \cdot \lambda_l \mu_1}, C_{l,1}' = (g^{b^2 \cdot ID_1} h_{\rho(l)}^b)^{\lambda_l \mu_1} \\
C_{1,2}^* &= g^{b \cdot \lambda_1 \mu_2}, \quad C_{1,2}' = (g^{b^2 \cdot ID_2} h_{\rho(1)}^b)^{\lambda_1 \mu_2} \\
&\qquad C_{l,2}^* = g^{b \cdot \lambda_l \mu_2}, C_{l,2}' = (g^{b^2 \cdot ID_2} h_{\rho(l)}^b)^{\lambda_l \mu_2} \\
&\qquad\qquad \cdots \\
C_{1,r}^* &= g^{b \cdot \lambda_1 \mu_r}, \quad C_{1,r}' = (g^{b^2 \cdot ID_r} h_{\rho(1)}^b)^{\lambda_1 \mu_r} \\
\cdots \\
&\qquad C_{l,r}^* = g^{b \cdot \lambda_l \mu_r}, C_{l,r}' = (g^{b^2 \cdot ID_r} h_{\rho(l)}^b)^{\lambda_l \mu_r}.
\end{aligned}
$$

d. Decrypt(CT, SK)

CT is the input ciphertext with access structure (M, ρ) and SK is a private key for a set S. Suppose that S satisfies the access structure and let $I \subset \{1, 2, ..., l\}$ be defined as $I = \{i : \rho(i) \in S\}$. Let $\{\omega_i \in \mathbb{Z}_p\}_{i \in I}$ be a set of constants such that if $\{\lambda_i\}$ are valid shares of any secret s according to M, then $\Sigma_{i \in I} \omega_i \lambda_i = s$. If the identity ID combined in the SK is not equal to the revocation identity ID_j in the ciphertext, we can perform

$$
\frac{e(C_0, K)}{\prod_{i \in I} (\prod_{j=1}^r [e(K_{\rho(i)}, C_{i,j}^*) \cdot e(L, C_{i,j}')]^{1/(ID-ID_j)})^{\omega_i}}
$$

$$
\begin{aligned}
&= e(g^{s\mu}, g^{\alpha} g^{b^2 t}) / \prod_{i \in I} (\prod_{j=1}^r [e((g^{b \cdot ID} h_{\rho(i)})^t, g^{b \cdot \lambda_i \mu_j}) \\
&\quad \cdot e(g^{-t}, (g^{b^2 \cdot ID_j} h_{\rho(i)}^b)^{\lambda_i \mu_j})]^{1/(ID-ID_j)})^{\omega_i} \\
&= e(g^{s\mu}, g^{\alpha}) \cdot e(g^{s\mu}, g^{b^2 t}) / \prod_{i \in I} (\prod_{j=1}^r [e(g^{b \cdot ID \cdot t}, g^{b \cdot \lambda_i \mu_j}) \\
&\quad \cdot e(h_{\rho(i)}^t, g^{b \cdot \lambda_i \mu_j}) \cdot e(g^{-t}, g^{b^2 \cdot ID_j \cdot \lambda_i \mu_j}) \\
&\quad \cdot e(g^{-t}, h_{\rho(i)}^{b \cdot \lambda_i \mu_j})]^{1/(ID-ID_j)})^{\omega_i} \\
&= e(g,g)^{\alpha s \mu} \cdot e(g,g)^{b^2 s \mu t} / \prod_{i \in I} (\prod_{j=1}^r [e(g^{b \cdot ID \cdot t}, g^{b \cdot \lambda_i \mu_j}) \\
&\quad \cdot e(g^{-t}, g^{b^2 \cdot ID_j \cdot \lambda_i \mu_j})]^{1/(ID-ID_j)})^{\omega_i} \\
&= e(g,g)^{\alpha s \mu} \cdot e(g,g)^{b^2 s \mu t} \\
&\quad \cdot 1 / \prod_{i \in I} (\prod_{j=1}^r [e(g,g)^{b^2 t \lambda_i \mu_j (ID-ID_j)}]^{1/(ID-ID_j)})^{\omega_i} \\
&= e(g,g)^{\alpha s \mu} \cdot e(g,g)^{b^2 s \mu t} / \prod_{i \in I} (\prod_{j=1}^r e(g,g)^{b^2 t \lambda_i \mu_j})^{\omega_i} \\
&= e(g,g)^{\alpha s \mu} \cdot e(g,g)^{b^2 s \mu t} / \prod_{i \in I} (e(g,g)^{(\Sigma_{j=1}^r \mu_j) b^2 t \lambda_i})^{\omega_i} \\
&= e(g,g)^{\alpha s \mu} \cdot e(g,g)^{b^2 s \mu t} / e(g,g)^{b^2 t \mu \Sigma_{i \in I} \lambda_i \omega_i} \\
&= e(g,g)^{\alpha s \mu}.
\end{aligned}
$$

4.6 DELEGATION

Based on the presented OIDR-CP-ABE and MIDR-CP-ABE scheme, we present a Delegation function in this section, which allows the TA to outsource its key generation function to multiple delegators. In this delegation approach, we assume each delegator can have the full privilege as the TA to generate new valid keys based on its allocated private keys from the TA and a user *must* derive his/her private keys from the same delegator. Thus, design goals of the delegation function focus on two main management features: (a) users can decrypt a ciphertext even if their private keys are generated from different delegators; (2) Using the previously presented OIDR-CP-ABE and MIDR-CP-ABE schemes, if a delegator's ID is revoked, then all users who generated their key from this delegator are also revoked.

When the TA allows an entity to serve as a delegator, it provides $\{h_x\}_{\forall x \in S}$ to the delegator. Here, $\{h_x\}$ is a set of random elements corresponding to each attribute. They are secrets owned by the TA in ODIR-CP-ABE and MDIR-CP-ABE schemes. After releasing these secrets to a delegator, then the delegator has the privilige to create new private keys based on its ID. Thus, the new private keys generation function by a delegator $KeyGen_{del}()$ is presented as follows.

$KeyGen_{del}(SK_{del}, S)$:

$$\tilde{SK} = (\tilde{K} = K \cdot g^{b^2 t'} = g^\alpha g^{b^2(t+t')};$$

$$\forall x \in S, \tilde{K}_x = K_x^{((t+t')/t)} = (g^{b \cdot ID} h_x)^{(t+t')} = K_x \cdot g^{b \cdot ID \cdot t'} \cdot h_x^{t'};$$

$$\tilde{L} = L).$$

In the function presented above, input SK_{del} is the private key derived from the TA based on MIDR-CP-ABE $KeyGen()$ function and S is the set of attributes, and t' is randomly selected by the delegator. $\tilde{SK}$ is the new private key for a user and it consists of three components: $\tilde{K}, \tilde{K}_x, \tilde{L}$. All these key elements can be easily computed based on t', $\{h_x\}$, and SK_{del}.

With this delegation approach, when the delegator's ID is included in the revocation list using the presented MIDR-CP-ABE scheme, all members whose keys are assigned by the delegator are also revoked. This is an effective approach to revoke multiple users in a batch. Anonymity is achieved in this revocation process because the delegation scheme only involves delegator's ID in the key generation function, and as a result, no individual's IDs are used.

4.7 PERFORMANCE EVALUATION

In this section, the two schemes presented in this chapter are evaluated in terms of their computation, storage, and communication performance. The evaluation is performed in three parts: Firstly, we analyze the performance complexity of presented revocation schemes compared to the original CP-ABE scheme. Secondly, we implement the IR-CP-ABE schemes based on a PBC library [154]. A computation performance evaluation is conducted including comparison with CP-ABE scheme. Finally, we present an experimental study to show the benefit of using IR-CP-ABE scheme for secure group construction.

4.7.1 COMPUTATION, STORAGE, AND COMMUNICATION COMPLEXITY ANALYSIS

A comparative analysis is carried out among the OIDR-CP-ABE scheme, the MIDR-CP-ABE scheme, and the original CP-ABE scheme. Under all the schemes, there are four functions to be evaluated, i.e., *Setup()*, *KeyGen()*, *Encrypt()*, and *Decrypt()*. The analysis is carried out corresponding to the same function in each scheme on computation cost, storage cost, and communication cost.

Computation Complexity Analysis

Table 4.1

Computation complexity comparison in the number of pairing operations

Function	CP-ABE	OIDR-CP-ABE	MIDR-CP-ABE
Setup()	1	1	1
KeyGen()	0	0	0
Encrypt()	0	0	0
Decrypt()	$2|I|+1$	$2|I|+1$	$2|I|r+1$

Table 4.2

Computation complexity comparison in exponentiation operations

Function	CP-ABE	OIDR-CP-ABE	MIDR-CP-ABE
Setup()	3	m+3	m+3
KeyGen()	$|S|+2$	$|S|+3$	$|S|+3$
Encrypt()	$3l+2$	$3l+2$	$3lr+2$
Decrypt()	$|I|+1$	$|I|+1$	$|I|r+1$

In these three schemes, there are mainly four types of operations that are time-consuming: Pairing, Exponentiation, Multiplication, and Inversion. According to [138], the most computation-intensive operations are *pairing* and *exponentiation*. Thus, in this section, we evaluate the number of *pairing* and *exponentiation* operations for each function as metrics for computation complexity. The complexity of all the schemes involved in a number of *parings* and *exponentiations* are presented in Table 4.1 and Table 4.2, respectively.

In *Setup()* function of all the three schemes, the number of pairing operations is 1. Pairing is only incurred in calculating the value of $e(g,g)^{\alpha}$. In CP-ABE, the number of exponentiation computations in *Setup()* function is 3. However, in both OIDR-CP-ABE and MIDR-CP-ABE, the number of exponentiation computations is $m + 3$, where m is the number of attributes defined globally. The increased number of exponentiation operations comes from the fact that each attribute is defined as a value h_x in CP-ABE, while it is defined as h_x^b in both OIDR-CP-ABE and MIDR-CP-ABE.

For all the schemes, there is no need for any pairing operation in the key generation process. Exponentiation operation is the key contributor to the cost in the *KeyGen()* function for all three schemes. In CP-ABE, the number of exponentiation computations needed is $|S| + 2$. In both OIDR-CP-ABE and MIDR-CP-ABE, this number is increased to $2|S| + 2$. This increase comes from the fact that not only a random value t but also the ID of the user is used as the exponent in the key component for each attribute.

For the *Encrypt()* function, the computation cost in terms of pairing is the same for CP-ABE, OIDR-CP-ABE, and MIDR-CP-ABE. The number of exponentiation operations is significant in differentiating the computation cost among these schemes. In both CP-ABE and OIDR-CP-ABE, it takes $3l + 2$ exponentiations for encryption operation. Here l is the number of attributes involved in the encryption process. Thus, the OIDR-CP-ABE scheme does not incur any more computation cost than the CP-ABE scheme. Comparatively, in MIDR-CP-ABE, the number is increased to $3lr + 2$, here r is the number of IDs that are revoked in the ciphertext.

In CP-ABE, the number of pairing needed for *Decrypt()* is $2|I| + 1$, where I is the set of attributes involved in the decryption process. It requires the same amount of pairing in OIDR-CP-ABE. However, $2|I|r + 1$ pairing operations are needed in MIDR-CP-ABE due to the fact that it conducts more computation for each of the IDs that are revoked. It shows that the computation overhead from exponentiation operations is less than pairing operations. The numbers of exponentiations in CP-ABE, OIDR-CP-ABE, and MIDR-CP-ABE are $|I| + 1$, $|I| + 1$, and $|I|r$, respectively.

Storage and Communication Cost Analysis

Table 4.3

Storage cost comparison

Function	CP-ABE	OIDR-CP-ABE	MIDR-CP-ABE
Setup()	m+5	m+6	m+6
KeyGen()	$|S| + 2$	$|S| + 2$	$|S| + 2$

The storage cost and communication cost are evaluated separately. From a storage perspective, the main overhead is from *Setup()* and *KeyGen()* functions, in which both functions create secret materials that need to be stored locally. For communication costs, the function *Encrypt()* is evaluated as results from this function constitute the ciphertext of transmitted messages. There is no additional storage or communication

Table 4.4

Communication cost comparison

Function	CP-ABE	OIDR-CP-ABE	MIDR-CP-ABE
Encrypt()	$2l+2$	$2l+2$	$2lr+2$

cost for the *Decrypt()* function as the result is directly used as plaintext. The storage for temporary variables that are normally used in computer memories are not considered. Only those needed for final results of each function are counted. Based on the implementation, which is further illustrated in the next section, each element is stored as an element data structure. Therefore, the number of elements is used as a metric for storage and communication cost analysis. Table 4.3 and Table 4.4 summarize the cost corresponding to each function in all three schemes.

In *Setup()* function, it takes $m+4$ elements for storing PK in all the three schemes. The storage cost for MSK is 1 in CP-ABE and 2 in the two schemes presented in this chapter. The cost difference is 1, which is not significant when the number of attributes is large. In *KeyGen()*, the required storage space is $|S|+2$ in all the three schemes. The cost for *Encrypt()* function equals the size of the ciphertext, which is $2l+2$ in both CP-ABE and OIDR-CP-ABE. The size of ciphertext in MIDR-CP-ABE is $2lr+2$. This difference is due to the fact that a separate pair of key components need to be generated for each revoked ID.

Based on the analysis presented, it can be seen that the presented schemes are more computation intensive. Between the two schemes, OIDR-CP-ABE performs better than MIDR-CP-ABE in computation, storage, and communication. The costs for OIDR-CP-ABE have the same order of complexity as a CP-ABE scheme, which means the presented solution does not incur significant overhead compared to CP-ABE. However, the new functional benefits for revoking users is useful in many applications.

4.7.2 IMPLEMENTATION AND TESTING RESULTS

The presented schemes are implemented in C language using the PBC library [154] on Ubuntu 14.04 64bit operating system. The hardware configuration for the machine that runs the experiment is: Intel i7 Quad-core CPU at 2.60GHz; 8GB memory. To test the relations between the amount of attributes involved and the time consumption, we fix the number of IDs revoked to 1 and increase the number of attributes that are involved in each of four functions *Setup()*, *KeyGen()*, *Encrypt()*, and *Decrypt()*. The time consumption of these functions is tested separately. Comparison is made for each function between the CP-ABE scheme and OIDR-CP-ABE scheme. For each attribute setting, the experiments are run ten times, and the average values are used as presented in Figure 4.5 and Figure 4.6.

As can be seen in the figure, the time consumption for all four functions are generally linear to the number of involved attributes. The difference in time consumption

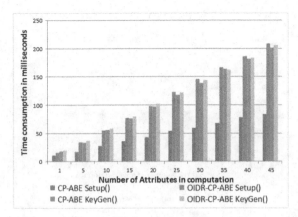

Figure 4.5: Relations between the amount of attributes and time consumption for key assignment.

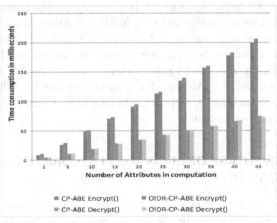

Figure 4.6: Relations between the amount of attributes and time consumption for communication.

for *KeyGen()* function is relatively small between two schemes. The most significant time difference happens with *Setup()* function. The time cost in OIDR-CP-ABE is about twice that of CP-ABE. In a real-world application scenario, this function is run one time by the TA and can be precomputed. Therefore, such computation cost differences do not significantly influence the overall performance of the entire cryptosystem. When 45 attributes are involved for the *Setup()*, *KeyGen()*, and *Encrypt()* function, the overall time cost is right over 200 milliseconds. The cost for *Decrypt()* is less than 100 milliseconds. The overall performance under this scenario is acceptable for real-world applications.

To further explore the influence on the time consumption from the number of IDs revoked, a second experiment is conducted with a fixed number of attributes and changing the number of revoked IDs. In this experiment, the number of attributes is

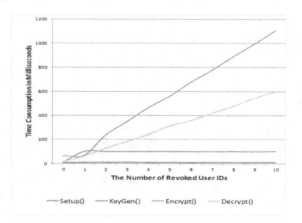

Figure 4.7: Relations between the amount of revoked IDs and time consumption.

set to 20 and the number of revoked IDs is gradually increased from 0 to 10. The evaluation result is shown in Figure 4.7 for MIDR-CP-ABE scheme. It can be seen that the time consumption of *Setup()* and *KeyGen()* are not sensitive to the number of IDs revoked. This is because both functions do not have the revoked ID list involved in their operations. The *Encrypt()* function is sensitive to the number of revoked IDs. Both *Encrypt()* and *Decrypt()* follow a linear trend in Figure 4.7. When the number of revoked IDs is greater than 9, with 20 attributes involved, the overall time cost for *Encrypt()* increases over 1 second.

4.7.3   THE ADVANTAGE OF IR-CP-ABE IN SECURE GROUP CONSTRUCTION

In this subsection, we did simulation to demonstrate that IR-CP-ABE supports much more ID groups than CP-ABE does. Assume there exists 5 attributes in the system and the number of identities increases from 10 to 30. In addition, the probability p for each attribute to be assigned to an identity belongs to $[20\%, 40\%, 60\%, 80\%, 100\%]$. For each number value of identities and p value, the simulation ran 10 times to retrieve the average number of different ID groups generated by CP-ABE and that generated by IR-CP-ABE.

Figure 4.8 illustrates that the ratio of the average number of different ID groups generated by CP-ABE to that generated by IR-CP-ABE is very small. The reason is that IR-CP-ABE could construct a very large number of different ID groups by combining all the IDs associated with the selected attributes first and then revoking any IDs, while CP-ABE can only generate ID groups based on attributes. Note that the ratio decreases when the number of identity increase, as the number of ID groups generated by IR-CP-ABE increases with the order of exponentiation. In addition, the ratio reaches its minimum when the number of identities is fixed and $p = 100\%$ because CP-ABE can only generate one ID group in this case. This implies IR-CP-ABE provides a more comprehensive solution in group construction than CP-ABE.

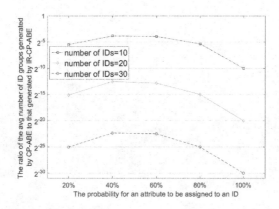

Figure 4.8: The ratio of the average number of ID groups generated by CP-ABE to that generated by IR-CP-ABE.

4.8 SUMMARY

In this chapter, we presented a new identity-revocable CP-ABE scheme to improve the group management capability of existing CP-ABE solutions. The presented research is a first solution to address NOT logic in a CP-ABE scheme, which can be used to construct all possible attribute-confined policy groups with any attributes' assignments to group members. We also presented a delegation approach to show how to use identity revocation to revoke a large group of users.

There are several research issues need that to be further investigated. First, the ID revocation scheme still needs to explicitly specify which users need to be revoked in the revocation list. Ideally, revoked users should not be known by any group users. Although the delegation scheme provides group-based anonymity, it is restricted by its delegation formation. Thus, privacy protection to revoke individual users is desired. Second, the delegation scheme requires to delegate all attributes and private key generation. A user can only get his/her attributes and private keys from one of the delegators. Thus, a federated delegation approach is desired, in which a user can use his/her attributes and private keys generated from different delegators.

5 Extended Identity-Revocable CP-ABE

This chapter focuses on how to extend Identity-Based Revocable CP-ABE (IR-CP-ABE) by considering the scalability of ABE revocation management solutions such as ABE key federation, interoperability, delegation, etc. The presented scheme is called EIR-CP-ABE. In particular, EIR-CP-ABE does not need a centralized access control infrastructure, where the authorization is done by incorporating security access control policies into ciphertext. In this way, protected data can be stored on even untrusted storage providers' servers and transmitted over untrusted networks, thus significantly improving the flexibility and usability of the ABAC model. Moreover, the access control policy is defined by the data owners, thus conforming to the data ownership policy. To implement EIR-CP-ABE, existing ABE-based ABAC solutions face challenges to realize important management features of access control such as delegation, federation, interoperability, and revocation, which prevent them from being widely deployed. In this chapter, we present a design of EIR-CP-ABE solution by incorporating users' private key generation procedure, which allows the ABAC solution to address all these access control management features, which make EIR-CP-ABE approaches practical. The performance evaluation demonstrates the solution is secure and efficient to establish a large-scale attribute-based access control framework.

5.1 INTRODUCTION

In order to achieve a more flexible ABE-based ABAC solution, several important access control management features need to be addressed. Since ABE-based ABAC relies on the cryptographic algorithm, these capabilities need to be addressed by the ABE algorithm itself.

Federation

According to dictionary.com's definition of federation, it is "a federated body formed by a number of nations, states, societies, unions, etc., each retaining control of its own internal affairs". Take a hospital use case as an example: if a data access control policy specifies that a doctor affiliating with *Hospital A* should be able to decrypt a ciphertex using the public-encrypting parameters issued by *Hospital B*, which means the crypto system parameters should be used and shared among different hospitals. For example, a heart MRI image encrypted by using attribute *att = MRI image* generated from hospital B is able to be decrypted by a doctor in hospital A who has the attribute *MRI image*. In this case, hospital A and B are federated. However, due to the administration barriers between hospitals, generating a common public parameter requires having a Trust Authority (TA), who has the private master key information for each hospital, and this may face a single-point failure issue. Moreover, organizations may not desire other organizations to be in possession of their root of trust.

To achieve the federation feature of ABAC, we need to design a coalition framework to generate common publicly-shared encryption parameters and derive master private keys for different administrative domains without the single-point failure issue. The federation approach is based on a secure multiple-party computation protocol running among a subset of selected participating organizations (say n) to generate system parameters and private keys for the root authority of each organization. The presented solution is resistant against $n - 1$ collusion in the coalition.

Delegation

ABE schemes require a Trusted Authority (TA) to generate users' private keys, where an ABE key generation authority can grant full or partial of its key-generation privileges to one or multiple delegators. Delegating TA's key-generation capabilities not only can relieve the single-point failure issue, but also can naturally follow the organization's hierarchical management structure. Following the hospital's hierarchical management structure, the delegation is flowing down to the bottom level above patients. Doctors, nurses, and other hospital employees can derive their attributes and the corresponding private keys at their registration delegator, or called parent delegator.

To achieve the delegation feature from a parent to a delegation node (or called child), we design a delegation protocol to allow a parent node to securely delegate its key generation capability to its children. A child can fully or partially inherit the parent's key generation capability based on secrets shared between them. Children delegators cannot collude to derive more capability beyond their delegated key-generation capabilities.

Interoperability

In practice, users need to interact with several different organizations that may not trust each other; a user also may own attributes belonged to different subdivisions of an organization. Moreover, at the organization level, we cannot easily find a root authority to generate private keys for all the users in these organizations. In the presented example, a doctor belongs to the Hospital A's emergency room and he also affiliates with Hospital B in the medical laboratory (imaging division), and thus he can derive two attributes from two delegators in the trust hierarchy and the doctor should be able to use them together, e.g., an access policy could be *att1=emergency room AND att2=MRI image*. Existing ABE algorithms require that all attributes of a user be given from the same delegated TA. The interoperability feature allows users to receive attribute private keys from different delegated TAs or federated TAs.

To achieve the interoperability feature, we need to rely on a closest common ancestor to assist two delegators to issue the private keys for a user. To follow the hierarchical delegation trust framework and prevent collusion issue where two delegators sharing secrets to generate secret keys for a user, the presented protocol incorporates a random exponent in each generated private keys to enforce the interoperability protocol must go through the closest ancestor.

Revocation

Revocation is an inherently difficult problem in public key cryptographic algorithms. Existing ABE schemes mainly incorporate a NOT logic gate to revoke an attribute when constructing an access policy tree, which may not provide the level of granularity or accuracy. For example, when revoking one specific doctor using a policy NOT(*emergency room*) cannot easily identify the revoked user, and more attributes need to be involved. As a result, using a unique identifier for a user or a group is a natural approach and can be easily adopted. Previous approaches [225] treat a group ID or a user's ID as an attribute, and then we can simply implement the NOT logic on the attribute in the ABE scheme to revoke a known group of entities or individuals.

However, this approach will significantly increase the size of the attribute set and make attributes management extremely complicated when the user-group is large. To address this issue, the solution is to incorporate users' and groups' IDs into their allocated private keys. In this way, we can directly revoke a user or a group (e.g., a group generated under one delegator) by revoking their IDs. The solution will be compatible with traditional attribute-based revocation, and a data owner can build an access policy with both attributes and a set of revoked identities.

In summary, the main focus of this chapter includes:

- a representative use-case that takes full advantage of the revolutionary capabilities of ABE-based ABAC. By incorporating users' unique identities into private keys, we can ease the revocation by revoking users' identities or delegators' identities (i.e., group-based revocation).
- a comprehensive approach through a set of ABE protocols to achieve the desired ABAC management features: delegation, federation, interoperability, and revocation. These features are essential for the EIR-CP-ABE solution.
- a comprehensive evaluation on computation, communication, storage, and security. The evaluation results show that the solutions can be applied in large-scale applications.

The rest of this chapter is arranged as follows: Section 5.2 presents the system and models that the EIR-CP-ABE solution is built on, and the EIR-CP-ABE scheme is presented in Section 5.3. Both performance analysis and numerical evaluations are presented in Section 5.4. Finally, Section 5.5 summarizes this chapter.

5.2 SYSTEM AND MODELS

5.2.1 EIR-CP-ABE TRUST FRAMEWORK

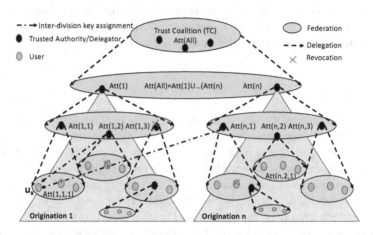

Figure 5.1: EIR-CP-ABE system and models.

The overall EIR-CP-ABE trust framework is presented in Figure. 5.1, which follows a decentralized trust architecture. At the top, the Trust Coalition (TC) is in charge of all attributes' registration $Att(All)$ and public security parameters management. We propose use TC to avoid the single point failure issue. Using a multi-party computation protocol, we can create a private key delegation model, where neither a single trusted authority nor a subset of trusted authorities can generate a valid private key for an attribute.

Delegation follows the hierarchical trust framework. Under TC, multiple organizations exist to manage their attributes, e.g., $Att(i)$, where $i = 1, 2, \cdots, n$, $|Att(All)| = |Att(1)| \cup \cdots \cup |Att(n)|$, and $|Att(i)| \cap |Att(k)|$ may not be empty ϕ. Each organization can form their own trust domain by establishing a hierarchical trust framework, where the organization's root node needs to register its managed attributes at the top level, and it can delegate its key generation capabilities to delegators at a lower level, e.g., $Att(1) \rightarrow Att(1, *)$ and $Att(1, 1) \rightarrow Att(1, 1, 1)$.

Federation in EIR-CP-ABE addresses the problem that different organizations can share attributes. Based on the protocol *Federation-Setup* and *Federation-Key-Generation*, a group of members which have benefit conflicts to conclude would work in a secure way to generate the system parameters as well as generate private keys for the root authorities of each organization.

Interoperability in EIR-CP-ABE addresses the problem that a user might be affiliated with two different delegators or even organizations. Interoperability in EIR-CP-ABE addresses the application scenario that a user can derive attributes and corresponding keys from different delegators. In the example shown in Figure 5.1, u_x may derive attributes and keys from $Att(1, 1)$, $Att(1, 2)$, and $Att(n, 1)$, where dot-dash lines represent the attributes, and keys are derived from non-parent delegators.

In order to make sure that the attributes and corresponding keys derived from different delegators can be used together to decrypt a message, it requires the common ancestor of these delegators to derive u_x's private key.

Revocation is another important feature in EIR-CP-ABE. A user's attributes may be revoked due to bad behaviors. Moreover, a user may transfer to a new trust domain that handled by a different delegator, e.g., transfer from $Att(1,1,1)$ to $Att(n,2,1)$, where the user's originally assigned attributes and private keys should be revoked. Revoking a user and revoking an attribute is different in two aspects: (a) an attribute may be shared with multiple users and thus revoking an attribute may revoke a group of users, which may not be known since the same attribute maybe assigned by multiple delegators; (b) a user's ID can uniquely identify a user, and thus revoking a user's ID is preferred to revoking an attribute when the application requires precise access control.

5.2.2 SECURITY MODEL

Compared to traditional access control model, where the data storage service provider is usually fully trusted, in the EIR-CP-ABE solution, the storage servers and communication network are assumed to be honest-but-curious. They might do some statistical analysis over the encrypted data or collude with some unauthorized data users to obtain access to protected data.

As for entities in the organizational structure, they are divided into four categories: *data owner*, *data consumer/user*, *authorities*, and *members of the trust coalition*. The data owner is fully trusted. The data consumer is assumed to be dishonest, i.e., they may collude to access data that they do not have access privileges for. In particular, data consumer A with private key SK_A linked with attribute set U_A and data consumer B with private key SK_B linked with attribute set U_B (both keys are generated by the protocol External Delegation and $U_A \neq U_B$) might collude together in order to gain access privilege extracted from attribute set $U_A \cup U_B$.

Within one organization, assume that there are two authorities DA_1 and DA_2 with the closest ancestor authority P. DA_1 and DA_2 have private delegation keys for generating private keys for attribute set Att_1 and Att_2, respectively ($Att_1 \neq Att_2$). These two domain authorities might want to obtain more access privileges without the authorization from their common ancestor domain authorities, say constructing access policies from attribute set $Att_1 \cup Att_2$. DA_1 and DA_2 could either be domain authorities within an organization or two root authorities of two different organizations. If they are two authorities within an organization, it is the protocol *Delegation-Internal* to guarantee the collusion security problem. If they are two root authors, we should ensure the *Federated-Key-Generation* protocol to be resistant against the collusion problem.

Assume that the members in the trusted coalition will not collude with the organizations but parts of the members of the trusted coalition might collude together aiming at controlling the whole structure, but not all of them. Assume that there are N members in the trusted coalition and fewer than $N-1$ members will collude together; the *Federated-Setup* protocol should resist against the collusion attack.

Each security problem will be analyzed in the security analysis section.

5.3   EIR-CP-ABE PROTOCOLS

As shown in Figure 5.1, there exist two structures among organizations, i.e., the inter and the inner. The inter level corresponds to the federation among multiple organizations which do not trust each other. The inner level corresponds to the organizational structure within each individual organization. The organizational structure is actually a tree structure. For clarity of statement, we define some terminologies related with the tree structure below.

In the presentation, *Root* denotes the top node in a tree, e.g., the root node of the organizational structure is *Att(all)*. *Child* denotes a node directly connected to another node, e.g., the node *Att(1, 1)* is the child of the node *Att(1)*. *Parent* is the converse notion of child, e.g., the node *Att(1)* is the parent of the node *Att(1, 1)*. *Ancestor* is a node reachable by repeated proceeding from child to parent, e.g., the node *Att(1)* and *Att(all)* are the ancestor of the node *Att(1, 1)*. *External node or leaf* is a node with no children. For example, an individual user in the organizational structure is dented by a leaf node. *Internal node* is a node with at least one child. *Level* is defined by 1+(the number of connections between the node and the root), e.g. the root authority of each organization is on Level-2 since there exists one connection between the root authority and the *Root*.

To realize the federation features of the presented ABE approach, we designed a secure multiple-party computation system setup protocol and a secure multiple-party computation key generation protocol to generate keys for the root authority of each individual organization. The selected group of organizations have benefit collisions, thus will not collude with each other. We name the selected group of organizations TC. The protocols run to enable the functionalities of the EIR-CP-ABE model are presented in the following subsections.

5.3.1   GLOBAL SETUP

In this phase, global parameters are negotiated between the trusted authorities. The global parameters include the universal set of attributes denoted by U, the pairing that will be used. The public parameters include the pairing e being used, the generator g of the group G_1 as well as the group elements of the attributes, denoted by h. $GP = \{U, g, e, \{h_x\}_{x \in U}\}$.

5.3.2   FEDERATED SETUP AND KEY GENERATION PROTOCOL

Each member in TC= $\{TC_1, \cdots, TC_N\}$ (N is the number of members in TC) will run the setup protocol of the ABE scheme to generate their share of the master secret key and the public parameter and then generate the system-wide master secret key and public parameters by running a secure multiple-party computation protocol. Figure 5.2 shows the workflows of system setup and private key generation for organizational root authorities. The two protocols are described as follows.

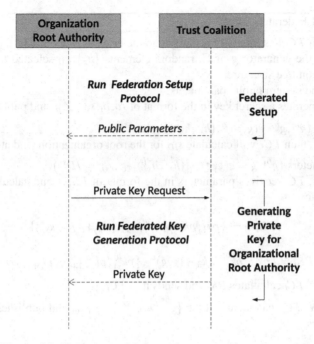

Figure 5.2: Flowchart of federated setup and key generation.

Federation Setup

Each TC member performs computation and communication as described in Protocol 5.1. When the protocol completes, the system-wide master secret key and public parameters will be as follows:

$$MSK = (\alpha, b, s_0),$$

$$PK = (g, g^b, g^{b^2}, e(g,g)^\alpha, \{h_x^b, h_x^{b^2}\}_{x \in \mathcal{U}}, \{g^{bsID_j^{-1}}, g^{sID_j^{-1}}\}_{ID_j \in Orgs}),$$

where $\alpha = \sum_{i=1}^N \alpha_i, b = \prod_{i=1}^N b_i, s_0 = \sum_{i=1}^N s_i$.

s_{ID_j} denotes the component generated for all the non-individual organizations or subdivisions in the tree structure. $sID_j = ID_j^{s_0}$ if ID_j is the root authority of an organization, otherwise, $s_{ID_{child}} = ID_{child}^{SID_{parent}}$. $Orgs$ denotes the set of all organizations and their subdivisions.

For statement simplicity, we focus on a single root organization use case. This could be extended to multiple organizations scenario in an easy way.

Federated Key Generation

The members of the trusted coalition work together to generate the private key for the root authority of an organization ID as shown in Protocol 5.2. The generated private

---

Protocol 5.1 Federated-Setup

Input: Each TC_i has inputs of security parameter λ, attribute set $\mathbf{U}$, a prime-order group $\mathbb{G}$, the generator g of $\mathbb{G}$, random elements $\{h_x\}_{x \in \mathbf{U}}$ selected from $\mathbb{G}$. The organization tree structure.

Output: The system public parameters.

1: TC_i generates a secret key in the format of $\alpha_i, b_i, s_i \in \mathbb{Z}_p$, and public key in the format of $\left(g^{b_i}, g^{b_i^2}, e(g,g)^{\alpha_i}, (h_x^{b_i}, h_x^{b_i^2})_{x \in U}\right)$.

2: If $i = 1$, then TC_1 will calculate s_{ID} for the root organization, and at the end got the parameters $\left(g^{b_1}, g^{b_1^2}, e(g,g)^{\alpha_1}, (h_x^{b_1}, h_x^{b_1^2})_{x \in U}, s_{ID} = ID^{s_1}\right)$.

3: If $i \neq 1$, TC_i receives parameters in the format of PK_{i-1} and calculates PK_i as shown below.

$$PK_{i-1} = \{pk_1, pk_2, pk_3, \{pk_{x1}, pk_{x2}\}_{x \in U}, pk_{s_{ID}}\}$$

$$PK_i = \{pk_1^{b_i}, pk_2^{b_i^2}, pk_3 \cdot e(g,g)^{\alpha_i}, \{pk_{x1}^{b_i}, pk_{x2}^{b_i^2}\}_{x \in U}, pk_{s_{ID}} \cdot ID^{s_i}\}$$

4: If $i \neq N$, TC_i calculates PK_i and sends it to TC_{i+1}.

5: If $i = N$, TC_N calculates all the $\{g^{bs_{ID_j}^{-1}}, g^{s_{ID_j}^{-1}}\}_{ID_j \in Orgs}$ and publishes the public parameters

$$PK = (g, g^b, g^{b^2}, e(g,g)^{\alpha}, \{h_x^b, h_x^{b^2}\}_{x \in U}, \{g^{bs_{ID_j}^{-1}}, g^{s_{ID_j}^{-1}}\}_{ID_j \in Orgs})$$

---

key is the form of SK.

$$SK = (K_0 = g^{\alpha} g^{b^2 t}, \ L_a = g^{-s_{ID}^{-1} t}, \ L_u = g^{-t}, \ K_a = (g^{bs_{ID}} h_x^b)^t, \ K_u = (g^{bID} h_x)^t,$$

$$gbsID = g^{bs_{ID}}, \ hx = h_x, \ gbt = g^{bt}, \ ht = h_x^t, \ hbt = h_x^{bt}, \ s_{ID})_{x \in U_{ID}},$$

where $t = \sum_{i=1}^{N} t_i$, $b = \prod_{i=1}^{N} b_i$, $s_{ID} = ID^s$, $s = \sum_{i=1}^{N} s_i$ and x is in the set of attributes managed by organization ID. With the private component s_{ID}, each root authority generates private components for the domain authorities with the rule $s_{child} = (ID_{child})^{s_{parent}} \cdot g^{bs_{child}^{-1}}$ and $g^{s_{child}^{-1}}$ are part of the system public parameters.

5.3.3 KEY GENERATION DELEGATION PROTOCOL

On the inner-organization level, the hierarchical structure naturally reflexes the internal organizations' authority and responsibility. For the internal nodes in an organizational structure, the key generation delegation privilege of the parent domain authority could be distributed to the child domain authority. For the external nodes (individual users) in an organizational structure, it is the internal nodes which are the parent of the external nodes to generate the private key for them. The workflow of the key generation delegation is shown in Figure 5.3.

Protocol 5.2 Federated-Key-Generation

Input: Each TC_i has public parameters, the secret share α_i, b_i, s_i, the attribute set U_{ID}, and the identity ID of the root organization.

Output: The private attribute key of the root organization.

1: TC_i generates $t_i \in \mathbb{Z}_p$.

2: TC_i calculates

$$SK_i = \{K_0 = g^{\alpha_i} g^{b^2 t_i}, \; L_a = g^{-s_{ID}^{-1} t_i}, \; L_u = g^{-t_i}, \; K_a = (g^{bs_{ID}} h_x^b)^{t_i}, \; K_u = (g^{bID} h_x)^{t_i},$$

$$gbsID = g^{bs_{ID}}, \; hx = h_x, \; gbt = g^{bt_i}, \; ht = h_x^{t_i}, \; hbt = h_x^{bt_i}, \; s_{ID}\}_{x \in U_{ID}}.$$

3: If $i = 1$, TC_1 calculates SK_1 and sends it to TC_2.

4: If $i \neq 1$, TC_i obtains $SK_{i-1} = \{K_0, L_a, L_u, K_a, K_u, gbsID, hx, gbt, ht, hbt\}_{x \in U_{ID}})$ from TC_{i-1} and calculates $K_0 \cdot g^{\alpha_i} \cdot (g^{b^2})^{t_{i+1}}, \; L_a \cdot (g^{-s_{ID}^{-1}})^{t_{i+1}}, \; L_u \cdot g^{-t_{i+1}}, \; K_a \cdot (g^{bs_{ID}} h_x^b)^{t_{i+1}}, \; K_u \cdot (g^{bID} h_x)^{t_{i+1}}, \; gbt \cdot (g^b)^{t_{i+1}}, \; ht \cdot (h_x)^{t_{i+1}}, \; hbt \cdot (h_x^b)^{t_{i+1}}.$

5: If $i \neq N$, TC_i sends the generated components to TC_{i+1}.

6: If $i = N$, TC_N sends the generated private key SK to the root authority of organization ID.

Delegation-Internal

Protocol 5.3 is run between a parent domain authority ID_{ia} and a child domain authority $ID_{(i+1)a}$ to generate the private key for the child domain authority on level $i+1$ based on that of the parent domain authority on level i. The detailed description is presented as follows.

Delegation-External

Protocol 5.4 is run by a domain authority to generate a private key for a user.

The components in the private key of the domain authority are updated as follows. The integer t' is a random integer selected by the domain authority.

$$g^{\alpha} g^{b^2 t_u} = g^{\alpha} g^{b^2 t_a} \cdot (g^{b^2})^{t'},$$

$$g^{s_{ja}^{-1} t_u} = g^{s_{ja}^{-1} t_a} \cdot (g^{s_{ja}^{-1}})^{t'},$$

$$g^{-t_u} = g^{-t_a} \cdot g^{-t'},$$

$$(g^{bs_{ja}} h_x^b)^{t_u} = (g^{bs_{ja}} h_x^b)^{t_a} \cdot (g^{bs_{ja}} \cdot h_x^b)^{t'},$$

$$g^{bt_u} = g^{bt_a} \cdot g^{bt'},$$

$$h_x^{t_u} = h_x^{t_a} \cdot h_x^{t'}.$$

Protocol 5.3 Delegation-Internal

Input: Parent DA with identity ID_P and attribute set U_{ID_P} has private key SK_{ia} shown below.

$$SK_{ia} = \{K_0 = g^\alpha g^{b^2 t_{ia}}, \ L_a = g^{-s_{ID}^{-1} t_{ia}}, \ L_u = g^{-t_{ia}}, \ K_a = (g^{bs_{ID_P}} h_x^b)^{t_{ia}}, \ K_u = (g^{bID} h_x)^{t_{ia}},$$

$$gbsID = g^{bs_{ID_P}}, \ hx = h_x, \ gbt = g^{bt_{ia}}, \ ht = h_x^{t_{ia}}, \ hbt = h_x^{bt_{ia}}, \ s_{ID_P}\}_{x \in U_{ID_P}}.$$

Output: The private attribute key of the child DA with the identity ID_C and attribute set U_{ID_C}.

1: The components in the private key of the child domain authority are updated in the following way, where $t' \in \mathbb{Z}_p$ and $x \in U_{ID_C}$

$$g^\alpha g^{b^2 t_{(i+1)a}} = g^\alpha g^{b^2 t_{ia}} \cdot (g^{b^2})^{t'},$$

$$g^{s_{ID_{(i+1)a}}^{-1} t_{(i+1)a}} = \left(g^{s_{ID_{ia}}^{-1} t_{ia}}\right)^{s_{ID_{ia}} \cdot s_{ID_{(i+1)a}}^{-1}} \cdot \left(1/g^{s_{ID_{(i+1)a}}}\right)^{t'}$$

$$g^{-t_{(i+1)a}} = g^{-t_{ia}} \cdot g^{-t'},$$

$$(g^{bs_{(i+1)a}} h_x^b)^{t_{(i+1)a}} = (g^{bt_{ia}} \cdot g^{bt'})^{s_{(i+1)a}} \cdot h_x^{bt_{ia}} \cdot h_x^{bt'},$$

$$(g^{bID} h_x)^{t_{(i+1)a}} = (g^{bt_{ia}} \cdot (g^b)^{t'})^{ID} \cdot h_x^{t_{ia}} \cdot h_x^{t'}$$

$$g^{bs_{(i+1)a}} = (g^b)^{s_{(i+1)a}},$$

$$g^{bt_{(i+1)a}} = g^{bt_{ia}} \cdot (g^b)^{t'},$$

$$h_x^{t_{(i+1)a}} = h_x^{t_{ia}} \cdot h_x^{t'},$$

$$h_x^{bt_{(i+1)a}} = h_x^{bt_{ia}} \cdot (h_x^b)^{t'}.$$

$$s_{(i+1)a} = ID^{s_{ia}}.$$

2: The parent DA sends the generated private attribute key to the subdivision.

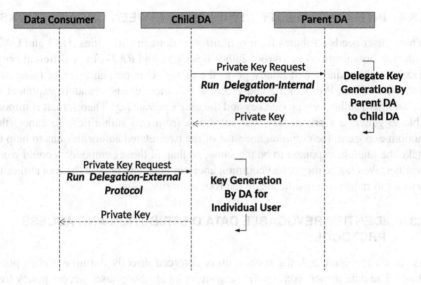

Figure 5.3: Flowchart of key generation delegation.

Protocol 5.4 Delegation-External

Input: Domain authority with identity ID_a has private key SK_a shown below.

$$SK_a = \{K_0 = g^\alpha g^{b^2 t_a},\, L_a = g^{-s_{ID}^{-1} t_a},\, L_u = g^{-t_a},\, K_a = (g^{bs_{ID_a}} h_x^b)^{t_a},\, K_u = (g^{bID} h_x)^{t_a},$$

$$gbsID = g^{bs_{ID_a}},\, hx = h_x,\, gbt = g^{bt_a},\, ht = h_x^{t_a},\, hbt = h_x^{bt_a},\, s_{ID_a}\}_{x \in U_{ID_a}}.$$

Output: The private attribute key of an individual user with identity ID_u and attribute set U_{ID_u}.

1: The components in the private key of the child domain authority are updated in the following way, where $t' \in \mathbb{Z}_p$ and $x \in U_{ID_u}$

$$g^\alpha g^{b^2 t_u} = g^\alpha g^{b^2 t_a} \cdot (g^{b^2})^{t'},$$

$$g^{s_{ID_a}{}^{-1} t_u} = g^{s_{ID_a}{}^{-1} t_a} \cdot (1/g^{s_{ID_a}})^{t'}$$

$$g^{-t_u} = g^{-t_a} \cdot g^{-t'},$$

$$(g^{bs_{ID_a}} h_x^b)^{t_u} = (g^{bt_a} \cdot g^{bt'})^{s_{ID_a}} \cdot h_x^{bt_a} \cdot h_x^{bt'},$$

$$(g^{bID_u} h_x)^{t_u} = (g^{bt_a} \cdot (g^b)^{t'})^{ID_u} \cdot h_x^{t_a} \cdot h_x^{t'}.$$

2: The DA sends the generated private attribute key to the user.

5.3.4 INTEROPERABILITY WITHIN AND BETWEEN ORGANIZATIONS

When a user needs attributes from two different domain authorities (DA1 and DA2) within an organization or two root authorities (RA1 and RA2) of two different organizations, the request will finally go to the closest common ancestor of these two authorities. From Protocol 5.2 and Protocol 5.3, there exists a random exponent in each domain authority's private key and the user's private key. Therefore, it is impossible to generate a private key with attributes from two authorities. To cancel the random exponent, the common ancestor of the two related authorities has to help to make the random exponent to be the same, so that all these components could work together. We choose the closest common ancestor authority but not the root authority in order to reduce the overheads on the root authority.

5.3.5 IDENTITY-REVOCABLE DATA DISTRIBUTION AND ACCESS PROTOCOL

As shown in Figure 5.4, the revocation is enforced directly during the encryption phase. The data owner would at first construct an attribute-based access policy tree and then kick out the undesired data consumers by adding their identities into a revocation identity set. Taking inputs of the access policy, revoked identity set as well as the plaintext data, the encryption algorithm outputs the ciphertext to be distributed. In this way, only data consumers whose attributes satisfy the access policy and are not revoked by the data owner can decrypt the ciphertext by running the decryption algorithm described below.

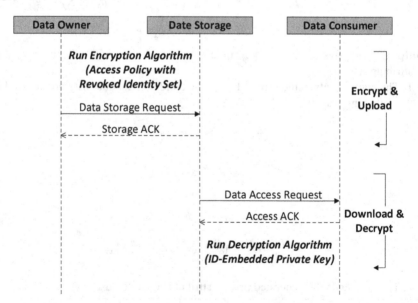

Figure 5.4: Flowchart of identity revocable access control.

Encrypt$(PK, (M, \rho), \mathcal{M}, \mathbf{ID})$

This is an algorithm revoking both multiple users and multiple domain authorities. The *encrypt* algorithm takes as inputs an LSSS access infrastructure (M, ρ), where M is an $l \times n$ matrix and the function ρ associates each row of M to corresponding attributes. $\mathbf{ID} = \mathbf{ID}_a \cup \mathbf{ID}_u$ and $|\mathbf{ID}_a| + |\mathbf{ID}_u| = r_a + r_u = r$. Denote the set of revoked domain authority identities as $\mathbf{ID}_a = \{(ID'_{a,j})\}_{j \in [1, r_a]}$. The set of revoked user identities is denoted by $\mathbf{ID}_u = \{ID'_{u,j}\}_{j \in [1, r_u]}$. The *encrypt* algorithm first chooses a random vector $v = (s, y_2, \cdots, y_n) \in \mathbb{Z}_p^n$. These values will be used to share an encryption exponent s. For $x \in [1, l]$, it calculates $\lambda_x = v \cdot M_x$. The *encrypt* algorithm chooses random $s \in \mathbb{Z}_p$. The algorithm chooses random μ_a, μ_u such that $\mu = \mu_a + \mu_u$, and $\mu'_1, \cdots, \mu'_{r_u} \in \mathbb{Z}_p$ such that $\mu_u = \mu'_1 + \cdots + \mu'_{r_u}$. Then, for message $\mathcal{M}$, the ciphertext is presented as follows:

$$CT = (C, C_0, \hat{C}_a, \hat{C}_u, \hat{C}_a', \mathbf{ID}), \text{ where}$$

$$C = \mathcal{M} e(g, g)^{\alpha s \mu},$$

$$C_0 = g^{s \mu},$$

$$\hat{C}_a = \left(C^*_{akj} = g^{bs \frac{1}{ID_{aj}} \lambda_k u_a}, C'_{ak} = (h_{\rho(k)}^{b^2})^{\lambda_k u_a} \right)_{k \in [1, l], ID_{aj} \notin ID_a}^{i \in \mathscr{I}_{nr}},$$

$$\hat{C}_u = \left(\{ C^*_{ukj} = g^{b \lambda_k u'_j} \}_{k \in [1, l], j \in [1, r_u]}, \right.$$

$$\left. \{ C'_{ukj} = (g^{b^2 \cdot ID_{u,j}} h_{\rho(k)}^b)^{\lambda_k u'_j} \}_{k \in [1, l], j \in [1, r_u]} \right).$$

Decrypt(CT, SK):

CT is the ciphertext with access structure (M, ρ), and SK is a private key for a set $\mathbf{S}$. Suppose that $\mathbf{S}$ satisfies the access structure, and let $\mathbf{I} \subset [1, l]$ be defined as $\mathbf{I} = \{i : \rho(i) \in \mathbf{S}\}$. Let $\{\omega_i \in \mathbb{Z}_p\}_{i \in \mathbf{I}}$ be a set of constants such that if $\{\lambda_i\}$ are valid shares of any secret s according to M, then $\Sigma_{i \in \mathbf{I}} \omega_i \lambda_i = s$. For the jth revoked user identity, denote the identity of the non-revoked domain authority administrating ID_u by $ID_{a,j}$. The decryption algorithm calculates $e(g, g)^{b^2 t s \mu'_j}$ as follows:

$$\begin{cases} (\prod_{i \in \mathbf{I}} [e(K_{\rho(i)u}, C^*_{uij}) \cdot e(C'_{uij}, L_u)]^{\omega_i})^{\frac{1}{(ID_u - ID_j)}}, ID_{u,j} \neq ID'_{u,j}, \\ \prod_{i \in \mathbf{I}} [e(K'_{\rho(i)aj}, C^*_{aij}) \cdot e(C'_{ai}, L_a)]^{\omega_i}, ID_a \notin ID'_a. \end{cases} \quad (5.1)$$

Then we could get $e(g, g)^{b^2 t s \mu_u}$ in the following way:

$$e(g, g)^{b^2 t s \mu_u} = \prod_{j \in [1, r_u]} e(g, g)^{b^2 t s \mu'_j}.$$

For the jth revoked domain authority, denote the identity of the domain authority on the h_j^{th} layer managing ID_u by $ID_{a,j}$. The decryption algorithm evaluates $e(g, g)^{b^2 t s u_j}$ as follows:

$$e(g,g)^{b^2 tsu_j} = \prod_{i \in \mathbf{I}} [e(K'_{a\rho(i)j}, C^*_{aij}) \cdot e(L_{aj}, C'_{aij})]^{\omega_i}.$$

Then we could get $e(g,g)^{b^2 ts\mu_a}$ from the authority component.

If SK's holder is not managed by any revoked domain authority and is not among the revoked users, then we can get $e(g,g)^{\alpha s\mu} = \frac{e(C_0, K)}{e(g,g)^{b^2 ts\mu_u} \cdot e(g,g)^{b^2 ts\mu_a}}$. Finally get the message $\mathcal{M}$ by evaluating $\frac{C}{e(g,g)^{\alpha s\mu}}$.

5.4 ANALYSIS AND EVALUATION

In this section, the ABE schemes presented in this chapter are evaluated in terms of their computation, storage, and communication performance.

5.4.1 COMPLEXITY ANALYSIS

We show the complexity analysis summarized in Table 5.1 of the designed scheme. There are four types of time-consuming operations: pairing, exponentiation, multiplication, and inversion, included in the schemes. Among them, the pairing and exponentiation operations are the dominant costs. Therefore, we utilize the number of pairing and exponentiation operations as metrics for computation complexity of each scheme. The main storage overhead comes from the setup algorithm and key generation algorithm.

Table 5.1
Complexity analysis

Overhead	Setup	KeyGen-RA	KeyGen-U
Computation (Pairing)	1	0	0
Computation(Exponent)	$2|\mathscr{R}\mathscr{I}| + 2|U| + 3$	$2|U_{ID_a}| + 5$	$2|U_{ID_u}| + 3$
Communication	$2|\mathscr{R}\mathscr{I}| + 2|U| + 4$	$2|U_{ID_a}| + 5$	$2|U_{ID_u}| + 3$
Storage	$2|\mathscr{R}\mathscr{I}| + 2|U| + 7$	$2|U_{ID_a}| + 5$	$2|U_{ID_u}| + 3$

Overhead	Encrypt	Decrypt
Computation (Pairing)	1	$2|\mathbf{I}|(r_u + 1)$
Computation(Exponent)	$x(3r_u l + 2)$ $+$ $y((|\mathscr{I}_{nr}| + 1)l + 2)$	$x(|\mathbf{I}|r_u) + y|\mathbf{I}|$
Communication	$x(2lr_u)$ $+$ $y(l|\mathscr{I}_{nr}| + 1) + 2$	0
Storage	0	0

Since the setup of the master secret key and system public parameters is performed by all the members of the Trust Coalition, we show the computation complexity for each of the TC member. Each member will perform one pairing and

$2|\mathscr{R}\mathscr{I}|+|U|+2$ exponents. Since each member will send the intermediate result to the next member, there will be $2|\mathscr{R}\mathscr{I}|+2|U|+3$ elements transmitted from one member to another. After the system setup, all the TC members will store both the public parameters and the share of the master secret key. Totally, the storage complexity will be $2|U|+2|\mathscr{R}\mathscr{I}|+7$.

There are two kinds of key generation, the first one is generating private key by the TC for the root authority of each organization. The second is the key generation by the domain authority within the organization for either child domain authority or individual users. Since generating delegation private keys for the domain authorities within an organization is system overall computation overhead, we exclude it here. The computation complexity of each TC when generating the private key for the root authority is $2|U_{ID}|+5$, where $|U_{ID}|$ is the number of attributes of the root authority. We assume that the height of the identity structure tree is 2, i.e., $H = 1$. The complexity of generating a private key for a user is $2|U_{ID_u}|+4$, where U_{ID_u} is the set of attributes assigned to the user.

One pairing computation is performed during the encryption. the number of exponents is $x(3r_u l + 2) + y((|\mathscr{I}_{nr}| + 1)l + 2)$. If only multiple users are revoked then $x = 1, y = 0$; if only multiple domain authorities are revoked then $x = 0, y = 1$; if there are both multiple users and multiple domain authorities revoked then $x = 1, y = 1$. The communication complexity of the encryption algorithm is $x(2lr_u) + y(r_g + l|\mathscr{I}_{nr}|) + 2$, where x and y is the same as above. The computation of decryption consists of $2|\mathbf{I}|(r_u + 1)$ pairings and $x(|\mathbf{I}|r_u) + y|\mathbf{I}|$ exponents.

Figures 5.6 to Figure 5.11 show the experimental performance evaluation of the algorithms. The algorithms are implemented in Python using PBC library on a MacOS 10.10.5 operating system. The example setting of the organization structure is described as in Figure 5.5.

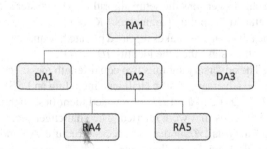

Figure 5.5: An example of the organization structure.

5.4.2 SECURITY ANALYSIS

Security Against Passive Attacks

By passive attacks, we mean statistical analysis by cloud servers or attacks from unauthorized individual users.

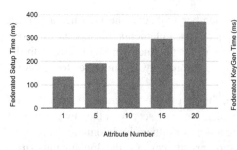

Figure 5.6: Federated setup.   Figure 5.7: Federated KeyGen.

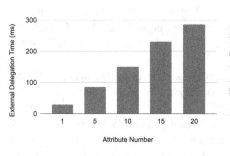
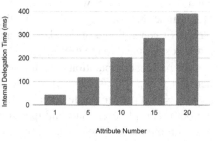

Figure 5.8: Internal delegation.   Figure 5.9: External delegation.

- **Init**: The adversary $\mathscr{A}$ commits to the challenge access structure $\mathbb{A}^*$ and the revoked identity set ID^* and sends this to the challenger.
- **Setup**: The challenger runs the setup algorithm. The master secret key MSK is kept secret, and the public parameters PK are given to the adversary $\mathscr{A}$.
- **Phase1**: The adversary $\mathscr{A}$ makes repeated private key queries $(\mathbf{U_i}, ID_i)_{i \in [1, q_1]}$, where if $\mathbf{S}_i$ satisfies $\mathbb{A}^*$ then the identity $ID_i = ID^*$.
- **Challenge**: The adversary submits two equal-length messages $\mathscr{M}_0$ and $\mathscr{M}_1$. In addition, the adversary gives a challenge including an LSSS access structure $\mathbb{A}^* = (M^*, \rho^*)$ and a set ID^* of revoked identities, such that ID^* must include all identities that were queried. The challenger picks up a random coin b, and encrypts $\mathscr{M}_b$ under the access structure $\mathbb{A}^*$ and the revoked identity set ID^*. Then the challenge ciphertext CT^* is sent to $\mathscr{A}$.
- **Phase2**: Repeat **Phase1** with the restriction that the queried sets of $(\mathbf{S_i}, ID_i)_{i \in [q_1+1, q]}$, where if $\mathbf{S}_i$ satisfies $\mathbb{A}^*$ then the identity $ID_i = ID^*$.
- **Guess**: The adversary outputs a guess bit b' of b. Define $\text{Adv}_{\mathscr{A}} = |\Pr[b' = b] - \frac{1}{2}|$ as the advantage of the adversary $\mathscr{A}$ winning the game.

Definition 5.1 (EIR-CP-ABE-Security). An EIR-CP-ABE scheme is secure if the advantage of any probabilistic polynomial time adversary $\mathscr{A}$ winning the above game is at most a negligible function of the security parameter.

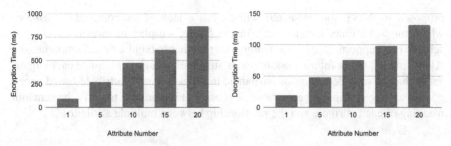

Figure 5.10: Data encryption.   Figure 5.11: Data decryption.

Security Against Colluding Users/Authorities

The proofs of security against colluding users and authorities are similar. Because of the limited space, here we just demonstrate the proof for resistance against colluding data consumers. Assume there are two individual users A and B with private key SK_A and SK_B as follows.

$$SK_A = (K_0 = g^\alpha g^{b^2 t_A}, \; L_a = g^{-s_{ID_A}^{-1} t_A}, \; L_u = g^{-t_A}, \; K_a = (g^{bs_{ID_A}} h_x^b)_A^t, \; K_u = (g^{bID_A} h_x)_A^t,$$

$$gbsID = g^{bs_{ID_A}}, \; hx = h_x, \; gbt = g^{bt_A}, \; ht = h_x^{t_A}, \; hbt = h_x^{bt_A}, \; s_{ID_A})_{x \in U_{ID_A}},$$

$$SK_B = (K_0 = g^\alpha g^{b^2 t_B}, \; L_a = g^{-s_{ID_B}^{-1} t_B}, \; L_u = g^{-t_B}, \; K_a = (g^{bs_{ID_B}} h_x^b)^{t_B}, \; K_u = (g^{bID_B} h_x)^{t_B},$$

$$gbsID = g^{bs_{ID_B}}, \; hx = h_x, \; gbt = g^{bt_B}, \; ht = h_x^{t_B}, \; hbt = h_x^{bt_B}, \; s_{ID_B})_{x \in U_{ID_B}}.$$

Since t_A and t_B are two different random integers. Even if A and B collude, i.e., put components of their private key together, they cannot produce a valid private key.

Security Against Colluding TC members

Federated efforts from members in the trust coalition are needed in two processes, i.e. federated setup and federated key generation for the root authority of each organization. As discussed, we assume the members in the trusted coalition have benefit collision and will not collude with each other. Hence the worst situation will be $N-1$ members collude. Since the final master secret key or the private key contains secret shares from all the members in the trust coalition, without the remaining honest member, the $N-1$ members cannot generate legitimate system parameters or private keys.

5.5 SUMMARY

In this chapter, to perform the ABE-based Attribute-Based Access Control model, we presented a new ABE scheme, which supports federation, delegation, interoperability, and identity-based revocation all at once. Compared with the ABAC model proposed by NIST, the presented scheme has a lack of environmental attributes, which might be time, location, *etc*. In the future, we plan to investigate an ABE scheme that supports all the nice features all together to build a more comprehensive ABAC model. Other future work includes attribute management. Since each organization has its own definition of attributes, in order to enable attribute-based access control among multiple organizations, it is of great importance to unify the semantic meaning of each attribute and the relationship between multiple attributes.

6 Search over ABE-Protected Data

Searchable encryption is a primitive, which not only protects data privacy of data owners but also enables data users to search over the encrypted data. Most existing searchable encryption schemes are in the single-user setting. There are few schemes in the multiple data users setting, i.e., encrypted data sharing. Among these schemes, most of the early techniques depend on a trusted third party with interactive search protocols or need cumbersome key management.

To remedy the defects, the most recent approaches borrow ideas from attribute-based encryption to enable Attribute-Based Keyword Search (ABKS). However, all these schemes incur high computational costs and are not suitable for mobile devices, such as mobile phones, with power consumption constraints. In this chapter, we develop new techniques that split the computation for the keyword encryption and trapdoor/token generation into two phases: a preparation phase that does the vast majority of the work to encrypt a keyword or create a token before it knows the keyword or the attribute list/access control policy that will be used. A second phase then rapidly assembles an intermediate ciphertext or trapdoor when the specifics become known. The preparation work can be performed while the mobile device is plugged into a power source, then it can later rapidly perform keyword encryption or token generation operations on the move without significantly draining the battery. We name the presented scheme as Online/Offline ABKS, in which it constructs an efficient multi-user searchable encryption scheme for mobile devices through moving the majority of the cost of keyword encryption and token generation into an offline phase.

6.1 INTRODUCTION

Storage services over cloud, e.g., Microsoft's Azure storage, Amazon's S3, and Apple iCloud, are a fundamental component of cloud computing, which allows the users to outsource their data to remote cloud servers. Data outsourcing relieves the users from maintaining their proprietary data, which is usually extremely costly. However, the cloud users would worry about their data privacy, since their private data are now placed on the cloud servers which are out of their trusted domains. Both malicious insiders, such as administrators, and outside attackers, such as hackers with root rights, may have full access to the server and consequently to users' data. Therefore, providing sufficient security and privacy protection on users' data is of great significance.

Encryption-before-outsourcing has been regarded as a fundamental means of protecting users' sensitive data against the untrusted cloud servers. Encryption hides all information about the plaintext, thus reducing security and privacy risks. However, it brings a new problem at the same time—search capabilities are removed from the data users. A naïve solution to solve this problem is to have a user download all encrypted data from the server and then search on them locally. That is, the user would indiscriminately download and decrypt all encrypted data, regardless of what data she is interested in. This solution is impractical since too much bandwidth, storage and computation resources are required and user devices have limited resources.

Searchable encryption (SE) [41] is an important enabling technique to solve the aforementioned problem. The schemes are built on the client/server model, where the untrusted cloud server stores encrypted data on behalf of one or more clients (i.e., the data owners or writers). To request content from the server, one or more clients (i.e., the data users or readers) are able to generate tokens for the server, which then searches on behalf of the client. This results in the following four SE architectures: single-owner/single-user (S/S); multi-owner/single-user (M/S); single-owner/multi-user (S/M); multi-owner/multi-user (M/M). Schemes in the S/S setting are called symmetric searchable encryption (SSE) [203], where a single client outsources his/her data and then searches over the data all by him/herself. Schemes in the M/S setting are called public key encryption with keyword search (PEKS) [31]. There exist several schemes on SSE and PEKS, however there are few approaches in the multiple data users' setting. In this chapter, we focus on the schemes in the M/M setting, i.e., encrypted data sharing. One motivating application for schemes in this setting is an encrypted file-sharing system. File owners store encrypted files on an untrusted cloud server and may want to share files only with particular users.

The M/M architecture was intensively researched between 2007 and 2008 [211, 212, 213, 214, 119] but seems to be out of interest until last year, when three schemes [198, 204, 239] were proposed in the literature. Earlier schemes (except for that in [119]) introduce a Trusted Third Party (TTP) for user authentication or re-encryption of the trapdoors. Assuming the existence of the trusted server is risky since the private data can be exposed by either software bugs and configuration errors at the trusted server or by a malicious administrator, who may give data access to unauthorized users (e.g., the competitor of a company) to make more profits. In addition, all these traditional methods [211, 212, 213, 214, 119] involve complicated key management, which incurs high overhead on the data owner end.

The most recent approaches [198, 204, 239] solve the aforementioned problems by encrypting keywords based on attributes, so that only users with the preset attributes can search over the encrypted data. Data owners don't need to perform cumbersome key management and can be offline after uploading the encrypted data. For sake of its nice properties and the increasing adoption of cloud computing, attribute-based keyword search will be further investigated in the literature and will be broadly applied in encrypted data sharing in the future. However, despite aforementioned nice properties, computation costs in the Keyword Encryption (KE) and Token Generation (TG) algorithm in all existing works scale with the number of attributes assigned to the keyword ciphertext and the size of the Boolean formula ascribed to a user's private key, respectively. In [204], the time consumption for index generation (i.e., keyword encryption) and the trapdoor generation is almost proportional to the number of attributes. In [239], the authors ran their ABKS scheme on a computer with Linux OS, 2.93GHz Intel Core™ Duo CPU, and 2GB RAM. From the experimental results, in the key-policy ABKS scheme, time consumption of token generation when there are 20 attributes is more than 0.6 second and is 1.5 seconds when the attributes number reaches 50. Time consumed in encrypting a keyword is more than 1 second when there are 20 attributes, and reaches more than 2.4 seconds when there are 50 attributes. These costs could impact several applications. For example, these schemes are not applicable on mobile devices with power consumption constraints. An exacerbating factor is that the cost for operations may vary widely between each ciphertext and token; thus, forcing a system to provision for load matching a worst-case scenario.

To address the aforementioned problem, we present an Online/Offline Attribute-Based Keyword Search (OO-ABKS) that splits the computation for the *KE* and *TG* algorithm into two phases: a preparation phase that does the vast majority of the work to encrypt a keyword or generates a trapdoor before it knows the keyword or the attribute list/access control policy that will be used. A second phase then rapidly assembles a keyword ciphertext or trapdoor when the specifics become known. To show how the presented solution can improve the online computation efficiency, we give a construction of the Online/Offline ABKS, scheme based on Zheng-Xu-Ateniese (ZXA) scheme [239] and compare the efficiency of the presented scheme with the basic one. Through simulation experiments, we show that by splitting keyword encryption and trapdoor generation into two phases, the offline phase does more than 97% of the work when the number of the attributes is greater than 10 and more than 99% when there are more than 50 attributes. Note that the total computation required between the offline and online phase is identical to the work required by the basic ZXA scheme.

The rest of this chapter is arranged as follows: Section 6.2 presents the related work. Section 6.3 describes the system and models that the searchable encryption scheme is built upon. Section 6.4 shows the syntax and security definition of an OO-ABKS scheme. Section 6.5 shows the construction of the proposed OO-ABKS scheme. Section 6.6 evaluates security of the proposed scheme. Section 6.7 provides performance evaluation. Section 6.8 concludes this chapter.

6.2   RELATED WORKS

In this section, we briefly review the relevant techniques in the research areas of searchable encryption solutions.

6.2.1   S/M SEARCHABLE ENCRYPTION

Curtmola et al. [64] proposed a generic combination of broadcast encryption and any S/S scheme to construct an S/M searchable encryption scheme. Recently, two provably secure schemes [182, 227], which are an example for the trade-off of security versus efficiency, were proposed. The search algorithm of Raykova et al. [182] is linear in the number of documents, but the scheme uses deterministic encryption and directly leaks the search pattern in addition to the access pattern. Yang et al. [227] achieve a higher level of security at the cost of efficiency. The search complexity is linear in the number of keywords per document. In addition, the schemes in this setting usually introduce a TTP for user authentication or re-encryption of the tokens and no security proof is provided.

6.2.2 M/M SEARCHABLE ENCRYPTION

Bellare et al. [24] propose an efficient scheme in the M/M setting by using deterministic encryption, at the cost of a weaker security model. Dong et al. [71] propose two schemes, where each user has its own unique key to encrypt, search, and decrypt data. In both schemes, a trusted key server is required to manage the keys. Bao et al. [22] propose a multi-user scheme, where the index generation and data encryption are interactive algorithms. Hwang and Lee [119] introduce the concept of multi-user public key encryption with conjunctive keyword search. They propose using multi-receiver public key encryption and randomness reuse to improve the computation and communication complexity. Wang et al. propose four searchable encryption schemes [211, 212, 213, 214] in the M/M setting. All these schemes either depend on a TTP (except for [119]) or incur high key management costs on the data user end.

Li et al. [144] propose a framework for authorized private keyword search over encrypted data and present two constructions. Their schemes depend on trusted third parties as well and the access control policies are not defined by data owners. The most recent schemes [198, 204, 239], as we introduced in Part I, borrow ideas from attribute-based encryption schemes to enable data owners to define the access policies. However, the keyword encryption and token generation time scales with the number of attributes assigned to the keyword ciphertext and the size of the Boolean formula ascribed to a user's private keys, respectively, thus making them unsuitable for mobile devices.

6.2.3 ATTRIBUTE-BASED ENCRYPTION

ABE, as a cryptographic means, is a popular method for enforcing access control policies. Basically, this technique allows entities with proper private keys to decrypt a ciphertext that is encrypted according to an access control policy. There are two variants of ABE schemes: KP-ABE (key-policy ABE) where the decryption key is associated to the access control policy [95], and CP-ABE (ciphertext-policy ABE) where the ciphertext is associated to the access control policy [220]. Hohenberger et al. [104] propose a new version of ABE to enhance key generation and encryption efficiency.

6.3 SYSTEM AND MODELS

The model of Online/Offline ABKS is pictured in Figure 6.1, which consists of the following participants: multiple data owners outsource their encrypted keywords and

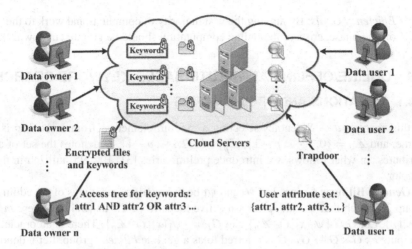

Figure 6.1: System model of attribute-based keyword search.

data to the cloud; multiple data users generate search tokens according to some interested keywords; the cloud server receives tokens from the users, conducts the search operations over outsourced encrypted data and returns the search results to the users. Note that, the data owners and users will encrypt keywords and generate tokens in an online/offline way.

6.3.1 THREAT MODEL

We consider the cloud servers to be honest-but-curious. This assumption is also employed by most previous works [41] on secure search over encrypted data. The cloud server honestly follows the designated protocol, but curiously infers additional private information based on the data available to him/her. Active attacks such as deleting and modifying the stored data or sending wrong results back to the data users are out of the scope of this chapter.

6.3.2 DESIGN GOALS

The presented online/offline attribute-based keyword search aims to achieve the following function and security goals.

- *Authorized Keyword Search*: The secure search system should enable data-owner-enforced search authorization, i.e., only users who meet the owner-defined access policy can obtain the valid search results.
- *Multiple Data Owners and Data Users*: The designed scheme should accommodate many data owners and data users. Specifically, each user is able to search over the encrypted data contributed by multiple data owners.
- *Security Goals*: We are mainly concerned with privacy requirements that are related with secure search and define them as follows: 1) secure against chosen-keyword attack, 2) keyword secrecy.

- *Efficiency Goals*: By moving the vast majority of computational work to the offline phase, costs of the online computation should be greatly cut down.

6.4 ONLINE/OFFLINE ATTRIBUTE-BASED KEYWORD SEARCH

6.4.1 NOTATIONS AND PRELIMINARIES

In this chapter, $a \leftarrow S$ denotes selecting a random element a from a set S. p is a prime, and $\mathbb{Z}_p = \{0, 1, \cdots, p-1\}$, $\mathbb{Z}_p^* = \{1, 2, \cdots, p-1\}$. U denotes the set of all attributes. In what follows, we introduce preliminaries [31, 104, 95] utilized in this chapter.

Generic **Bilinear** *Group*: Let ψ_0 and ψ_1 be two random encodings of the additive group $\mathbb{Z}_p^*$, such that ψ_0 and ψ_1 are injective maps from $\mathbb{Z}_p^*$ to $\{0, 1\}^m$, where $m > 3log(p)$. Let $G = \{\psi_0(x) | x \in \mathbb{Z}_p\}$ and $G_T = \{\psi_1(x) | x \in \mathbb{Z}_p\}$. There is an oracle to compute $e : G \times G \rightarrow G_T$. G is referred to as a generic *bilinear* group. Let g denote $\psi_0(1)$, g^x denote $\psi_0(x)$, $e(g, g)$ denote $\psi_1(1)$, and $e(g, g)^y$ denote $\psi_1(y)$.

Access Tree: An access tree represents access control policies. In an access tree, a leaf is associated with an attribute and an inner node represents a threshold gate. Let num_v be the number of children of node v, and label the children from the left to the right as $1, \cdots, num_v$. Let k_v, $1 \le k_v \le num_v$, be the threshold value associated with node v, where $k_v = 1$ represents the OR gate and $k_v = num_v$ represents the AND gate. Let $parent(v)$ denote the parent of node v, $ind(v)$ denote the label of node v, $attr(v)$ denote the attribute associated to leaf node v, lvs(T) denote the set of leaves of access tree T, and T_v denote the subtree of T rooted at node v.

Secret Distribution: Given an access tree T, we denote the algorithm for distributing a secret s according to T by: $\{q_v(0) | v \in \text{lvs(T)}\} \leftarrow \text{Share(T, } s)$. This algorithm generates a polynomial q_v of degree $k_v - 1$ for each node v in a top-down fashion (for each leaf node $k_v = 1$):

- If v is the root of T (i.e., $v = $ root), set $q_v(0) = s$ and randomly pick $k_v - 1$ coefficients for polynomial q_v.
- If v is a leaf of T, set $q_v(0) = q_{parent(v)}(ind(v))$.
- If v is an inner node (but not the root), set $q_v(0) = q_{parent(v)}(ind(v))$ and randomly select $k_v - 1$ coefficients for polynomial q_v

When the algorithm halts, each leaf v is associated with a value $q_v(0)$, which is the secret share of s at node v.

Secret Reconstruction: Given an access tree T and a set of values $\{E_{u_1}, \cdots, E_{u_m}\}$, where $u_1, \cdots, u_m$ are the leaves of T and attributes $attr(u_1), \cdots, attr(u_m)$ satisfy the access control policy represented by T. $E_{u_j} = e(g, h)^{q_{u_j}(0)}$ for $1 \le j \le m$, $g, h \in G$, e is a *bilinear* map, and $q_{u_1}(0), \cdots, q_{u_m}(0)$ are secret shares of s according to T, we denote the algorithm for reconstructing $e(g, h)^s$ by

$$e(g, h)^s \leftarrow \text{Combine(T, } \{E_{u_1}, \cdots, E_{u_m}\}).$$

This algorithm executes the following steps with respect to node v in a bottom-top fashion according to T:

- If attributes $attr(u_1), \cdots, attr(u_m)$ do not satisfy the access control policy represented by tree T_v, then set $E_v = \perp$
- If attributes $attr(u_1), \cdots, attr(u_m)$ satisfy the access control policy represented by tree T_v, then execute the following:
  - If v is a leaf, set $E_v = E_{u_j}(0) = e(g,h)^{q_{u_j}(0)}$, where $v = u_j$ for some j.
  - If v is an inner node (including the root), for v's children nodes $\{v_1, \cdots, v_{num_v}\}$, there exists a set of indices S such that $|S| = k_v, j \in S$, and attributes $attr(u_1), \cdots, attr(u_m)$ satisfy the access control policy represented by tree T_{v_j}. Set $E_v = \prod_{j \in S} E_{v_j}^{\Delta_{v_j}} = e(g,h)^{q_v(0)}$, where $\Delta_{v_j} = \prod_{l \in S, l \neq j} \frac{-j}{l-j}$.

When the algorithm halts, the root of T is associated with the reconstructed secret $E_{root} = e(g,h)^{q_{root}(0)} = e(g,h)^s$.

6.4.2 DEFINITIONS AND SECURITY

In this subsection, we show the definitions and security of the presented Online/Offline ABKS, schemes. Let S represent a set of attributes and T an access tree. For generality, we will define (I_{key}, I_{enc}) as the inputs to the key extraction and keyword encryption algorithm, respectively. In a KP-ABKS scheme $(I_{key}, I_{enc}) := (T, S)$, while in a CP-ABKS scheme, we have $(I_{key}, I_{enc}) := (S, T)$. We define the function f as follows:

$$f(I_{key}, I_{enc}) := \begin{cases} 1, & \text{if S satisfies the access control} \\ & \text{policy represented by T} \\ 0, & \text{otherwise.} \end{cases} \quad (6.1)$$

Definition 6.1. An Online/Offline ABKS (OO-ABKS) scheme consists of the following algorithms:

- Setup(1^ℓ)$\rightarrow$ (PK, MK). This algorithm initializes the public parameter PK and generates a master key MK.
- Extract(MK, I_{key}) $\rightarrow$ SK. This algorithm generates private key SK for a user according to I_{key}.
- Offline.Encrypt(PK, I_{enc}) $\rightarrow$ IC. This algorithm outputs an intermediate cipheretext IC.
- Online.Encrypt(PK, IC, w, I_{enc}) $\rightarrow$ CT$_w$. This algorithm encrypts keyword w to obtain ciphertext CT$_w$.
- Offline.TokenGen(SK, PK) $\rightarrow$ IT. This algorithm outputs an intermediate search token IT.
- Online.TokenGen(PK, IT, w') $\rightarrow$ TK$_{w'}$. This algorithm allows a data user to generate a search token TK$_{w'}$ for the keyword w'.

- Search(CT_w, $TK_{w'}$) $\rightarrow$ $\{0,1\}$. This algorithm is run by the cloud server and returns 1 if (i) $f(I_{key}, I_{enc}) = 1$ and (ii) $w = w'$; returns 0 otherwise.

According to the presented threat model, the data owners and the authorized data users are fully trusted, while the cloud server may attempt to infer private information based on the information available to him. Therefore, security means that the cloud server learns nothing beyond the search results. Specifically, for the cloud server modeled by a Probabilistic Polynomial Time (PPT) adversary $\mathscr{A}$, an OO-ABKS scheme is secure if the following two security requirements are satisfied.

Selective Security Against Chosen-Keyword Attack (SCKA): An OO-ABKS scheme is secure against SCKA if it satisfies the following requirement: If the adversary $\mathscr{A}$ does not obtain any matching search token, he would not infer any information about the keyword in the selective security model, where I_{enc} the adversary intends to attack must be determined before the system is bootstrapped. We formalize this security property via the following selective chosen-keyword attack game.

Setup: $\mathscr{A}$ chooses a non-trivial challenge I_{enc}^* (a trivial challenge I_{enc}^* is one that can be satisfied by any data user who does not have any secret key), and sends it to the challenger, who then runs Setup(1^{ℓ}) to generate the public parameter PK and the master key MK.

Phase 1: The challenger keeps a keyword list L_{kw}, which is initially empty. The adversary $\mathscr{A}$ can query the following oracles for polynomial times.

- $\mathscr{O}_{Extract}(I_{key})$: If $f(I_{key}, I_{enc}^*) = 1$, then abort, otherwise, the challenger returns to $\mathscr{A}$ the private key SK corresponding to I_{key}.
- $\mathscr{O}_{TokenGen}(I_{key}, w)$: The challenger generates private key SK with I_{key} and returns to $\mathscr{A}$ a search token TK_w = Online.TokenGen(PK, Offline.TokenGen(SK, PK), w). If $f(I_{key}, I_{enc}^*) = 1$, the challenger appends w to L_{kw}.

Challenge phase: The adversary $\mathscr{A}$ selects two keywords w_0 and w_1. The challenger selects $\lambda \leftarrow \{0,1\}$, computes $CT_{w_\lambda} \leftarrow$ Online.Encrypt(PK, Offline.Encrypt(PK, I_{enc}^*), w_λ, I_{enc}^*), and delivers CT_{w_λ} to $\mathscr{A}$. Note that we require $w_0, w_1 \notin L_{kw}$ in order to prevent the adversary $\mathscr{A}$ from trivially guessing λ with tokens received from $\mathscr{O}_{TokenGen}$.

Phase 2: $\mathscr{A}$ continues to query the oracles as in **Phase 1**. The restriction is that he is forbidden to query $\mathscr{O}_{TokenGen}$ with (I_{key}, w_0) or (I_{key}, w_1), if $f(I_{key}, I_{enc}^*) = 1$.

Guess: $\mathscr{A}$ outputs a bit λ', and wins if $\lambda' = \lambda$. Let $|\Pr[\lambda = \lambda'] - \frac{1}{2}|$ be the advantage of $\mathscr{A}$ winning the SCKA game.

Definition 6.2. An OO-ABKS scheme is selectively secure against chosen-keyword attack if the advantage of any PPT $\mathscr{A}$ winning the SCKA game is negligible in the security parameter.

Note that, the security definition above is in the selective setting. To be fully secure, the adversary will select the non-trivial challenge I_{enc}^* in the **Challenge**

Phase, with the constraint that the query I_{key} in **Phase 1** and **Phase 2** satisfies $f(I_{key}, I_{enc}^*) = 0$ and the challenge ciphertext is $CT_{w_\lambda} \leftarrow$ Online.Encrypt(PK, Offline.Encrypt(PK, I_{enc}^*), w_λ, I_{enc}^*).

Keyword secrecy: In the public-key setting, it is impossible to protect the search tokens against the keyword guessing attack [41]. Therefore, a weaker security notion called keyword secrecy [239] is used to assure that the probability $\mathscr{A}$ learns the keyword from the ciphertexts and search tokens is negligibly more than the probability of correct random keyword guess. This notion is formalized via the following game.

Setup: The challenger runs Setup(1^ℓ) to generate the public parameter PK and the master key MK.

Phase 1: The adversary $\mathscr{A}$ can query the following two types of oracles for polynomial times:

- $\mathscr{O}_{Extract}(I_{key})$: The challenger returns to $\mathscr{A}$ the private key SK corresponding to I_{key} and then adds I_{key} to the list L_{key}, which is initially empty.
- $\mathscr{O}_{TokenGen}(I_{key}, w)$: The challenger generates the private key SK with I_{key}, and returns to $\mathscr{A}$ a search token $TK_w =$ Online.TokenGen(PK, Offline.TokenGen(SK, PK), w).

Challenge Phase: $\mathscr{A}$ selects a non-trivial I_{enc}^* and sends it to the challenger, who selects w^* from the message space uniformly at random and selects I_{key}^* such that $f(I_{key}^*, I_{enc}^*) = 1$. The challenger runs $CT_{w^*} \leftarrow$ Online.Encrypt(PK, Offline.Encrypt(PK, I_{enc}^*), w^*, I_{enc}^*) and $TK_{w^*} \leftarrow$ Online.TokenGen(PK, Offline. TokenGen(SK, PK), w^*) and delivers (CT_{w^*}, TK_{w^*}) to $\mathscr{A}$. We require that $\forall I_{key} \in L_{key}, f(I_{key}, I_{enc}^*) = 0$.

Phase 2: The adversary $\mathscr{A}$ continues to query the key extraction and token generation oracle with the following restriction: the queried I_{key} must satisfy $f(I_{key}, I_{enc}^*) = 0$. $\mathscr{A}$ may also use some background knowledge or whatever feasible approaches and guesses q distinct keywords.

Guess: $\mathscr{A}$ outputs w', and wins the game if $w' = w^*$.

Definition 6.3. An OO-ABKS scheme achieves keyword secrecy if the probability that $\mathscr{A}$ wins the keyword secrecy game is at most $\frac{1}{|\mathscr{M}|-q} + \varepsilon$, where $\mathscr{M}$ is the keyword space, q is the number of distinct keywords that the adversary has attempted, and ε is a negligible in security parameter ℓ.

6.5 ABKS SCHEMES WITH ONLINE/OFFLINE ENCRYPTION AND TRAPDOOR GENERATION

In order to illustrate that the online/offline approach enhances efficiency of the ABKS schemes, in this section, we give a specific construction of the OO-ABKS scheme. Currently, there exist three ABKS schemes [198, 204, 239] in the literature: the scheme proposed in [204] only supports limited authorization policies with AND gates; the scheme of [198] is constructed on composite order *bilinear* groups, which is much more computation costly than prime order *bilinear* groups [84] and is less suitable for mobile devices with power consumption constraints. Therefore, we choose to extend the ZXA scheme proposed in [239]. In what follows, we assume a bound P on the maximum number of attributes that can be used to encrypt a keyword or generate a trapdoor.

The basic idea of the ABKS scheme is described as follows. There are two parts for each keyword ciphertext and search token, one for the keyword and the other for the access policy. Only when the attributes satisfy the access policy, can a user determine whether the keyword ciphertext and the search token are associated with the same keyword. In the following we will describe how this is enforced in the key-policy and ciphertext-policy ABE scheme respectively.

6.5.1 ONLINE/OFFLINE KP-ABKS

Let $H_1 : \{0,1\}^* \to G$ denote a hash function which is modeled as a random oracle, and $H_2 : \{0,1\}^* \to \mathbb{Z}_p$ denote an one-way hash function. g denotes the generator of group G. Select at random $t \leftarrow \mathbb{Z}_p$. The credentials of a data user are $A_v = g^{q_v(0)}H_1(att(v))^t, B_v = g^t$ for each leaf node v, where $q_v(0)$ is the share of leaf $v's$ secret ac according to the access tree T. The keyword ciphertext and search token are generated in the following way [239].

- Keyword w is encrypted into two parts: one is to "blend" w with randomness $r_1, r_2 \leftarrow \mathbb{Z}_p$ by letting $W' = g^{cr_1}$, $W = g^{a(r_1+r_2)}g^{bH_2(w)r_1}$ and $W_0 = g^{r_2}$, and the other is associated to attribute set $Atts$ in the way of setting $W_j = H_1(at_j)^{r_2}$ for each $at_j \in Atts$. r_2 ties these two parts together.
- Search token for a keyword w could be generated based on a set of credentials as follows. One part $tok_1 = (g^a g^{bH_2(w)})^s, tok_2 = g^{cs}$ where $s \leftarrow \mathbb{Z}_p$ is associated to the keyword w, the other part $A'_v = A^s_v$, $B'_v = B^s_v$ for each $v \in lvs(T)$ is associated with the credentials. The random component s ties these two components together.

The key idea is that when the attribute set $Atts$ satisfies the access tree T, the party running the search algorithm could use A'_v, B'_v and W_0, W_j to recover $e(g,g)^{acr_2s}$.

- Setup(1^ℓ) $\to$ (PK, MK). Select a *bilinear* map $e : G \times G \to G_T$, where G and G_T are cyclic groups of order p, which is an ℓ-bit prime. Let $H_1 : \{0,1\}^* \to G$ be a hash function modeled as random oracle and $H_2 : \{0,1\}^* \to \mathbb{Z}_p$ be a one-way hash function, select $a,b,c, \leftarrow \mathbb{Z}_p$ and $g \leftarrow G$ and set PK = $(H_1, H_2, e, g, p, g^a, g^b, g^c, G, G_T)$, MK = (a, b, c).
- Extract(MK, T) $\to$ SK. Execute Share(T, ac) to obtain secret share $q_v(0)$ of ac for each leaf $v \in lvs(T)$ on the access tree T. For each leaf $v \in lvs(T)$, pick $t \leftarrow \mathbb{Z}_p$, and compute $A_v = g^{q_v(0)}H_1(att(v))^t$ and $B_v = g^t$. Set SK = (T, $\{(A_v, B_v)|v \in lvs(T)\}$).
- Offline.Encrypt(PK) $\to$ IC. Select $r_1, r_2 \leftarrow \mathbb{Z}_p$, and compute $W' = g^{cr_1}$, $W = g^{a(r_1+r_2)}$ and $W_0 = g^{r_2}$. For each $at_j \in U$, compute $W_j = H_1(at_j)^{r_2}$. Set IC = $(r_1, U, W', W, W_0, \{W_j|at_j \in U\})$.

- Online.Encrypt(PK, IC, w, S) $\rightarrow CT_w$. The online encryption algorithm takes as input the public parameters, intermediate ciphertext IC, a set of attributes S $= (A_1, A_2, \cdots, A_{k \leq P})$ and a keyword w. It computes $W = IC.W \cdot g^{br_1 H_2(w)}$ and sets the ciphertext as $CT_w = (S, IC.W', W, IC.W_0, \{IC.W_j | at_j \in S\})$.
- Offline.TokenGen(SK, PK) $\rightarrow$ IT. Select $s \leftarrow \mathbb{Z}_p$, and compute $A'_v = A^s_v$, $B'_v = B^s_v$ for each $v \in lvs(SK.T)$. Compute $tok_1 = g^{as}$ and $tok_2 = g^{cs}$. Set IT $= (s, SK.T, tok_1, tok_2, \{(A'_v, B'_v) | v \in lvs(SK.T)\})$.
- Online.TokenGen(PK, IT, w') $\rightarrow TK_{w'}$. Compute $\cdot tok'_2 = IT.tok_2 \cdot g^{bsH_2(w')}$ and set the search token as $TK_{w'} = (IT.(SK.T), IT.tok_1, tok'_2, \{(IT.A'_v, IT.B'_v) | v \in lvs(IT.(SK.T))\})$.
- Search(CT_w, $TK_{w'}$) $\rightarrow \{0,1\}$. Given attribute set S specified in CT_w, select an attribute set AS satisfying the access tree T=IT.(SK.T) specified in $TK_{w'}$. If AS does not exist, return 0; otherwise, for each $at_j \in AS$, compute $E_v = e(IT.A'_v, IC.W_0)/e(IT.B'_v, IC.W_j) = e(g,g)^{sr_2 q_v(0)}$, where $attr(v) = at_j$ for $v \in lvs(T)$. Compute $e(g,g)^{sr_2 q_{root}(0)} \leftarrow$ Combine(T, $\{E_v | attr(v) \in AS\}$) so that $E_{root} = e(g,g)^{acsr_2}$. Return 1 if $e(IC.W', IT.tok_1)E_{root} = e(W, tok'_2)$, and 0 otherwise.

6.5.2 ONLINE/OFFLINE CP-ABKS

Different from the KP-ABKS scheme, in the CP-ABKS scheme, the access policy is enforced on the ciphertext.

- Setup(1^ℓ): Select $e : G \times G \rightarrow G_T$, where G and G_T are cyclic groups of order p, which is an ℓ-bit prime. $H_1 : \{0,1\}^* \rightarrow G$ is a hash function modeled as random oracle and $H_2 : \{0,1\}^* \rightarrow \mathbb{Z}_p$ is an one-way hash function, select $a,b,c \leftarrow \mathbb{Z}_p$ and $g \leftarrow G$, set PK $= (H_1, H_2, e, g, p, g^a, g^b, g^c, G, G_T)$, MK $= (a, b, c)$.
- Extract(MK, S): Select r, $r_j \leftarrow \mathbb{Z}_p$, compute $A = g^{\frac{ac-r}{b}}$ $A_j = g^r H_1(at_j)^{r_j}$ and $B_j = g^{r_j}$. Set SK $= (S, A, \{(A_j, B_j) | at_j \in S\})$.
- Offline.Encrypt(PK, T): Select $r_1, r_2 \leftarrow \mathbb{Z}_p$, and compute $W = g^{cr_1}$, $W_0 = g^{a(r_1+r_2)}$ and $W' = g^{br_2}$. Compute secret shares of r_2 for each leave of access tree T as $\{q_v(0) | v \in lvs(T)\} \leftarrow$ Share(T, r_2). For each $v \in lvs(T)$, compute $W_v = g^{q_v(0)}$ and $D_v = H_1(att(v))^{q_v(0)}$. Set intermediate ciphertext IC $= (r_1, T, W, W_0, W', \{(W_v, D_v) | v \in lvs(T)\})$.
- Online.Encrypt(PK, IC, w): Compute $W'_0 = W_0 \cdot g^{br_1 H_2(w)}$, and set the ciphertext $CT_w = (IC.T, IC.W, W'_0, IC.W', \{(IC.W_v, IC.D_v) | v \in lvs(IC.T)\})$.
- Offline.TokenGen(SK, PK): Select $s \leftarrow \mathbb{Z}_p$, and compute $tok_1 = g^{as}$, $tok_2 = g^{cs}$ and $tok_3 = A^s = g^{(acs-rs)/b}$. For each $at_j \in$ Atts, compute $A'_j = A^s_j$ and $B'_j = B^s_j$. Set intermediate token IT $= (s, SK.S, tok_1, tok_2, tok_3, \{(A'_j, B'_j) | at_j \in SK.S\})$.
- Online.TokenGen(PK, IT, w'): Compute $tok'_1 = IT.tok_1 \cdot g^{bsH_2(w)}$, and set the token as $TK_{w'} = (IT.(SK.S), tok'_1, IT.tok_2, IT.tok_3, \{(IT.A'_j, IT.B'_j) | at_j \in IT.(SK.S)\})$.

- Search(CT_w, $TK_{w'}$): Given attribute set IT.(SK.S) as specified in $TK_{w'}$, select an attribute set AS that satisfies the access tree T=IC.T specified in CT_w. If AS does not exist, return 0; otherwise, for each $at_j \in AS$, compute $E_v = e(IT.A'_j, IC.W_v)/e(IT.B'_j, IC.D_v) = e(g,g)^{rsq_v(0)}$, where $attr(v) = at_j$ for $v \in lvs(T)$. Compute $e(g,g)^{rsq_{root}(0)} \leftarrow$ Combine(T, $\{E_v|attr(v) \in AS\}$) and $E_{root} = e(g,g)^{rsr_2}$. Return 1 if $e(W'_0, IT.tok_2) = e(IC.W, tok'_1)E_{root}$ $e(IT.tok_3, IC.W')$, and 0 otherwise.

Note that the online/offline technique can be used to enhance efficiency in other attribute-based searchable encryption schemes [198, 204], even for future constructed schemes, whose computation costs scale with the number of the attributes.

6.6   SECURITY ANALYSIS

6.6.1   DEFINITION FOR ABKS

Definition 6.4. An ABKS scheme consists of the following four algorithms [239]:

- Setup(1^ℓ)$\rightarrow$ (PK, MK). This algorithm initializes the public parameter PK and generates a master key MK.
- Extract(MK, I_{key}) $\rightarrow$ SK. This algorithm generates private key SK for a user according to I_{key}.
- Encrypt(PK, I_{enc}, w) $\rightarrow$ CT_w. This algorithm outputs keyword ciphertext CT_w.
- TokenGen(SK, PK, w') $\rightarrow$ $TK_{w'}$. This algorithm outputs a search token $TK_{w'}$ for keyword w'.
- Search(CT_w, $TK_{w'}$) $\rightarrow$ $\{0,1\}$. This algorithm is run by the cloud server and returns 1 if (i) $f(I_{key}, I_{enc}) = 1$ and (ii) $w = w'$; returns 0 otherwise.

6.6.2   SELECTIVE SECURITY AGAINST CHOSEN-KEYWORD ATTACK

The selectively chosen-keyword attack game SCKA for ABKS is defined as follows [239]:

Setup: Adversary $\mathscr{A}$ chooses a non-trivial challenge I^*_{enc} (a trivial challenge I^*_{enc} is one that can be satisfied by any data user who does not have any secret key), and sends it to the challenger, who then runs Setup(1^ℓ) to generate the public parameter PK and the master key MK.

Phase 1: The challenger keeps a keyword list L_{kw}, which is initially empty. The adversary $\mathscr{A}$ can query the following oracles for polynomial times.

- $\mathscr{O}_{Extract}(I_{key})$: If $f(I_{key}, I^*_{enc}) = 1$, then abort, otherwise, the challenger returns to $\mathscr{A}$ the private key SK corresponding to I_{key}.
- $\mathscr{O}_{TokenGen}(I_{key}, w)$: The challenger generates private key SK with I_{key} and returns to $\mathscr{A}$ a search token $TK_w \leftarrow$ TokenGen(SK, PK, w). If $f(I_{key}, I^*_{enc}) = 1$, the challenger appends w to L_{kw}.

Challenge phase: The adversary $\mathscr{A}$ selects two keywords w_0 and w_1. The challenger selects $\lambda \leftarrow \{0,1\}$, computes $CT_{w_\lambda} \leftarrow$ Encrypt(PK, I_{enc}, w_λ) and delivers CT_{w_λ} to $\mathscr{A}$. Note that we require $w_0, w_1 \notin L_{kw}$ in order to prevent the adversary $\mathscr{A}$ from trivially guessing λ with tokens got from $\mathscr{O}_{TokenGen}$.

Phase 2: $\mathscr{A}$ continues to query the oracles as in **Phase 1**. The restriction is that he is forbidden to query $\mathscr{O}_{TokenGen}$ with (I_{key}, w_0) or (I_{key}, w_1), if $f(I_{key}, I_{enc}^*) = 1$.

Guess: $\mathscr{A}$ outputs a bit λ', and wins if $\lambda' = \lambda$. Let $|\Pr[\lambda = \lambda'] - \frac{1}{2}|$ be the advantage of $\mathscr{A}$ winning the SCKA game.

Definition 6.5. An ABKS scheme is selectively secure against chosen-keyword attack if the advantage of any PPT(probabilistic polynomial time) $\mathscr{A}$ winning the above SCKA game is negligible in the security parameter.

6.6.3   KEYWORD SECRECY

The authors of [239] did not introduce details of the "**Guess**" phase. Here, we refine the keyword secrecy definition of an ABKS scheme and show how the adversary guesses the q distinct keywords. In particular, the keyword secrecy of an ABKS scheme is defined by the following game:

Setup: The challenger runs Setup(1^ℓ) to generate the public parameter PK and the master key MK.

Phase 1: The adversary $\mathscr{A}$ can query the following two types of oracles for polynomial times:

- $\mathscr{O}_{Extract}(I_{key})$: The challenger returns to $\mathscr{A}$ the private key SK corresponding to I_{key} and then adds I_{key} to the list L_{key}, which is initially empty.
- $\mathscr{O}_{TokenGen}(I_{key}, w)$: The challenger generates the private key SK with I_{key}, and returns to $\mathscr{A}$ a search token $TK_w = $ TokenGen(SK, PK, w).

Challenge Phase: $\mathscr{A}$ selects a non-trivial I_{enc}^* and sends it to the challenger, who selects w^* from the message space uniformly at random and selects I_{key}^* such that $f(I_{key}^*, I_{enc}^*) = 1$. The challenger runs $CT_{w^*} \leftarrow$ Encrypt(PK, I_{enc}^*, w^*) and $TK_{w^*} \leftarrow$ TokenGen(SK, PK, w^*) and delivers (CT_{w^*}, TK_{w^*}) to $\mathscr{A}$. We require that $\forall I_{key} \in L_{key}, f(I_{key}, I_{enc}^*) = 0$.

Phase 2: The adversary $\mathscr{A}$ continues to query the key extraction and token generation oracle with the following restriction: the queried I_{key} must satisfy $f(I_{key}, I_{enc}^*) = 0$. $\mathscr{A}$ may also use some background knowledge or whatever feasible approaches and guesses q distinct keywords.

Guess: $\mathscr{A}$ outputs a keyword w', and wins the game if $w' = w^*$ holds.

Definition 6.6. An ABKS scheme achieves keyword secrecy if the probability that $\mathscr{A}$ wins the keyword secrecy game is at most $\frac{1}{|\mathscr{M}|-q} + \varepsilon$, where $\mathscr{M}$ is the keyword space, q is the number of distinct keywords that the adversary has attempted, and ε is a negligible function of security parameter ℓ.

6.6.4 CRYPTOGRAPHIC ASSUMPTION

Given $(g, f, h, f^{r_1}, g^{r_2}, Q)$ where $g, f, h, Q \leftarrow G$, $r_1, r_2 \leftarrow \mathbb{Z}_p$, DL assumption says that any probabilistic polynomial-time algorithm $\mathscr{A}$ can determine whether $Q = h^{r_1 + r_2}$ holds at most with a negligible advantage in security parameter ℓ, where "advantage" is defined as

$$|\Pr[\mathscr{A}(g, f, h, f^{r_1}, g^{r_2}, h^{r_1 + r_2}) = 1] - \Pr[\mathscr{A}(g, f, h, f^{r_1}, g^{r_2}, Q) = 1]|.$$

6.6.5 SECURITY OF ZXA

Theorem 6.1. *[239] Given the DL assumption and one-way hash function H_2, the KP-ABKS scheme is selectively secure against chosen-keyword attack in the random oracle model.*

Proof of Theorem 6.1. *[239] If there is a PPT adversary $\mathscr{A}$, who wins the SCKA game with advantage μ, then a challenger algorithm that solves the DL problem with advantage $\frac{\mu}{2}$ could be constructed.*

The DL instance is $(g, h, f, f^{r_1}, g^{r_2}, Q)$, where g, f, and hQ are random elements from G, and r_1, r_2 are randomly chosen from $\mathbb{Z}_p$. The challenger simulates the SCKA game in the following way.

- **Setup:** *The challenger sets $g^a = h$ and $g^c = f$, where a and c are unknown, selects $d \rightarrow \mathbb{Z}_p$ and computes $g^b = f^d = g^{cd}$ by implicitly defining $b = cd$. Let H_2 be an one-way hash function and $pm = (e, g, p, h, f^d, f)$ and $mk = (d)$. The adversary $\mathscr{A}$ selects an attribute set $Atts^*$ and gives it to the challenger. The random oracle $0_{H_1}(at_j)$ is defined as follows:*
  - *If at_j has not been queried previously,*
    - *if $at_j \in Atts^*$, select $\beta_j \leftarrow \mathbb{Z}_p$, add $(at_j, \alpha_j = 0, \beta_j)$ to O_{H_1}, and return g^{β_j};*
    - *otherwise, select $\alpha_j, \beta_j \leftarrow \mathbb{Z}_p$, add $(at_j, \alpha_j, \beta_j)$ to O_{H_1}, and return $f^{\alpha_j} g^{\alpha_j}$.*
  - *otherwise, retrieve (α_j, β_j) from O_{H_1} and return $f^{\alpha_j} g^{\beta_j}$.*
- **Phase 1:** *$\mathscr{A}$ adaptively queries the following oracles for polynomial times, and the challenger keeps a keyword list L_{kw}, which is empty initially. The following two procedures are defined to determine the polynomial for each node of T.*
  - *PolySat$(T_v, Atts^*, \lambda_v)$: Given secret λ_v, this procedure determines the polynomial for each node of T_v rooted at v when $F(Atts^*, T_v) = 1$. It works as follows: Suppose the threshold value of node v is k_v, it sets $q_v(0) = \lambda_v$ and picks $k_v - 1$ coefficients randomly to fix the polynomial q_v. For each child node v' of v, recursively call PolyUnsat$(T_{v'}, Atts^*, \lambda_{v'})$ where $\lambda_{v'}(Index(v'))$.*
  - *PolyUnsat$(T_v, Atts^*, g^{\lambda_v})$: Given element $g^{\lambda_v} \in G$, where the secret λ_v is unknown, this procedure determines the polynomial for each node of*

T_v rooted at v where $F(Atts^*, T_v) = 0$ as follows. Suppose the threshold value of the node v is k_v. Let V be the empty set. For each child node v' of v, if $F(Atts, T_{v'}) = 1$, then set $V = V \cup \{v'\}$. Because $F(Atts, T_v) = 0$, then $|V| < k_v$. For each node $v' \in V$, it selects $\lambda_{v'} \leftarrow \mathbb{Z}_p$, and sets $q_v(Index(v')) = \lambda_{v'}$. Finally it fixes the remaining $k_v - |V|$ points of q_v randomly to define q_v and makes $g^{q_v(0)} = g^{\lambda_v}$. For each child node v' of v,

- if $F(Atts^*, T_{v'}) = 1$, then run $PolySat(T_{v'}, Atts^*, q_v(Index(v')))$, where $q_v(Index(v'))$ is known to the challenger.
- otherwise, call $PolyUnsat(T_{v'}, Atts^*, g^{\lambda_{v'}})$, where $g^{\lambda_{v'}} = g^{q_v(Index(v'))}$ is known to the challenger.

With the above two procedures, the challenger runs $PolyUnsat(T, Atts^*, g^a)$, by implicitly defining $q_{root}(0) = a$. Then for each $v \in lvs(T)$, the challenger gets $q_v(0)$ if $att(v) \in Atts^*$, and gets $g^{q_v(0)}$ otherwise. Because $cq_v(0)$ is the secret share of ac, due to the linear property, the challenger generates credentials for each $v \in lvs(T)$ as follows:

- If $att(v) = at_j$ for some $at_j \in Atts^*$: Select $t \leftarrow \mathbb{Z}_p$, set $A_v = f^{q_v(0)} g^{\beta_j t} = g^{cq_v(0)} H_1(att(v))^t$ and $B_v = g^t$;
- If $att(v) \notin Atts^*$ (assuming $att(v) = at_j$): Select $t' \leftarrow \mathbb{Z}_p$, set $A_v = (g^{q_v(0)})^{\frac{-\beta_j}{\alpha_j}} (f^{\alpha_j} g^{\beta_j})^{t'}$ and $B_v = g^{q_v(0) \frac{-1}{\alpha_j}} g^{t'}$. Note that (A_v, B_v) is a valid credential.

Eventually, the challenger returns $sk = \{(A_v, B_v) | v \in lvs(T)\}$ to $\mathscr{A}$. $O_{TokenGen}(T, w)$: The challenger runs $O_{KeyGen}(T)$ to get $sk = (T, \{A_v, B_v | v \in lvs(T)\})$, computes $tk \leftarrow TokenGen(sk, w)$, and returns tk to $\mathscr{A}$. If $F(Atts, T) = 1$, the challenger adds w to the keyword list L_{kw}.

- **Challenge phase**: $\mathscr{A}$ chooses two keywords w_0 and w_1 of equal length, such that $w_0, w_1 \notin L_{kw}$. The challenger outputs cph^* as:
  - Select $\lambda \leftarrow \{0, 1\}$
  - For each $at_j \in Atts^*$, set $W_j = (g^{r_2})^{\beta_j}$.
  - Set $W' = f^{r_1}$, $W = Q(f^{r_1})^{dH_2(w_\lambda)}$, and $W_0 = g^{r_2}$.
  - Set $cph^* = (Atts^*, W', W, W_0, \{W_j | at_j \in Atts^*\})$ and returns cph^* to $\mathscr{A}$.

- **Phase 2**: $\mathscr{A}$ continues to query the oracles as in Phase 1. The only restriction is that (T, w_0) and (T, w_1) cannot be the input to $O_{TokenGen}$ if $F(Atts^*, T) = 1$.

- **Guess**: Finally, $\mathscr{A}$ outputs a bit λ' and gives it to the challenger. If $\lambda' = \lambda$, then the challenger outputs $Q = h^{r_1 + r_2}$, otherwise, it outputs $Q \neq h^{r_1 + r_2}$

This completes the simulation. In the challenge phase, if $Q = h^{r_1 + r_2}$, then cph^* is a valid ciphertext of w_λ, so the probability of $\mathscr{A}$ outputting $\lambda = lambda'$ is $\frac{1}{2} + \mu$. If Q is an element randomly selected from G, then cph^* is not a valid ciphertext of w_λ. The probability of $\mathscr{A}$ outputting $\lambda = \lambda'$ is $\frac{1}{2}$ since W is an random element in G. Therefore, the probability of the challenger correctly guessing whether Q is equal to $h^{r_1 + r_2}$ with the DL instance $(g, h, f, f^{r_1}, g^{r_2}, Q)$ is $\frac{1}{2}(\frac{1}{2} + \mu + \frac{1}{2}) = \frac{1}{2} + \frac{\mu}{2}$. That is, the *challenger solves the DL problem with advantage $\frac{\mu}{2}$ if $\mathcal{A}$ wins the SCKA game with an advantage μ.*

Theorem 6.2. *[239] Given the one-way hash function H_2, the KP-ABKS scheme achieves keyword secrecy in the random oracle model.*

Proof of Theorem 6.2. *[239] A challenger that exploits the keyword secrecy game is constructed as follows.*

- **Setup:** *The challenger selects $a, b, c \leftarrow \mathbb{Z}_p$, $f \leftarrow G$. Let H_2 be an one-way hash function and $pm = (e, g, g^a, g^b, g^c, f)$ and $mk = (a, b, c)$.*
  The random oracle $O_{H_1}(at_j)$ is simulated as follows: If at_j has not been queried before, the challenger selects $\alpha_j \leftarrow \mathbb{Z}_p$, adds (at_j, α_j) to O_{H_1}, and returns g^{α_j}; otherwise, the challenger retrieves α_j from O_{H_1} and returns g^{α_j}.

- **Phase 1:** *$\mathcal{A}$ can adaptively query the following oracles for polynomial times.*
  - *$O_{KeyGen}(T)$: The challenger generates $sk \leftarrow KeyGen(T, mk)$ and returns sk to $\mathcal{A}$. It adds T to the list L_{KeyGen}, which is initially empty.*
  - *$O_{TokenGen}(T, w)$: The challenger runs $O_{KeyGen}(T)$ to obtain $sk = (T, \{A_v, B_v | v \in lvs(T)\})$, computes $tk \leftarrow TokenGen(sk, w)$, and returns tk to $\mathcal{A}$.*

- **Challenge Phase:** *$\mathcal{A}$ selects an attribute set $Atts^*$. The challenger chooses an access control policy that is represented as T^*, such that $F(Atts^*, T^*) = 1$, computes $sk^* \leftarrow KeyGen(mk, T^*)$. By taking as input $Atts^*$ and sk^*, it selects w^* from keyword space uniformly at random, and computes cph^* and tk^* with Enc and $TokenGen$. $Atts^*$ should satisfy the requirement defined in the keyword secrecy game.*

- **Guess:** *Finally, $\mathcal{A}$ outputs a keyword w' and gives it to the challenger. The challenger computes $cph' \leftarrow Enc(Atts, w')$ and if $Search(tk^*, cph') = 1$, then $\mathcal{A}$ wins the game.*
  This finishes the simulation. Suppose $\mathcal{A}$ has already attempted q distinct keywords before outputting w', we can see that the probability of $\mathcal{A}$ winning the keyword secrecy game is at most $\frac{1}{|\mathcal{M}| - q} + \varepsilon$. This is because the size of the remaining keyword space is $|\mathcal{M}| - q$, and as the H_2 is an one way secure hash function, meaning deriving w^* from $H_2(w^*)$ is at most a negligible probability ε. Therefore, given q distinct keywords $\mathcal{A}$ has attempted, the probability of $\mathcal{A}$ winning the keyword secrecy game is at most $\frac{1}{|\mathcal{M}| - q} + \varepsilon$.

Theorem 6.3. *[239] Given the one-way hash function H_2, the CP-ABKS scheme is selectively secure against chosen-keyword attack in the generic bilinear group model.*

Proof of Theorem 6.3. *[239] The following shows the ZXA CP-ABKS scheme is selectively secure against chosen-keyword attack in the generic bilinear group model, where H_1 is modeled as a random oracle and H_2 is a one-way hash function.*

In the SCKA game, $\mathscr{A}$ attempts to distinguish
$g^{a(r_1+r_2)}g^{br_1H_2(w_0)}$ *from* $g^{a(r_1+r_2)}g^{br_1H_2(w_1)}$. *Given* $\theta \leftarrow \mathbb{Z}_p$, *the probability of distinguishing* $g^{a(r_1+r_2)}g^{br_1H_2(w_0)}$ *from* g^θ *is equal to that of distinguishing* g^θ *from* $g^{a(r_1+r_2)}g^{br_1H_2(w_1)}$. *Therefore, if* $\mathscr{A}$ *has advantage* ε *in breaking the SCKA game, then it has advantage* $\frac{\varepsilon}{2}$ *in distinguishing* $g^{a(r_1+r_2)}g^{br_1H_2(w_0)}$ *from* g^θ. *Thus, let us consider a modified game where* $\mathscr{A}$ *can distinguish* $g^{a(r_1+r_2)}$ *from* g^θ. *The modified SCKA game is described as follows:*

- **Setup:** *The challenger chooses* $a,b,c \leftarrow \mathbb{Z}_p$ *and sends public parameters* (e,g,p,g^a,g^b,g^c) *to* $\mathscr{A}$. $\mathscr{A}$ *chooses an access tree* T^*, *which is sent to the challenger.* $H_1(at_j)$ *is simulated as follows: If* at_j *has not been queried before, the challenger chooses* $\alpha_j \leftarrow \mathbb{Z}_p$, *adds* (at_j,α_j) *to* O_{H_1} *and returns* g^{α_j} ; *otherwise the challenger returns* g^{α_j} *by retrieving* α_j *from* O_{H_1}.
- **Phase 1:** $\mathscr{A}$ *can query* O_{KeyGen} *and* $O_{TokenGen}$ *as follows:*
  - $O_{KeyGen}(Atts)$: *The challenger selects* $r^{(t)} \leftarrow \mathbb{Z}_p$ *and computes* $A = g^{(ac+r^{(t)})/b}$. *For each attribute* $at_j \in Atts$, *the challenger chooses* $r_j^{(t)} \leftarrow \mathbb{Z}_p$, *computes* $A_j = g^{r^{(t)}}g^{\alpha_j r_j^{(t)}}$ *and* $B_j = g^{r_j^{(t)}}$, *and returns* $(Atts,A,\{(A_j,B_j)|a_tj \in Atts\})$.
  - $O_{TokenGen}(Atts,w)$: *The challenger queries* $O_{KeyGen}(Atts)$ *to get* $sk = (Atts,A,\{(A_j,B_j)|at_j \in Atts\})$ *and returns* $tk = (Atts,tok_1,tok_2,tok_3,\{(A_j',B_j')|at_j \in Atts\})$ *where* $tok1 = (g^a g^{bH_2(w)})^s$, $tok_2 = g^{cs}$, $tok_3 = A^s$, $A_j' = A_j^s$ *and* $B_j' = B_j^s$ *by selecting* $s \leftarrow \mathbb{Z}_p$. *If* $F(Atts,T^*) = 1$, *the challenger adds* w *to the keyword List* L_{kw}.
- **Challenge phase:** *Given two keywords* w_0, w_1 *of equal length where* $w_0,w_1 \in L_{kw}$, *the challenger chooses* $r_1,r_2 \leftarrow \mathbb{Z}_p$, *and computes secret shares of* r_2 *for each leaves in* T^*. *The challenger selects* $\lambda \leftarrow \{0,1\}$. *If* $\lambda = 0$, *it outputs*

$$W = g^{cr_1}, W_0 = g^\theta, W' = g^{br_2}, \{(W_v = g^{q_v(0)}, D_v = g^{\alpha_j q_v(0)})|v \in lvs(T^*), att(v) = at_j\}$$

by selecting $\theta \in \mathbb{Z}_p$; *otherwise it outputs*

$$W = g^{cr_1}, W_0 = g^{r_1+r_2}, W' = g^{br_2}, \{(W_v = g^{q_v(0)}, D_v = g^{\alpha_j q_v(0)})|v \in lvs(T^*), att(v) = at_j\}$$

- **Phase 2:** *This is the same as in the SCKA game.*

We can see that if $\mathscr{A}$ can construct $e(g,g)^{\delta a(r_1+r_2)}$ for some g^δ that can be composed from the oracle outputs he has already queried, then $\mathscr{A}$ can use it to distinguish g^θ from $g^{a(r_1+r_2)}$. Therefore, we need to show that $\mathscr{A}$ can construct $e(g,g)^{\delta a(r_1+r_2)}$ for some g^δ with a negligible probability. That is, $\mathscr{A}$ cannot gain non-negligible advantage in the SCKA game.

In the generic group model, ψ_0 and ψ_1 are random injective maps from $\mathbb{Z}_p$ into a set of p^3 elements. Then the probability of $\mathscr{A}$ guessing an element in the image of ψ_0

and ψ_1 is negligible. Recall that $G = \{\psi_0(x) | x \in \mathbb{Z}_p\}$ and $GT = \{\psi_1(x) | x \in \mathbb{Z}_p\}$. Hence, let us consider the probability of $\mathscr{A}$ constructing $e(g,g)^{\delta a(r_1+r_2)}$ for some $\delta \in \mathbb{Z}_p$ from the oracle outputs he has queried.

We list all terms that can be queried to the group oracle G_T in the Table 6.1. Let us consider how to construct $e(g,g)^{\delta a(r_1+r_2)}$ for some δ. Because r_1 only appears in the term cr_1, δ should contain c in order to construct $e(g,g)^{\delta a(r_1+r_2)}$. That is, let $\delta = \delta'c$ for some δ' and $\mathscr{A}$ wishes to construct $e(g,g)^{\delta'a(r_1+r_2)}$. Therefore, $\mathscr{A}$ needs to construct $\delta'acr_2$, which will use terms br_2 and $(ac+r^{(t)})/b$. Because $(br_2)(ac+r^{(t)})/b = acr_2 + r^{(t)}r_2$, $\mathscr{A}$ needs to cancel $r^{(t)}r_2$, which needs to use the terms α_j , $r^{(t)} + \alpha_j r_j^{(t)}$, $q_v(0)$ and $\alpha_j q_v(0)$ because $q_v(0)$ is the secret share of r_2 according to T^*. However, it is impossible to construct $r^{(t)}r_2$ with these terms because $r^{(t)}r_2$ only can be reconstructed if the attributes corresponding to $r_j^{(t)}$ of $r^{(t)} + \alpha_j r_j^{(t)}$ satisfies the access tree T^*. Therefore, we can conclude that $\mathscr{A}$ gains a negligible advantage in the modified game, which means that $\mathscr{A}$ gains a negligible advantage in the SCKA game. This completes the proof.

Table 6.1

Possible terms for querying group oracle

a	$r_j^{(t)}$	$s(ac+r^{(t)})/b$	cr_1
b	$r^{(t)} + \alpha_i r_j^{(t)}$	$s(r_j^{(t)})$	$q_v(0)$
c	$(ac+r^{(t)})/b$	$s(r^{(t)} + \alpha_j r_j^{(t)})$	$\alpha_j q_v(0)$
α_j	cs	$s(a+bH_2(w))$	br_2

Theorem 6.4. *[239] Given the one-way hash function H_2, the CP-ABKS scheme achieves keyword secrecy in the random oracle model.*

Proof of Theorem 6.4. *Theorem 6.4 could be proved in the same way as Theorem 6.3.*

6.6.6   SECURITY OF OO-ABKS

The security of the Online/Offline KP-ABKS scheme is based on the security of the ZXA scheme in the key policy setting and the proof of the following two theorems is provided in the Appendices.

Theorem 6.5. *The above Online/Offline KP-ABKS scheme is selectively secure against chosen-keyword attack in the random oracle model assuming that the ZXA KP-ABKS scheme is selectively secure against chosen-keyword attack in the random oracle model.*

Proof of Theorem 6.5. *Assume there exists an adversary $\mathcal{B}$ that wins the SCKA game of the OO-ABKS schemes with non-negligible probability, then we can construct an adversary $\mathcal{A}$ that succeeds in attacking the basic ABKS schemes. The main idea is let $\mathcal{A}$ personate the OO-ABKS challenger for adversary $\mathcal{B}$ by interacting with the ABKS challenger and take advantage of $\mathcal{B}'s$ capability in attacking OO-ABKS schemes to attack the ABKS schemes.*

- *In the* **Setup** *phase, adversary $\mathcal{B}$ chooses a non-trivial I^*_{enc} and sends it to adversary $\mathcal{A}$, who then sends it to the ABKS challenger. The challenger runs Setup(1^ℓ) to generate the public parameter PK and the master key MK. Adversary $\mathcal{A}$ sends the public parameter PK to $\mathcal{B}$.*
- *During* **Phase 1**, *$\mathcal{A}$ sends $\mathcal{B}'s$ queries to the challenger who generates the private keys and tokens as in the algorithms of the ABKS schemes. It is easy to verify that the private key and tokens constructed in the ABKS scheme is the same as that in the online/offline way. Therefore, when receiving the results from $\mathcal{A}$, $\mathcal{B}$ will consider them as generated by an OO-ABKS challenger.*
- *In the* **Challenge phase**, *$\mathcal{B}$ selects two keywords w_0 and w_1 and sends them to $\mathcal{A}$. The keywords are sent to the challenger, who encrypts one of them w_λ ($\lambda = 0$ or 1), and sends the ciphertext CT_{w_λ} to $\mathcal{A}$.*
- *$\mathcal{B}$ gets the challenge ciphertext from $\mathcal{A}$ and continues to send queries to $\mathcal{A}$. $\mathcal{A}$ responds $\mathcal{B}$'s queries as in* **Phase 1** *but with more constraints as stated in* **Phase 2**. *When $\mathcal{B}$ finishes querying, it outputs a guess bit for λ, denoted by λ'. $\mathcal{A}$ outputs the same bit as its guess for λ.*

It is clear that if $\mathcal{B}$ gets non-negligible advantage in winning the SCKA game defined for the OO-ABKS schemes, then $\mathcal{A}$ succeeds in attacking the ABKS schemes with non-negligible probability, which contradicts Theorem 6.1 and Theorem 6.3. Therefore, we can conclude that there exists no PPT adversary $\mathcal{B}$ can win the SCKA game of OO-ABKS with non-negligible probability and the constructed OO-ABKS scheme OO-ABKS scheme is selectively secure against chosen-keyword attack.

Theorem 6.6. *The above Online/Offline CP-ABKS scheme is selectively secure against chosen-keyword attack in the generic* bilinear *group model with respect to Definition 6.2 assuming that the ZXA CP-ABKS scheme is selectively secure against chosen-keyword attack in the generic* bilinear *group model.*

Proof of Theorem 6.6. *Theorem 6.6 can be proven in the same way as shown in Proof 6.5.*

Theorem 6.7. *The above Online/Offline KP-ABKS scheme achieves keyword secrecy in the random oracle model if the ZXA KP-ABKS scheme achieves keyword secrecy in the random oracle model.*

Proof of Theorem 6.7. *Assume there exists an adversary $\mathcal{B}$ that wins the keyword secrecy game of the OO-ABKS schemes with non-negligible probability, then we can* construct an adversary $\mathcal{A}$ that succeeds in attacking the basic ABKS schemes in term of keyword secrecy.

- In the **Setup** phase, the ABKS challenger runs $Setup(1^{\ell})$ to generate the public parameter PK and the master key MK. Adversary $\mathcal{A}$ sends the public parameter PK to $\mathcal{B}$.
- During **Phase 1**, $\mathcal{A}$ sends $\mathcal{B}'s$ queries to the challenger who generates the private keys and tokens as in the algorithms of the ABKS schemes.
- In the **Challenge phase**, $\mathcal{B}$ selects a non-trivial I_{enc}^{*} and sends it to $\mathcal{A}$, who sends it to the challenger. The challenger chooses the challenge keyword w^{*} and sends the ciphertext $CT_{w^{*}} \leftarrow Encrypt(PK, I_{enc}^{*}, w^{*})$ to adversary $\mathcal{A}$. $\mathcal{A}$ then sends it to $\mathcal{B}$.
- In the second query phase **Phase 2**, $\mathcal{A}$ continues queuing the key extraction and token generation oracle with the restriction that $f(I_{key}, I_{enc}^{*}) = 0$.
- After **guess**ing q distinct keywords, $\mathcal{B}$ outputs a keyword w' as its guess of w^{*}. The adversary $\mathcal{A}$ outputs the same keyword.

If the probability $\mathcal{B}$ wins the keyword secrecy game in the OO-ABKS schemes is non-negligibly more than $\frac{1}{|\mathcal{M}|-q}$, the same will $\mathcal{A}$ in the ABKS schemes, which contradicts Theorem 6.2 and Theorem 6.4. Therefore, Theorem 6.7 and Theorem 6.8 hold and the constructed OO-ABKS schemes achieve keyword secrecy.

Theorem 6.8. *The above Online/Offline CP-ABKS scheme achieves keyword secrecy in the random oracle model if the underlying CP-ABKS scheme achieves keyword secrecy in the random oracle model.*

Proof of Theorem 6.8. *Theorem 6.8 can be proven in the same way as shown in Proof 6.7.*

6.7 PERFORMANCE EVALUATION

In order to show online computation efficiency enhancement after splitting the keyword encryption and token generation into two phases, in this section, we compare the Online/Offline ABKS scheme with the basic ZXA scheme. Since the *bilinear* operations are the dominate costs, we ignore minor factors such as arithmetic in $\mathbb{Z}_p$. A basic problem to answer: how much pre-processing is needed for an ABKS encryption and token generation before the data requester knows the keyword that he/she wants to encrypt/search or the access structure that he/she wants to encrypt under? Based on the presented solution, almost all of the work can be done offline, which can significantly reduce the overhead to be executed in real-time. Let E donate an exponentiation, M denote a multiplication, and H_1 denote hash function as described in the scheme. S is the number of a data user's attributes and N is the number of attributes that are involved in data owner's access control policy. Table 6.2 and Table 6.3 describe the asymptotic complexities of the KE and TG algorithm, respectively.

Table 6.2

Complexity of KP-ABKS and CP-ABKS in keyword encryption algorithms

Encryption Algorithms	Complexity
ZXA.KP-ABKS	$(S+4)E + SH_1 + M$
KP-Offline	$(S+3)E + SH_1$
KP-Online	$E + M$
ZXA.CP-ABKS	$(2N+4)E + NH_1 + M$
CP-Offline	$(2N+3)E + NH_1$
CP-Online	$E + M$

Table 6.3

Complexity of KP-ABKS and CP-ABKS in token generation algorithm

Token Generation Algorithms	Complexity
ZXA.KP-ABKS	$(2N+3)E + M$
KP-Offline	$(2N+2)E$
KP-Online	$E + M$
ZXA.CP-ABKS	$(2S+4)E + M$
CP-Offline	$(2S+3)E$
CP-Online	$E + M$

From Table 6.2 and Table 6.3, we can see that only one exponentiation and one multiplication are done during the online phase in the presented scheme. Therefore, in the online phase only a fixed amount of computation is needed, i.e., compared with the basic scheme, the computation costs in encrypting keywords and generating tokens do not scale with the number of attributes, which incurs low efficiency of ABKS schemes. Additionally, the overall computation costs are the same as that in the ZXA scheme.

To evaluate the performance of the Online/Offline ABKS, schemes, we ran experiments on a client machine with MAC OS X Yosemite system, 1.4 GHz, and 2GB RAM. We varied N, the number of attributes, from 1 to 100 with step length 10. All the data reported below are averaged over 1000 randomized runs. The percentage of online computation costs decrease quickly with the increasing attributes number. We show the experimental results for KP-ABKS in Figure 6.2 and Figure 6.3, for CP-ABKS in Figure 6.4 and Figure 6.5.

In both KP-ABKS and CP-ABKS, when the number of attributes is greater than 30, over 99% of the work in keyword encryption could be shifted to the offline phase; when the number of attributes is greater than 50, less than 1% of the work in token generation needs to be done in the online phase. Note that the total computation

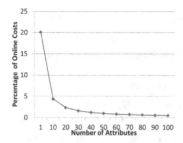

Figure 6.2: Percentage of online costs in *TG* of KP-ABKS.

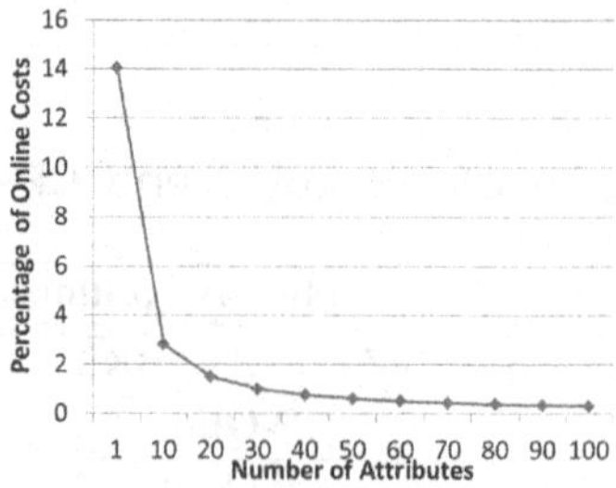

Figure 6.3: Percentage of online costs in *KE* of KP-ABKS.

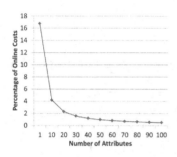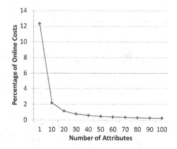

Figure 6.4: Percentage of online costs in *TG* of CP-ABKS.

Figure 6.5: Percentage of online costs in *KE* of CP-ABKS.

required between the offline and online phases is identical to the work required by the basic scheme. Thus, the total work remains the same, but the vast majority of the computational work can be shifted in time to a moment when the device is least busy or has access to a power source. Therefore, the Online/Offline ABKS, scheme is very useful for mobile devices with power consumption constraints.

6.8 SUMMARY

We developed a new technique for ABKS that splits the computational work into two phases. The offline phase does the vast majority of the work to encrypt a keyword or generate a trapdoor. The online phase then rapidly assembles an ABKS ciphertext or trapdoor when the specifics become known. We provide efficient constructions for both key-policy and ciphertext-policy ABKS systems. We provide performance estimates that show a large majority of the computational work can be moved to the offline phase. We expect that this technology could reduce battery consumption on mobile devices. Overall, it helps reduce the cost of bringing attribute-based keyword search into practice.

Note that I_{enc} is included in the offline phase of keyword encryption in the definition of Online/Offline ABKS; it's better that this could be deleted from the inputs of the algorithm, which means that the encryption can be totally independent of both keywords and access structure. In the construction of the Online/Offline KP-ABKS scheme, the attribute list S is unnecessary when encrypting keywords offline, while in the Online/Offline CP-ABKS construction, the access tree T is taken as the input in the offline keyword encryption algorithm. This is decided by the basic scheme ZXA we use. As future work, we plan to construct online/offline CP-ABKS schemes that do not need access control policy as input in the offline phase of keyword encryption. Additionally, in this chapter, we present validation of efficiency of the presented scheme via simulation and get approximate experimental results.

7 Attribute-Based Signature with Policy-and-Endorsement Mechanisms

From Chapter 1 to Chapter 6, we present new extended features based on ABE schemes. In this chapter, we focus on attribute-based signature solutions, which demonstrate a rich type of cryptosystem, called Adaptive-Policy Attribute-Based Cryptosystem (AP-ABC), that is motivated by real-life challenges in access control management. At a high level, one can consider a natural access control system, called Fine-Grained Access Control (FGAC), in which each user is associated with a subset of attributes that specify which type of resources the user can access; each resource is also associated with an access policy that specifies which type of users can access the resource.

The AP-ABC is such a cryptographic system to implement FGAC due to the fact that this cryptosystem can deceptively change the protected resources according to access policy. To realize the AP-ABE, we presented a Policy-Endorsing Attribute-Based Signature (PE-ABS) solution to allow a signer to announce his endorsement using his claim without having to reveal the identity of signer. In this case, we consider it a special role signature with anonymity, which is a method for allowing a member of a group to anonymously sign a message on behalf of the specified role (known as a set of attributes), e.g., title, position, or time-bound in social organization. This functionality becomes more attractive in the case where it is practically infeasible for the signature holder and the verifier to know all possible signers in large-size systems.

7.1 INTRODUCTION

In this chapter a general framework for constructing attribute-based cryptosystems (ABC) with encryption, signature, and authentication is developed. Attribute-based systems are a natural fit for settings where the roles of the users depend on the combination of attributes they possess. In such systems, the users obtain multiple attributes from one or more attribute authorities, and a user's capability in the system (e.g., sending messages, accessing resources) depends on his attributes. We start with an informal description of the framework as follows.

- Let $\mathbb{A}$ be a finite set of attributes, and everyone can know this set;
- Let $\mathcal{R}$ be a finite set of roles, where each role is a subset of attributes, i.e., $\rho \in \mathcal{R}$ and $\rho \subseteq \mathbb{A}$. Each member has a role in $\mathcal{R}$ and can obtain a private key $K^{(\rho)}$ corresponding to its role;

- Let $\mathscr{P}$ be a finite set of policies, where each policy can be expressed as a logical function on attributes, $f_\pi(X)$ for any $\pi \in \mathscr{P}$ and $X \subseteq \mathbb{A}$. Roughly speaking, a message can be encrypted or signed to any policy π in $\mathscr{P}$; and
- We allow for an arbitrary predicate called open on the set $\mathscr{P} \times \mathscr{R}$ that specifies which roles in $\mathscr{R}$ can open what policies in $\mathscr{P}$.

In an encryption case, a key $K^{(\rho)}$ can decrypt ciphertexts encrypted for policy π, if and only if the role ρ opens the policy π, i.e., open open$(\pi, \rho) = f_\pi(\rho)$ is true. This kind of encryption is also called ciphertext-policy attribute-based encryption (CP-ABE), which can be considered a kind of spatial encryption [40].

In a signature case, everyone can also use policy π to verify signature for key $K^{(\rho)}$ if and only if the role ρ opens the policy π, i.e., open(π, ρ) is true. This kind of message authentication differs from that offered by traditional digital signatures due to the fact that it is a policy-and-endorsement mechanism that supports the claims of the form: "a single user, whose attributes satisfy the predicate, endorsed this message." As a simple example, suppose Alice wishes to fill out a form with the following claim to endorse her finance application: (*Warrantor is a Professor AND in school of (Computer Science OR Electronic Engineering*)). To give credibility to this form, she needs to find the proper warrantor to sign her form. It is a reasonable requirement that any referendary must be adaptable to validate Alice's form in a secure way even if she does not know her warrantor. That is, this kind of signature needs to provide anonymity for signers.

In this simple example, the set of attributes is defined as $\mathbb{A} = \{$Professor, Computer Science, Electronic Engineering$\}$, and they are divided into two categories, {*Faculty, Dept .*}, respectively. The policy about *Warrantor* can be defined as *Warrantor := ((Faculty == Professor) AND ((Dept. == Computer Science) OR (Dept. == Electronic Engineering)))*. Given an assignment of roles *Louis := Professor, Computer Science*, the open function returns true, that is *open(Warrantor, Louis) == true*.

This signature can also be efficiently converted into an entity authentication protocol: given a public policy π, the verifier can interactively check the availability of guarantees generated by the key $K^{(\rho)}$ of the prover if and only if open(π, ρ) is true. This means that this kind of identification can utilize contextual attributes to achieve anonymous authentication. Thus, it has the marked difference from the traditional approaches that require principals to present personal identity information in order to obtain access to various services.

In the rest of this chapter, Section 7.2 presents the related work. In Section 7.3, we formalize the model of AP-ABC and the precise definition of the security using the framework of Boneh and Hamburg [40]. We also briefly discuss the mechanics of the work with AP-ABC and the features that they can provide. In Section 7.4, a Policy-Endorsing Attribute-Based Signature (PE-ABS) approach is presented to realize a policy-and-endorsement mechanism. Another challenge, also an important part of the security requirement, is the provable security of the proposed PE-ABS scheme. In Section 7.5, we present the proof of two security requirements: selfless anonymity and existential unforgeability. The security of scheme is based on the

Strong Diffie-Hellman (SDH) assumption [38] in the random oracle model [222]. We also need the Decision Linear assumption that has been proved useful for constructing short group signature. Further, we analyze the performance of our scheme by constructing a practical AP-ABC in Section 7.6. Finally, a summary of this chapter is presented in Section 7.7.

7.2 RELATED WORKS

Attribute-based cryptosystem provides a fine-grained access control mechanism by means of encryption, signature, authentication, and identification, which depends on the match between access policies embedded into the resources and identity attributes ascribed into user's private key. Since the first ABE scheme was introduced by Sahai and Waters in 2005 [186], ABC has received much attention and many schemes have been proposed in the literature [25, 95, 193, 94, 220, 196]. According to the structure of access policy, these schemes can be roughly divided into two categories: single threshold structure and hierarchical threshold structure.

ABC schemes with single threshold structure generally use techniques from secret-sharing schemes as a core component of these schemes. For example, the scheme of Sahai and Waters [186], called fuzzy identity-based encryption, allows for a threshold attribute-based decryption of encrypted data. Messages can be encrypted by specifying a set of decryptor attributes ρ' during encryption. Such a ciphertext can then be decrypted by any user with the attribute set ρ such that $|\rho \cap \rho'| \geq t$. It is well known that secret-sharing technique can be used to express monotonic access structures, that is, $\forall \rho', \rho'' \in \mathscr{R}$, if open$(\pi, \rho)$ = true. So Goyal et al. [95] subsequently increased the expressibility of ABE systems by allowing the private key to express any monotonic access structure over attributes. This scheme is also called key-policy ABE (KP-ABE), where attributes are used to annotate the ciphertexts, and formulas over these attributes are ascribed to users' secret keys. Single threshold structure is also fit for an attribute-based signature scheme, in which a signer has a set of attributes ρ, and the verifier specifies a verification attribute set ρ'. A signature is verified as valid if $|\rho \cap \rho'| \geq t$, where t is fixed during the setup time. For example, Shahandashti and Safavi-Nain [193] proposed threshold attribute-based signatures (t-ABS), in which, signers are associated with a set of attributes and verification of a signed document against a verification attribute set succeeds if the signer has a threshold number t of attributes in common with the verification attribute set.

Another kind of ABC has hierarchical threshold structure, in which the policy is transformed into access tree with threshold gates: AND gates can be constructed as n-of-n threshold gates and OR gates as 1-of-n threshold gates. This structure has greater flexibility than a single-threshold structure since the latter can be obtained as a special case of the former. Bethencourt, Sahai, and Waters first gave a construction for this structure in the form of ciphertext-policy ABE (CP-ABE) [186]. In their construction, the roles of the ciphertexts and keys are reversed in the sense that attributes are used to describe the features of a key holder, and an encryptor will associate an access policy with the ciphertext. Since then, some cryptographically stronger CP-ABE constructions [94, 220] that allowed reductions to the

Decisional Bilinear Diffie-Hellman (DBDH) problem have been proposed in recent years. For example, Goyal et al. [94] presented bounded CP-ABE in the standard model. Waters [220] proposed the first fully expressive CP-ABE in the standard model. Hierarchical threshold structure can also be used to construct more flexible signature schemes. For example, Khader [196] proposed a signature scheme, called attribute-based group signatures, based on Boneh's group signature [38]. This scheme, however, lacks a corresponding encryption scheme to construct a complete attribute-based cryptosystem.

Techniques similar to ABC were proposed for many applications like Attribute-Based Access Control (ABAC, used in service-oriented architecture (SOA)) [235], Property-Based Broadcast Encryption (used in DRM) [15, 14], Hidden Credentials [125], as well as dynamic communication in vehicular ad hoc networks [112].

7.3 PRELIMINARIES

In this section we propose the formal definition of adaptive-policy attribute-based cryptosystem (AP-ABC) for fine-grained access control. Based on an attribute-based cryptographic infrastructure, this cryptosystem specifies two fundamental cryptographic tools: ciphertext-policy attribute-based encryption (CP-ABE) and policy-endorsing attribute-based signature (PE-ABS). In addition, we describe two adversary's attack models for PE-ABS.

7.3.1 ATTRIBUTE-BASED CRYPTOSYSTEM

Let $\mathbb{A}$ be the universe of possible attributes in FGAC. Given an access policy π over $\mathbb{A}$, we assume that there exists a function open(π, ρ) = true, where x is associated with attributes of $\mathbb{A}$. An attribute set $\rho \subseteq \mathbb{A}$ is said to satisfy a claim-predicate π if open$(\pi, \rho) = true$. In terms of these notations, we define an adaptive-policy attribute-based cryptosystem as follows.

Definition 7.1 (Adaptive-Policy Attribute-Based Cryptosystem). An attribute-based cryptosystem consists of the following three procedures, which are parameterized by a universe of attributes $\mathbb{A}$:

Infrastructure: generates the parameters and user's keys of the cryptosystem:

Setup(κ, A): takes as input the security parameter κ and the set A. It outputs a manager-key gmk and a public-key gpk;

Join(gmk, i, ρ_i): takes as input the manager key gmk, a user counter i, and the user's role $\rho_i \subseteq \mathbb{A}$. It outputs a user's private-key $gsk[i]$, where $\rho_i \in gsk[i]$.

Encryption: realizes the ciphertext-policy attribute-based encryption:

Encrypt(gpk, π, M): takes as input the public key gpk, the access policy π, and the plaintext $M \in \{0, 1\}^*$. It outputs a ciphertext C.

Decrypt$(gsk[i], C)$: takes as input a ciphertext C and a private key $gsk[i]$. If open(π, ρ) = true, this algorithm outputs the plaintext M.

Signature: realizes the policy-endorsing attribute-based signature:

Sign($gsk[i], \pi, M$): takes as input a private key $gsk[i]$, a policy π, and a message $M \in \{0,1\}^*$. It returns a signature σ and $\pi \in \sigma$.

Verify(gpk, σ, M): takes as input the public key gpk, a message M, and a purported signature σ on M. It returns a Boolean value, valid or invalid.

In this definition, CP-ABE and PE-ABS are intimately integrated into a complete system by a cryptographic key management infrastructure, which supports dynamically joining new users. As there already exist schemes on CP-ABE, we will design the AP-ABC system based on these existing schemes in this work. Hence, we will not consider CP-ABE scheme but direct our attention to the construction of PE-ABS in this chapter. Note that, the existing CP-ABEs still need to make the necessary changes to construct AP-ABC.

7.3.2 ACCESS POLICY AND ATTRIBUTE TREE

We now discuss how to construct the cryptographic function open(π, ρ) with secrecy and correctness for any $\pi, \rho \subseteq \mathbb{A}$. Without loss of generality, we assume that any access policy π over $\mathbb{A}$ can be expressed as a Boolean function $f_\pi(\cdot)$, whose inputs are associated with attributes of A. With the help of gpk or gsk, it is possible to use $f_\pi(\cdot)$ to hide a secret value, such that there merely exist some appointed values that the verifier can later accept as legal "opening".

Given an arbitrary access policy π with Boolean function $f_\pi(\cdot)$, we use hierarchical threshold structure to implement the above-mentioned approach, as follows: firstly transform $f_\pi(\cdot)$ into an attribute tree $\mathscr{T}$, which consists of *AND* and *OR* threshold gates, then use threshold cryptographic techniques to generate a random function, e.g., secret sharing schemes, and finally realize Boolean function under the Decisional Bilinear Diffie-Hellman (DBDH) assumption.

An attribute tree is a tree in which each interior node is a threshold gate and the leaves are linked with attributes. An $m - of - n$ threshold gate means that the secret of the parent node can be recovered if and only if at least *mofn* children is available. We note that AND gates can be constructed as $n - of - n$ threshold gates and OR gates as $1 - of - n$ threshold gates. Attribute tree also implements some power logic function, e.g., integer comparisons [25], and satisfaction of a leaf is achieved by owning an attribute.

7.3.3 CORRECTNESS AND SECURITY DEFINITIONS

Now we define the security requirements of an attribute-based cryptosystem. Since the security of CP-ABE has already been widely studied in previous work, we will only formulate the security of PE-ABS and pay attention to the construction of the PE-ABS scheme based on the existing CP-ABE scheme in this work.

Firstly, we must ensure that our scheme is correct. The correctness property of PE-ABS scheme is that honestly-generated signatures should pass the verification check:

Definition 7.2 (Correctness). A PE-ABS scheme is correct if for all attribute sets $\mathbb{A}$, $\forall (gpk, gmk) \, \mathrm{Setup}(\kappa, \mathbb{A})$, $\forall i \in \mathbb{N}$, $\forall gsk[i] \leftarrow \mathrm{Join}\,(gmk, i, \rho_i)$, $\forall M \in \{0,1\}^*$, all claim-predicates π with open $(\pi, \rho_i) = true$,

$$\mathrm{Verify}(gpk, \mathrm{sign}(\pi, gsk[i], M), M) = valid,$$

with probability 1 over the randomness of all the algorithms.

Secondly, we present two formal definitions, selfless anonymity and existential unforgeability, that together capture the desired notions of the security of PE-ABS, as follows:

Definition 7.3 (Selfless Anonymity). An n-user PE-ABS scheme is $(t, n, q_h, q_s, \varepsilon) -$ *selfless* anonymity if the success probability of any polynomial-time adversary in the following experiment:

- $(gpk, gmk) \leftarrow \mathrm{Setup}(\kappa, \mathbb{A})$ and give gpk to the adversary;
- the adversary is given access at most q_h times to oracle $Hash(\cdot)$ and q_s times to oracle $\mathrm{Sign}(\pi, gsk[i], \cdot)$ as well as at most $n - 1$ times to oracle $Join(gmk, \cdot, \cdot)$ to request the private key of the user $i \in [1, n]$;
- the adversary outputs (M, i_0, i_1) and the system returns a challenge $\sigma^* = \mathrm{Sign}\,(gpk, gsk\,[i_b], M)$, where $b \leftarrow_R \{0, 1\}$;
- the adversary outputs a bit b' as the guess of b.

We say the adversary succeeds if $b' = b$, and i_0, i_1 were never queried to the corruption oracle Join at either index.

In the selfless-anonymity game, the adversary's goal is to determine which of the two keys generated the signature. The adversary is not given access to either key but he is allowed to corrupt the private keys of other users.

Definition 7.4 (Existential unforgeability). A PE-ABS scheme is $(t, q_h, q_s, \varepsilon)$-existentially unforgeable under a chosen message attack if the success probability of any polynomial-time adversary in the following experiment is at most in time at most t:

- $\mathrm{run}(gpk, gmk) \leftarrow \mathrm{Setup}(\kappa, \mathbb{A})$ and give gpk to the adversary;
- the adversary is given access at most q_h times to oracle $Hash(\cdot)$ and q_s times to oracle $\mathrm{Sign}(\pi, gsk[i], \cdot)$, respectively;
- at the end the adversary outputs (M', π', σ').

We say the adversary succeeds if (M', π') was never queried to the Sign oracle, and $\mathrm{Verify}(gpk, \sigma', M') = valid$.

The security proof for our scheme is in the random oracle model and the extra parameter q_h in the security definitions denotes the number of random oracle queries that the adversary issues.

7.4 THE CONSTRUCTION FOR PE-ABS

Now we construct a fully secure attribute-based signature scheme with selfless anonymity and existential unforgeability in the random oracle model. This scheme is constructed on ciphertext-policy attribute-based encryption (in short BSW scheme) in [26].

7.4.1 BILINEAR GROUP SYSTEM

Let G_1, G_2, and G_T be three cyclic groups of prime order p. G_1 and G_2 are two additive groups and G_T is a multiplicative group. There exists an efficiently computable homomorphism ψ from G_2 to G_1, but there exists no efficiently computable homomorphism from G_1 to G_2.

Definition 7.5 (Bilinear map). Let e be an efficiently computable *bilinear* map $e : G_1 \times G_2 \to G_T$ with the following properties: for all $G \in G_1, H \in G_2$ and all $a, b \in \mathbb{Z}_p$,

- Bilinearity: $e([a]G, [b]H) = e(G, H)^{ab}$;
- Non-degeneracy: $e(G, H) \neq 1$ unless G or H is the identity of G_1 or G_2;
- Computability: $e(G, H)$ is efficiently computable.

In addition, for all $S, T \in G_2$ and an efficient homomorphism $\psi : G_2 \leftarrow G_1$, we have $e(\psi(S), T) = e(\psi(T), S)$. On this basis, we will use the following *bilinear* group system to construct PE-ABS scheme.

Definition 7.6. (Bilinear group system). We call $S = (p, G_1, G_2, G_T, e(\cdot, \cdot))$ a *bilinear* group system, if there exists an efficiently computable *bilinear* map $e : G_1 \times G_2 \to G_T$ and the operations in G_1, G_2, G_T are efficient.

7.4.2 HASHING FUNCTIONS

Our scheme makes use of two hash functions:

- $H_0 : \{0, 1\}^* \to G_2$ mapping an arbitrary string to an element of G_2;
- $H_1 : \{0, 1\}^* \to \mathbb{Z}_p$, mapping an arbitrary string to an element of $\mathbb{Z}_p$.

7.4.3 POLICY-ENDORSING ATTRIBUTE-BASED SIGNATURE

Consider a security parameter κ, a *bilinear* group system $S = (p, G_1, G_2, G_T, e(\cdot, \cdot))$ with homomorphism $\psi : G_2 \to G_1$ and $\lfloor \log p \rfloor = O(\kappa)$. The scheme employs hash functions H_0 and H_1 with ranges G_2 and $\mathbb{Z}_p$ respectively, treated as random oracles. We propose a PE-ABS scheme as follows.

Setup$(\kappa, \mathbb{A})$. Takes as inputs the security parameter κ, the *bilinear* group system S and the set of all attributes $A = \{attr_1, \ldots, attr_m\}$, where each $attr_i$ denotes an attribute and m is the total number of attributes. It proceeds as follows:

1. selects two generators G and h of $\mathbb{G}_1$ and $\mathbb{G}_2$ uniformly at random such that $e(G,h) \neq 1$;
2. selects two random integers $\alpha, \beta \in_R \mathbb{Z}_p$;
3. sets $g = \lfloor \beta \rfloor G \in \mathbb{G}_1$ and $\zeta = e(G,h)^\alpha \in \mathbb{G}_T$;
4. outputs the management key and public key:

$$gmk = (\alpha, \beta, G), \tag{7.1}$$

$$gpk = (\mathbb{S}, \mathbb{A}, g, h, \zeta). \tag{7.2}$$

The manager publishes gpk and keeps gmk secret.

$Join(gmk, i, \rho_i)$. Takes as inputs the management key gmk, a member counter $i \in \mathbb{Z}$, and a set of attributes $\rho_i = \{attr_{i_1}, \dots, attr_{i_l}\} \subseteq \mathbb{A}$. It proceeds as follows:

1. selects a fresh $r_i \in \mathbb{Z}_p$ and computes $D_i = \left[\frac{\alpha + r_i}{\beta}\right] h \in \mathbb{G}_2$;
2. picks a random integer $r_j \in_R \mathbb{Z}_p$ for each $attr_j \in \rho_i$, and computes

$$\begin{cases} A_{i,j} = [r_i]G + [r_j]\psi(H_0(attr_j)) \in \mathbb{G}_1, \\ A'_{i,j} = [r_j]\psi(h) \in \mathbb{G}_1; \end{cases} \tag{7.3}$$

3. outputs the private key:

$$gsk[i] = \left(D_i, \left(A_{i,j}, A'_{i,j}\right)_{attr_j \in \rho_i}\right). \tag{7.4}$$

The system manager sends $gsk[i]$ to this member. Obviously, nobody should be allowed to possess r_i except the system manager.

$Sign(gsk[i], \pi, M)$. Takes as inputs a private key $gsk[i]$, the policy π, and a message $M \in \{0,1\}^*$, and returns a signature σ. It proceeds as follows:

1. picks a random integer $t \in_R \mathbb{Z}_p$ and two random generators $u \in_R \mathbb{G}_1, v \in_R \mathbb{G}_2$, and computes

$$E_i = [t]u \in \mathbb{G}_1, \quad T_i = D_i + [t]v \in \mathbb{G}_2; \tag{7.5}$$

2. selects a random integer $s \in \mathbb{Z}_p$, set $Q_i = [s]g \in \mathbb{G}_1$, and then creates access policy tree $\mathscr{T}$ and invokes the Algorithm 7.1 (disperse algorithm) with which the signer can compute:

$$\{\Delta_{\mathscr{T}}(attr_j)\}_{attr_j \in \mathscr{T}} = Disperse(s, \mathscr{T}; \tag{7.6}$$

As a result, the following values can be computed for each $attr_j \in \rho_i \cap \mathscr{T}$,

$$\begin{cases} B_{i,j} = [\Delta_{\mathscr{T}}(attr_j)]A_{i,j} \in \mathbb{G}_1 \\ B'_{i,j} = [\Delta_{\mathscr{T}}(attr_j)]A'_{i,j} \in \mathbb{G}_1; \end{cases} \tag{7.7}$$

3. picks two blinding values $r_s, r_t \in_R \mathbb{Z}_p$ and computes

$$V_1 = [r_t]u \in \mathbb{G}_1, \quad V_2 = \zeta^{r_s} \cdot e(Q_i, v)^{r_t} \in \mathbb{G}_T; \tag{7.8}$$

4. computes a challenge $c \in \mathbb{Z}_p$ by using $c \leftarrow H_1\left(gpk, M, u, v, T_i, E_i, Q_i, V_1, V_2\right)$, and then computes the following parameters:

$$s_s = r_s + c \cdot s, \quad s_t = r_t + c \cdot t. \tag{7.9}$$

The final signature is

$$\sigma = \left(\mathscr{T}, u, v, T_i, E_i, Q_i, c, s_s, s_t, \left\{\left(B_{i,j}, B'_{i,j}\right)\right\}_{attr_j \in \rho_i \cap \mathscr{T}}\right). \tag{7.10}$$

Verify(gpk, σ, M). Takes as inputs the group public key gpk and a purported signature σ on a message M. It returns either valid or invalid. It can check whether σ is a valid signature as follows:

1. for each attribute $attr_j \in \sigma$, computes $C_j = h$, $C'_j = H_0\left(attr_j\right)$, and

$$S_{i_j} = \frac{e\left(B_{i,j}, C_j\right)}{e\left(B'_{i,j}, C'_j\right)} = e(G, h)^{r_i \cdot \Delta_{\mathscr{T}}\left(attr_j\right)}, \tag{7.11}$$

to obtain the set $\left\{S_{i_1}, \ldots, S_{i_l}\right\}$;

2. invokes the aggregate Algorithm 7.2 to get

$$S = e(G, h)^{r_i \cdot s} = Aggregate\left(\mathscr{T}, \left\{S_{i_1}, \ldots, S_{i_l}\right\}\right), \tag{7.12}$$

which may be regarded as the inverse process of the Algorithm 7.1 (disperse algorithm), and then computes

$$V'_1 = [s_t] u - [c] E_i, \quad V'_2 = \frac{\zeta^{s_s} \cdot S^c \cdot e\left(Q_i, v\right)^{s_t}}{e\left(Q_i, T_i\right)^c}; \tag{7.13}$$

3. checks whether the challenge c is correct:

$$c \stackrel{?}{=} H_1\left(gpk, M, u, v, T_i, E_i, Q_i, V'_1, V'_2\right). \tag{7.14}$$

If so, outputs valid; otherwise, outputs invalid.

7.4.4 DISPERSE AND AGGREGATE ALGORITHMS

We propose a pairwise algorithm: disperse algorithm and aggregate algorithm to fix and open the access policy $\pi \in \mathscr{P}$, as follows.

Disperse Algorithm 7.1. Given an integer s and a policy tree $\mathscr{T}$ generated from a logical function $f_\pi(X)$ for $\pi \in \mathscr{P}$ and $X = \{attr_i\} \subseteq \mathbb{A}$, the disperse algorithm shares the secret s in X and returns a set of values $\{\Delta_{\mathscr{T}}(attr_i)\}$ for $\forall attr_i \in X$. In Figure 7.1, a simple example is provided to explain the attribute tree, in which the variant Time is denoted as $Time_3 \| Time_2 \| Time_1$ in a binary format. In the attribute tree, the nodes are divided into two categories: leaf nodes and internal nodes (or non-leaf nodes). The attributes are assigned into the leaf nodes and each internal node denotes a logical relation (AND/OR).

Algorithm 7.1 Disperse$(s, \mathcal{T})$

Require: a policy tree $\mathcal{T}$ and a secret s;
Ensure: a set of secrets of attributes in $\mathcal{T}$;
1: Each node t_i is assigned a secret value $s_i = 0$, an index value $ind_i \in \mathbb{Z}_p$, and a polynomial $q_i(x) = 0$; and sets $s_1 = s$ for the root node t_1.
2: **while** $\mathcal{T}$ is not empty **do**
3:     Finds an unprocessed node t_i from top to bottom;
4:     Computes the secret $s_i = q_{\text{parent}(i)}(ind_i)$
5:     **if** t_i is an internal node **then**
6:        Finds its children $\{t_{i_1}, \ldots, t_{i_l}\}$
7:        **if** $\{t_{i_1}, \ldots, t_{i_l}\}$ has AND relation **then**
8:           Generates $q_i(x) = s_i + \sum_{j=0}^{l-1} a_{i,j} x^j \pmod{p}$ where any $a_{i,j} \in_R \mathbb{Z}_p^*$;
9:        **end if**
10:    **end if**
11:    **if** t_i is a leaf node **then**
12:       Outputs $s_i = q_{\text{parent}(i)}(ind_i)$ as the secret of corresponding attribute $\Delta_{\mathcal{T}}(attr_j)$ which is also denoted $\Delta_s(attr_j)$
13:    **end if**
14: **end while**

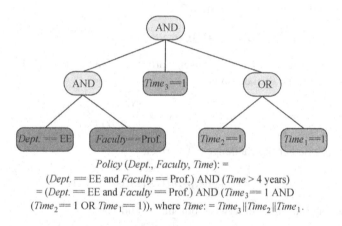

Policy $(Dept., Faculty, Time)$: $=$
$(Dept. == EE$ and $Faculty == Prof.)$ AND $(Time > 4$ years$)$
$= (Dept. == EE$ and $Faculty == Prof.)$ AND $(Time_3 == 1$ AND
$(Time_2 == 1$ OR $Time_1 == 1))$, where $Time: = Time_3 \| Time_2 \| Time_1$.

Figure 7.1: Example for policy tree.

We adopt the secret sharing method with hierarchical polynomials to realize the disperse algorithm: firstly, assign a random index ind_i for each node t_i in $\mathcal{T}$; secondly, choose a polynomial $q_i(x)$ for each internal node t_i, where $q_i(x) = \sum_{j=0}^{k-1} a_j x^j$ is a polynomial with $k-1$ degree for each AND node with k branches, or a constant polynomial for each OR node; third, set $q_i(0) = q_{\text{parent}(i)}(ind_i)$ and set $q_1(0) = s$ for the root node t_1; and then, execute the algorithm; finally, assign a value $q_{\mathcal{T}}(attr_i) = q_{\text{parent}(k)}(ind_k)$ and $k = Node(attr_i)$ for the corresponding $attr_i$.

Aggregate Algorithm 7.2. The aggregate algorithm can be considered as the inverse operation of disperse algorithm, but our scheme merely needs the "commitment" of secret s rather than the original secret s. The aggregate algorithm takes as input the set of secret of attributes $\left\{ e(G,h)^{r_i \cdot \Delta_{\mathcal{T}}(attr_j)} \right\}_{attr_j \in \sigma}$ and $\mathcal{T}$, output $e(G,h)^{r_i \cdot s}$ or failure. In this algorithm, the nodes in the policy tree are reduced from bottom to top, but a successful match requires only one efficient path to resume $e(G,h)^{r_i \cdot s}$. So, in order to find this efficient path, we need to mark the state of each node and check each path repeatedly. Here, we divide the state of nodes into three categories: *undecided, processed but unavailable,* and *available.* Otherwise, in the 8th step of aggregate algorithm, the equation holds since

$$
\begin{aligned}
S_k &= \prod_{t_{k_j} \in W_k} S_{k_j}^{\lambda_{W_k}(t_{k_j})} \\
&= \prod_{t_{k_j} \in W_k} e(G,h)^{r_i \cdot \Delta_T(attr_j) \cdot \lambda_{W_k}(t_{k_j})} \\
&= e(G,h)^{r_i \cdot \Sigma_{t_{k_j} \in W_k} \Delta_T(attr_{k_j}) \cdot \lambda_{W_k}(t_{k_j})} \\
&= e(G,h)^{r_i \cdot \Delta_T(attr_k)},
\end{aligned}
\tag{7.15}
$$

where $\lambda_{W_k}(t_{k_j}) = \prod_{t_{k_i} \in W_k, t_{k_i} \neq t_{k_j}} \frac{ind_{k_i}}{ind_{k_i} - ind_{k_j}}$ is Lagrange interpolation coefficient.

Algorithm 7.2 Aggregate $\left(\mathcal{T}, \{S_{i_1}, \ldots, S_{i_l}\} \right)$

Require: a policy tree $\mathcal{T}$ and a set of commitments of secrets $\left\{ S_{i_j} = e(G,h)^{r_i \cdot \Delta_T(attr_j)} \right\}_{attr_j \in \sigma}$ in σ.

Ensure: $e(G,h)^{r_i \cdot s}$ or failure;

1: for $\forall attr_j \in \sigma$, lets $t_k = \text{Node}(attr_j)$ and sets $S_k = e(G,h)^{r_i \cdot \Delta_{\mathcal{T}}(attr(j))}$ as the value of leaf node t_k and its status $st_k = 2$ (available), otherwise $st_k = 0$ (undecided).

2: **while** the set $\{t_k\}_{st_k} = 0$ is not empty **do**

3:     Find an undecided node t_k from down to up;

4:     **if** its children $W_k = \{t_{k_1}, \ldots, t_{k_l}\}$ has OR relation and $\exists t_{k_j} \in W_k, st_{k_j} = 2$ **then**

5:         Sets the secret value $S_k = S_{k_j}$ and $st_k = 2$;

6:         Sets $st_j = 0$ for $\forall st_j = 1$ (processed but unavailable);

7:     **else if** its children $W_k = \{t_{k_1}, \ldots, t_{k_l}\}$ has AND relation and $\forall t_{k_j}, st_{k_j} = 2$ **then**

8:         Computes $S_k = \prod_{t_{k_j} \in W_k} S_{k_j}^{\lambda_{W_k}(t_{k_j})}$ and $st_k = 2$;

9:         Sets $st_j = 0$ for $\forall st_j = 1$;

10:     **else**

11:         Sets $st_k = 1$ (processed but unavailable);

12:     **end if**

13:     **if** t_k is the root node and $sk_k = 2$ **then**

14:         Outputs S_k and halts;

15:     **end if**

16: **end while**Halts and outputs "the attributes are unsatisfied".

7.5   SECURITY PROOF OF PE-ABS SCHEME

In this section, we shall analyze the security of our scheme in three aspects: correctness, selfless anonymity, and existential unforgeability. Firstly, we prove the correctness of the scheme as follows.

Theorem 7.1. *The PE-ABS scheme constructed in Section 7.3 is correct.*

Proof of Theorem 7.1. *In order to validate the policy in signature σ, the proposed scheme involves the verification of three aspects: the match of attributes, the consistency of policy, and the correctness of signature.*

1. Match Attributes: it can hold since

$$
\begin{aligned}
S_{i_j} &= \frac{e(B_{i,j},C_j)}{e\left(B'_{i,j},C'_j\right)} = \frac{e([\Delta_s(attr_j)]A_{i,j},C_j)}{e\left([\Delta_s(attr_j)]A'_{i,j},C'_j\right)} \\
&= e([r_i \cdot \Delta_s(attr_j)]G,h) \\
&= e(G,h)^{r_i \cdot \Delta_s(attr_j)}.
\end{aligned}
\tag{7.16}
$$

2. Consistency with Policy: it can hold by invoking the aggregate algorithm 7.2 to compute $S = e(G,h)^{r_i s}$. From the algorithm, we know that the user will obtain the value S if and only if his attributes satisfy the access policy.

3. Verify Signature : it can hold since for each

$$
\begin{aligned}
V'_1 &= [s_t]u - [c]E_i = [r_t]u = V_1, \\
V'_2 &= \frac{\zeta^{s_s} \cdot S^c \cdot e(Q_i,v)^{s_t}}{e(Q_i,T_i)^c} \\
&= \frac{e(G,h)^{\alpha s_s} \cdot e(G,h)^{r_i \cdot c \cdot s} \cdot e(g,v)^{s_t \cdot s}}{e([s]g,D_i+[t]v)^c} \\
&= \frac{e(G,h)^{\alpha s_s} \cdot e(G,h)^{r_i \cdot c \cdot s} \cdot e(g,v)^{s_t \cdot s}}{e([s]g,D_i)^c \cdot e([s]g,[t]v)^c} \\
&= \frac{e(G,h)^{\alpha(r_s+cs)} \cdot e(G,h)^{r_i \cdot c \cdot s} \cdot e(g,v)^{(r_t+ct)\cdot s}}{e\left([s\beta]G,\left[\frac{\alpha+r_i}{\beta}\right]h\right)^c \cdot e(g,v)^{t \cdot s \cdot c}} \\
&= \frac{e(G,h)^{\alpha(r_s+cs)} \cdot e(G,h)^{r_i \cdot c \cdot s} \cdot e(g,v)^{(r_t+ct)\cdot s}}{e(G,h)^{(\alpha+r_i)s \cdot c} \cdot e(g,v)^{t \cdot s \cdot c}} \\
&= e(G,h)^{\alpha r_s} \cdot e(g,v)^{r_t s} = \zeta^{r_s} \cdot e(Q_i,v)^{r_t} = V_2.
\end{aligned}
\tag{7.17}
$$

Since $V'_1 = V_1$ and $V'_2 = V_2$, then we have $c' = c$. Hence, the attribute-based signature scheme is correct.

7.5.1   SELFLESS ANONYMITY SECURITY

For arbitrary generators P,Q and R of $\mathbb{G}_2$, consider the following decision linear problem: for all $a,b,c \in \mathbb{Z}_p$, given $(P,Q,R,[a]P,[b]Q,[c]R) \in \mathbb{G}_2^6$, decide whether

$c = a + b$ (mod p). One can easily show that an algorithm for solving decision linear in $\mathbb{G}_2$ gives an algorithm for solving DDH in $\mathbb{G}_2$. The converse is believed to be false. That is, it is believed that decision linear is a hard problem even in *bilinear* groups where decisional Diffie-Hellman problem is easy.

Definition 7.7 (Decision Linear Assumption). The $(t, \varepsilon) - decision$ linear assumption is said to hold in $\mathbb{G}_2$ if no t-time algorithm $\mathscr{A}$ has advantage at least ε in solving the decision linear problem in $\mathbb{G}_2$, i.e.,

$$\left| \begin{array}{l} \Pr[\mathscr{A}(P,Q,R,[a]P,[b]Q,[a+b]R) = 1] \\ - \Pr[\mathscr{A}(P,Q,R,[a]P,[b]Q,Z) = 1] \end{array} \right| \geq \varepsilon,$$

where $a, b \in \mathbb{Z}_p, P, Q, R, Z \in \mathbb{G}_2$, and the probability is over the random choices of the parameters and of the coin tosses of $\mathscr{A}$.

Boneh, Boyen, and Shacham [38] show that the decision linear assumption holds in generic *bilinear* groups.

Theorem 7.2. *The n-user PE-ABS scheme in* $(\mathbb{G}_1, \mathbb{G}_2)$ *has* $(t, q_h, q_s, n, \varepsilon)$ *selfless anonymity in the random oracle model, assuming the* (t, ε') *decision linear assumption holds in the group* $\mathbb{G}_2$ *for* $\varepsilon' = \frac{\varepsilon}{2}(\frac{1}{n^2} - \frac{q_s q_h}{p}) \approx \frac{\varepsilon}{2n^2}$.

Proof of Theorem 7.2. *Suppose Algorithm* $\mathscr{A}$ *breaks the selfless anonymity of the n-user PE-ABS signature scheme. We build an algorithm* $\mathscr{B}$ *that breaks the decision linear assumption in* $\mathbb{G}_2$. *Algorithm* $\mathscr{B}$ *is given as input of a 6-tuple* $(u_0, u_1, w, h_0 = [a]u_0, h_1 = \lfloor b \rfloor u_1, Z) \in \mathbb{G}_2^6$, *where* $u_0, u_1, w \in \mathbb{G}_2, a, b \in_R \mathbb{Z}_p$, *and either* $Z = [a+b]w \in \mathbb{U}_2$ *or Z is random in* $\mathbb{G}_2$. *Algorithm* $\mathscr{B}$ *decides which Z was given by interacting with* $\mathscr{A}$ *as follows.*

Setup. $\mathscr{B}$ *simulates Setup as follows:*

- $\mathscr{B}$ *selects a random* $G = \psi(u_0) \in \mathbb{G}_1, h = [a]w \in \mathbb{G}_2$ *(unknown) as generators of* $\mathbb{G}_1$ *and* $\mathbb{G}_2$, *respectively;*
- $\mathscr{B}$ *picks a random integer* $\alpha \in_R \mathbb{Z}_p, \beta = a$ *(unknown) and computes,*

$$\begin{aligned} g &= [\beta]G = [a]\psi(u_0) = \psi(h_0) \in \mathbb{G}_1 & (7.18) \\ \zeta &= e(G,h)^{\alpha} = e(\psi(u_0),[a]w)^{\alpha} & (7.19) \\ &= e(\psi(h_0),w)^{\alpha} \in \mathbb{G}_T; \end{aligned}$$

- $\mathscr{B}$ *picks a random* $e \in_R \mathbb{Z}_p^*$ *and defines* $G' = [1/e]\psi(h_0) = [a/e]\psi(u_0) \in \mathbb{G}_1$ *and* $n = \lfloor e \rfloor w \in \mathbb{G}_2$, *which replace G and h in attributes recovery, since* $e(G,h) = e(\psi(u_0),[a]w) = e(G',h')$;

- $\mathcal{B}$ *picks two random users* $i_0, i_1 \in_R [1,n]$. *For all* $i \in [1,n]$ *except* $i_0, i_1, \mathcal{B}$ *selects* $r_i, r_j \in_R \mathbb{Z}_p^*$ *and computes*

$$D_i = \left[\frac{\alpha + r_i}{\beta}\right] h = [\alpha + r_i] w \in \mathbb{G}_2, \tag{7.20}$$

$$A_{i,j} = [r_i] G' + [r_j] \psi (H_0(attr_j)) \tag{7.21}$$
$$= [r_i/e] \psi (h_0) + [r_j] \psi (H_0(attr_j)) \in \mathbb{G}_1,$$
$$A'_{i,j} = [r_j] \psi (h') = [r_j e] \psi(w) \in \mathbb{G}_1, \tag{7.22}$$

as the private key $\left(D_i, A_{i,j}, A'_{i,j}\right)$ *due to*

$$\frac{e(A_{i,j}, h')}{e\left(A'_{i,j}, H_0(attr_j)\right)} = e\left([r_i] G', h'\right) = e\left([r_i] G, h\right). \tag{7.23}$$

- *for* i_0 *and* i_1, $\mathcal{B}$ *picks a random* $r \in_R \mathbb{Z}_p$ *and defines* $D_{i_0} = \left[\frac{\alpha + ar}{\beta}\right] h = [\alpha + ar] w \in \mathbb{G}_2$ *and* $D_{i_1} = [r]Z + [\alpha - br] w \in \mathbb{G}_2$, *which is unknown since it knows neither a nor b. Observe that if* $Z = [a+b]w$, *then*

$$D_{i_0} = [\alpha + ar] w = [ar + br] w + [\alpha] w - [br] w \tag{7.24}$$
$$= [r]Z + [\alpha] w - [br] w = D_{i_1}.$$

Hence, users i_0 *and* i_1 *have the same private key in this case.*

Hash Queries. At any time, $\mathcal{A}$ *can query the hash functions* H_0, H_1. $\mathcal{B}$ *responds with random values with consistency.*

Phase 1. $\mathcal{A}$ *can request signing queries and corruption queries. If* $i \neq i_0, i_1$, *then* $\mathcal{B}$ *uses the secret key of i to respond to the query as usual. If* $i = i_0, i_1, \mathcal{B}$ *responds as follows.*

Signing Query. $\mathcal{B}$ *generates a signature for M using* D_{i_0} *or* D_{i_1}:

- *for user* $i = i_0, \mathcal{B}$ *picks random* $x, k, l \in_R \mathbb{Z}_p$, *sets* $t = (ar+x)/k$ *and computes:*

$$u = [k] \psi(u_0) \in \mathbb{G}_1, \tag{7.25}$$
$$v = [kl]u_0 - [k]w \in \mathbb{G}_2, \tag{7.26}$$
$$E_i = [t]u = [ar+x] \psi(u_0) = [r] \psi(h_0) + [x] \psi(u_0) \in \mathbb{G}_1, \tag{7.27}$$
$$T_i = D_{i_0} + [t]v = [\alpha - x]w + [rl]h_0 + [xl]u_0 \in \mathbb{G}_2; \tag{7.28}$$

- *for user* $i = i_1, \mathcal{B}$ *picks random* $x, k, l \in_R \mathbb{Z}_p$, *sets* $t = (br+x)/k$ *and computes:*

$$u = [k] \psi(u_1) \in \mathbb{G}_1, \tag{7.29}$$
$$v = [kl]u_1 + [k]w \in \mathbb{G}_2, \tag{7.30}$$
$$E_i = [t]u = [br+x] \psi(u_1) = [r] \psi(h_1) + [x] \psi(u_1) \in \mathbb{G}_1, \tag{7.31}$$
$$T_i = D_{i_1} + [t]v = [r]Z + [\alpha+x]w + [lr]h_1 + [x]]u_1 \in \mathbb{G}_2. \tag{7.32}$$

*Any way, $E_i = [t]u \in G_1$ and $T_i = D_i + \lfloor t \rfloor v \in G_2$ for some random $t \in Z_p$
and independent $u \in G_1, v \in G_2$. Since $r_i = ar$ and $\beta = a$, algorithm $\mathscr{B}$ selects
a random integer $y \in Z_p$, sets $s = y/a \in Z_p$ and computes $[r_i s] G = [ary/a] u_0 =
[yr] u_0$ and $Q_i = [s\beta]G = [ay/a]u_0 = [y]u_0$ and $Q_i = [s\beta]G = [ay/a]u_0 = [y]u_0$.
In terms of $e(G,h) = e(G', h')$, $\mathscr{B}$ sets $[r_i s]G' = [yr/e] \psi(h_0) \in G_1$. Moreover,
since $[\Delta_s(j)]([r_i]G') = [\Delta_{r_i s}(j)]G'$ (Let $\Delta_s(x) = s + \sum_{i=1}^{t} a_i x^i$. Then $(\Delta_s(x)]g =
[s]g + \sum_{i=1}^{t} [a_i x^i] g$. When $[s]g = [r_i s]G' = [yr/e] \psi(h_0), g = [1/e] \psi(h_0)$, we have
$[\Delta_{yr}(x)] G' = [yr/e] \psi(h_0) + \sum_{i=1}^{t} [a_i x^i/e] \psi(h_0)$), $\mathscr{B}$ sets $r_i S = ary/a = yr, r_j \in_R Z_p$,
and computes the following equation in terms of policy tree $\mathscr{T}$,*

$$B_{i,j} = [\Delta_{yr}(attr_j)] G' + [\Delta_s(attr_j) r_j] \psi(H_0(attr_j)) \tag{7.33}$$
$$= [\Delta_{yr}(attr_j)][1/e] \psi(h_0) + [\Delta_s(attr_j) r_j] \psi(H_0(attr_j)),$$
$$B'_{i,j} = [\Delta_s(attr_j) r_j] \psi(h') = [\Delta_s(attr_j) er_j] \psi(w). \tag{7.34}$$

$\mathscr{B}$ assures that the equation

$$\frac{e(B_{i,j}, h')}{e\left(B'_{i,j}, H_0(attr_j)\right)} = e([\Delta_{r_i s}(attr_j)] G', h') \frac{e([\Delta_s(attr_j) r_j] \psi(H_0(attr_j)), h')}{e(\Delta_s(attr_j)[r_j] \psi(h'), H_0(attr_j))}$$
$$= e([\Delta_s(attr_j)]([r_i] G'), h') \tag{7.35}$$
$$= e([\Delta_{r_i s}(attr_j)] G, h).$$

*So, $e([r_i s]G, h)$ can be recovered, and $e([r_i s]G, h) = e([yr] \psi(u_0), [a]w) =
e([ayr/e] \psi(u_0), [e]w) = e([yr/e] \psi(h_0), [e]w) = e([yr]G', h')$. Next, $\mathscr{B}$ picks $s_t, s_s \in_R
Z_p$ and computes the corresponding V_1, V_2:*

$$V_1 = [s_t]u - [c]E_i, \tag{7.36}$$
$$V_2 = \frac{\zeta^{s_s} \cdot e([r_i s]G, h)^c \cdot e(Q_i, v)^{s_t}}{e(Q_i, T_i)^c}. \tag{7.37}$$

*In the unlikely event $\mathscr{A}$ has already issued a hash query for $c =
H_1(gpk, M, I_i, E_i, Q_i, V_1, V_2)$. $\mathscr{B}$ reports failure and terminates. This happens with
probability at most q_h/p. Otherwise, $\mathscr{B}$ defines $c = H_1(gpk, M, T_i, E_i, Q_i, V_1, V_2)$.
Algorithm $\mathscr{B}$ then computes the signature $\sigma = (\mathscr{T}, u, v, T_i, E_i, Q_i, c, s_s, s_t,
\left\{ B_{i,j}, B'_{i,j} \right\}_{attr_j \in \rho_i \cap \mathscr{T}})$, and gives σ to $\mathscr{A}$.*

*Corruption Query. If $\mathscr{A}$ issues a corruption query for user i_0 or i_1, then $\mathscr{B}$ reports
failure and aborts.*

*Challenge. $\mathscr{A}$ outputs a message M, and two users, i_0^* and i_1^* where it wishes to
be challenged. If $\{i_0^*, i_1^*\} \neq \{i_0, i_1\}$, then $\mathscr{B}$ reports failure and aborts. Otherwise, $\mathscr{B}$
picks a random $b \in_R \{0,1\}$ and generates a signature σ^* under user i_b's key for M
using the same method responding to signing queries in Phase 1. It gives σ^* as the
challenge to $\mathscr{A}$.*

Phase 2. $\mathscr{A}$ issues restricted queries. $\mathscr{B}$ responds as in Phase 1.

*Output. Eventually, $\mathscr{A}$ outputs its guess $b' \in \{0,1\}$ for b. If $b = b'$, then $\mathscr{B}$ outputs
0 (meaning that Z is random in G_2); otherwise, $\mathscr{B}$ outputs 1 (meaning that $Z =
\lfloor a + b \rfloor w$).*

Suppose $\mathcal{B}$ does not terminate in the simulation. When Z is random in $\mathbb{G}_2$, $\mathcal{B}$ emulates the selfless-anonymity game perfectly. Hence, $\Pr\lfloor b = b' \rfloor > \frac{1}{2} + \varepsilon$. Then, when $Z = [a+b]w$, the private keys for user i_0 and i_1 are identical and hence the challenge signature σ^* is independent of b. If follows that $\Pr\lfloor b = b' \rfloor = \frac{1}{2}$. Therefore, assuming $\mathcal{B}$ does not abort, it has an advantage of at least $\varepsilon/2$ in solving the given linear challenge $(u_0, u_1, w, h_0, h_1, Z) \in \mathbb{G}_2^6$.

$\mathcal{B}$ does not abort if it correctly guesses the values i_0^* and i_1^* during the setup phase and none of the signing queries causes it to abort. The probability that a given signature query causes $\mathcal{B}$ to abort is at most q_s/p and therefore the probability that $\mathcal{B}$ aborts as a result of $\mathcal{A}$'s signature is at most $q_h q_s/p$. As long as $\mathcal{B}$ does not abort in Phase 1, $\mathcal{A}$ gets no information about i_0, i_1. So the probability that the query during Phase 1 and the choice of challenge do not cause $\mathcal{B}$ to abort is $1/\binom{n}{2}$, greater than $1/n^2$. It now follows that $\mathcal{B}$ can solve the given linear challenge with advantage at least $\frac{\varepsilon}{2}\left(\frac{1}{n^2} - \frac{q_s q_h}{p}\right)$, as required. This completes the proof of Theorem 7.2.

7.5.2 EXISTENTIAL UNFORGEABILITY UNDER CHOSEN MESSAGE ATTACKS

In the *bilinear* group pair $(\mathbb{G}_1, \mathbb{G}_2)$, q-strong Diffie-Hellman (SDH) problem is stated as follows: given a $(q+2)$-tuple of elements $\left(P, Q, [\gamma]Q, [\gamma^2]Q, \ldots, [\gamma^q]Q\right) \in \mathbb{G}_1 \times \mathbb{G}_2^{q+1}$, output a pair $\left(x, \left[\frac{1}{\gamma+x}\right]Q\right)$ for a freely chosen value $x \in \mathbb{Z}_p \backslash \{-\gamma\}$, where P, Q are generators in $\mathbb{G}_1$ and $\mathbb{G}_2$, respectively. We give a definition of the SDH assumption as follows.

Definition 7.8 (Strong Diffie-Hellman Assumption). (q, t, ε)-SDH assumption holds in $(\mathbb{G}_1, \mathbb{G}_2)$ if no t-time algorithm $\mathcal{A}$ has advantage at least ε in solving q-strong Diffie-Hellman on $(\mathbb{G}_1, \mathbb{G}_2)$, i.e.,

$$\Pr\left[\mathcal{A}\left(P, Q, [\gamma]Q, \ldots, [\gamma^q]Q\right) = \left(x, \left[\frac{1}{\gamma+x}\right]Q\right)\right] \geq \varepsilon,$$

where the probability is over the random choices of generators (P, Q) in $\mathbb{G}_1 \times \mathbb{G}_2$, of γ in $\mathbb{Z}_p$ and of the random bits of $\mathcal{A}$, $P = \psi(Q)$ for an efficient homomorphism ψ.

SDH assumption was proposed by Boneh and Boyen to construct a short signature scheme without random oracle. To gain confidence in the assumption they proved that it holds in generic groups and it has similar properties to the strong-RSA assumption.

Theorem 7.3. *Suppose (q, t', ε)-SDH assumption holds in $(\mathbb{G}_1, \mathbb{G}_2)$. Then PE-ABS scheme above is (t, q_s, ε)-secure against existential forgery under a weak chosen message attack provided that $q_s \leq q$ and $t \leq t' - \Theta\left(q^2 T\right)$, where T is the maximum time for an exponentiation in $\mathbb{G}_1, \mathbb{G}_2$, and $\mathbb{Z}_p$.*

Proof of Theorem 7.3. *Suppose that there is a* (t, q_s, ε)*-forger algorithm* $\mathscr{A}$ *that breaks the signature scheme. We construct an algorithm* $\mathscr{B}$*, by interacting with the forger* $\mathscr{A}$*, which can solve the q-SDH problem in time* t' *with advantage* ε*.* $\mathscr{B}$ *is given a random instance* $(P, Q, [\gamma]Q, \ldots, [\gamma^q]Q)$ *of the q-SDH problem in* $(\mathbb{G}_1, \mathbb{G}_2)$*, where* $P = \psi(Q)$*.*

Setup. Let f *be the univariate polynomial defined by* $f(X) = \prod_{i=1}^{q}(X + x_i)$ *and* $f_i(X) = \frac{f(X)}{X + x_i}$*. Expand* f *and write* $f(X) = \sum_{i=0}^{q} a_i X^i$ *where* $a_0, \ldots, a_q \in \mathbb{Z}_p$ *are the coefficients of the polynomial* f*. Let* $\alpha = f(\gamma)$*,* $\beta = \gamma, G = P$*, then the simulator* $\mathscr{B}$ *chooses a random integer* $\xi \in \mathbb{Z}_p^*$ *and sets*

$$gpk = \begin{cases} g &=& [\beta]G = [\gamma]P = \psi([\gamma]Q), \\ h &=& [\xi]Q, \\ \zeta &=& e(G, h)^\alpha = e(P, [\xi]Q)^{f(\gamma)} \\ &=& e(P, [f(\gamma)]Q)^\xi = e(P, Q)^{\xi f(\gamma)}. \end{cases} \tag{7.38}$$

Note that, gmk $= (\alpha, \beta, G) = (f(\gamma), \gamma, P)$ *is unknown.*

For each $r_i = -x_i(f_i(\gamma) + \gamma)$*, compute*

$$sk_i = \begin{cases} D_i &=& \left[\frac{\alpha + r_i}{\beta}\right]h = \left[\frac{f(\gamma) - x_i(f_i(\gamma) + \gamma)}{\gamma}\right]h = [\xi(f_i(\gamma) - x_i)]Q, \\ A_{i,j} &=& [r_i]G + [r_j]\psi(H_0(attr_j)) \\ &=& [-x_i(f_i(\gamma) + \gamma)]P + [r_j]\psi(H_0(attr_j)), \\ A'_{i,j} &=& [r_j]\psi(h). \end{cases} \tag{7.39}$$

Queries. The simulator $\mathscr{B}$ *must respond with at most q signatures on the respective messages from* $\mathscr{A}$*.* $\mathscr{B}$ *chooses* $t = x_i$*, and sets*

$$E_i = [t]u = [x_i]P, \tag{7.40}$$

$$T_i = D_i + [t]v = [f_i(\gamma) - x_i]h + [x_i]h = [\xi f_i(\gamma)]Q, \tag{7.41}$$

$$Q_i = [s]g = [\gamma s]G = [\gamma s]P, \tag{7.42}$$

$$B_{i,j} = [-x_i(f_i(\gamma) + \gamma)\Delta_s(attr_j)]P + [r_j\Delta_s(attr_j)]\psi(H_0(attr_j)), \tag{7.43}$$

$$B'_{i,j} = [r_j\Delta_s(attr_j)]\psi(h). \tag{7.44}$$

then selects $c, s_s, s_t \in \mathbb{Z}_p$*, and computes*

$$V_1 = [s_t]u - [c]E_i, \tag{7.45}$$

$$V_2 = \frac{\zeta^{s_s} e([r_i]G, h)^s e(Q_i, v)^{s_t}}{e(Q_i, T_i)^c}, \tag{7.46}$$

$$c = H_1(gpk, M, T_i, E_i, Q_i, V_1, V_2). \tag{7.47}$$

Output. After receiving the signature $(M, \tilde{\sigma} = (\mathscr{T}, \tilde{u}, \tilde{v}, \tilde{T}_i, \tilde{E}_i, Q_i, \tilde{c}, \tilde{s}_s, \tilde{s}_t,$ $\left\{\left(B_{i,j}, B'_{i,j}\right)\right\}_{attr_j \in \mathscr{T}})$*),* $\mathscr{B}$ *attempts to extract t, the discrete logarithm of* E_i*. Using this value,* $\mathscr{B}$ *can compute* $\tilde{x}_i$*. To achieve this value,* $\mathscr{B}$ *checks*

$$e([\gamma]G + \tilde{E}_i, \tilde{T}_i) = e\left([\gamma + \tilde{x}_i]G, \left[\frac{f(\gamma)}{\gamma + \tilde{x}_i}\right]h\right) = e(P, Q)^{\xi f(\gamma)} = \zeta. \tag{7.48}$$

If the equality holds, $\mathcal{B}$ runs the above process again with the same state as before but a different hash challenge $\bar{c} \in \mathbb{Z}_p$, and then obtains response $\left(\overline{M}, \overline{\sigma} = \left(\mathcal{T}, \bar{u}, \bar{v}, \overline{T}_i, \overline{E}_i, \overline{Q}_i, \bar{c}, \bar{s}_s, \bar{s}_t, \left\{ \left(\overline{B}_{i,j}, \overline{B}'_{i,j} \right) \right\}_{attr_j \in \mathcal{T}} \right) \right)$.

- *If $\tilde{E}_i = \overline{E}_i, \tilde{T}_i = \overline{T}_i$, and $\tilde{V}_1 = \overline{V}_1$, $\mathcal{B}$ computes $\tilde{x}_i = \bar{x}_i = \frac{\tilde{s}_t - \bar{s}_t}{\bar{c} - \bar{c}} \bmod p$ due to that $\tilde{s}_t = \tilde{r}_t + \tilde{c} \cdot \tilde{x}_i$, $\bar{s}_t = \bar{r}_t + \bar{c} \cdot \bar{x}_i$, and $\tilde{r}_t = \bar{r}_t$.*
- *If $\tilde{E}_i \neq \overline{E}_i, \tilde{T}_i \neq \overline{T}_i$, or $\tilde{V}_1 \neq \overline{V}_1$, (E_i, V_1, c, s_t) is a zero-knowledge scheme, then there exists a knowledge extractor that extracts $t = x_i$ from E_i.*

Now, $\mathcal{B}$ gets a pair $\left(\tilde{x}_i, \tilde{T}_i \right) = \left(\tilde{x}_i, \left[\tilde{f}_i(\gamma) \right] h \right) = \left(\tilde{x}_i, \left[\frac{f(\gamma)}{\gamma + \tilde{x}_i} \right] h \right)$. Write $f(x) = \left(\sum_{l=0}^{q-1} a'_l x^l \right) (x + \tilde{x}_i) + r$. Thus $\mathcal{B}$ knows $a'_0, \ldots, a'_{q-1}$ and r. According to

$$\left[\frac{f(\gamma)}{\gamma + \tilde{x}_i} \right] h = \left[\sum_{l=0}^{q-1} a'_l \gamma^l + \frac{r}{\gamma + \tilde{x}_i} \right] h, \tag{7.49}$$

$\mathcal{B}$ computes

$$\left[\frac{1}{\gamma + \tilde{x}_i} \right] Q = \left[\frac{1}{r\xi} \right] \left(\left[\frac{f(\gamma)}{\gamma + \tilde{x}_i} \right] h - \left[\sum_{l=0}^{q-1} a'_l \gamma^l \right] h \right) \tag{7.50}$$

$$= \left[\frac{1}{r\xi} \right] \left(T_i - \sum_{l=0}^{q-1} a'_l \cdot \left[\gamma^l \right] h \right).$$

Finally, $\mathcal{B}$ outputs $\left(\tilde{x}_i, \left[\frac{1}{\gamma + \tilde{x}_i} \right] Q \right)$ as the solution to the submitted instance of the SDH problem, which contradicts with SDH assumption.

The claimed bound $\mathfrak{t} \leq \mathfrak{t}' - \Theta \left(q^2 T \right)$ is obvious by the construction of the Algorithm $\mathcal{B}$.

7.6 PERFORMANCE ANALYSIS

In this section, we analyze the efficiency of the above-mentioned scheme. Firstly, we analyze the computation cost of the disperse algorithm, aggregate algorithm, and all phases in the signature scheme. The basic operation of our scheme is the computation of a multiple elliptic point in elliptic curve (Mul), namely, $[k]P$, where k is a positive integer and P is an elliptic curve point. We neglect the computation costs of an addition of elliptic points and simple modular arithmetic operations because they run fast enough. The important operations are the computation of a *bilinear* map $e(\cdot, \cdot)$ between two elliptic points (BM), Hash functions (Hashes), and exponential operation in $\mathbf{G}_T$ (EXP). For clearance, we give Table 7.1 to present them. In this table, we assume that the aggregate algorithm executes n times Equation (1) and that each internal node has $logn$ children, where n is the number of attributes.

Here we assume the pairing takes the form $e : E \left(\mathbb{F}_{p^m} \right) \times E \left(\mathbb{F}_{p^{km}} \right) \to \mathbb{F}^*_{p^{km}}$ ((we give here the definition from), where p is a prime, m is a positive integer, and k is the

Table 7.1
Performance analysis for PE-ABS

Phase\ Item	Mul	Hash	BM	EXP
Setup	2	0	1	1
Join	$1+3n$	n	0	0
Sign	$3+2n$	1	1	2
Verify	3	$n+1$	$2n+2$	4
Disperse	0	0	0	0
Aggregate	0	0	0	$n\log n$

embedding degree (or security multiplier). Without loss of generality, let the security parameter κ be 80 bits, we need the elliptic curve domain parameters over $\mathbb{F}_p$ with $|p| = 160$ bits and $m = 1$ in our experiments. This means that the length of integer is $l_0 = 2\kappa$ in $\mathbb{Z}_p$. Similarly, we have $l_1 = 4\kappa$ in $\mathbb{G}_1$, $l_2 = 24\kappa$ in $\mathbb{G}_2$, and $l_T = 24\kappa$ in $\mathbb{G}_T$ for the embedding degree $k = 6$. Hence, for PE-ABS scheme, the communication overhead of sign/interact is $3l_0 + (3+2n)l_1 + 2l_2 + |\mathscr{T}| = 66\kappa + 9n\kappa$ bits, where n is the number of attributes and $|T|$ denotes the length of policy tree (assume $|\mathscr{T}| = n\kappa$). This means that 1 KB can store a signature with more than 100 attributes.

7.7 SUMMARY

The endorsement of claim is an important signature form, widely used for accounting, banking, legal, business, insurance and other services. In order to implement a cryptographic signature for the endorsement of claim, we propose an attribute-based signature scheme with policy-and-endorsement mechanism. Depending on the match between access policies of signature and identity attributes of private key, this scheme can meet various requirements to authenticate identity of signers. Without doubt, the security proof of signature scheme is a challenging task for random oracle model in comparison to standard model. Based on strong Diffie-Hellman assumption and decision linear assumption, we describe the security of our scheme from two aspects: selfless anonymity and existential unforgeability. In addition, the performance analysis shows that our scheme has lower computational overheads and shorter signature length for a highly complex policy. Future studies in this area are two-fold: first, to make access policy more refined, we will try to produce Boolean values using relational comparisons; second, based on policy-and-endorsement mechanism, a group-oriented authentication scheme will be investigated to implement offline identification in an information sharing system.

Part II

Applications of Attribute-Based Encryption

Gartner predicted that by 2020, 70% of enterprises will use ABAC as the dominant mechanism to protect critical assets, up from less than 5% since 2014 [86]. However, as of today, ABAC still has not gained momentum. Infrastructure-centric application of ABAC does not provide enough additional capabilities to justify the cost and risk of migrating from widely adopted AC models such as RBAC. Most organizations simply do not have the complex access control policies to fully take advantage of ABAC. RBAC itself is flexible enough to support complex policies, even though they may be better expressed under ABAC. Furthermore, ABAC presents several disadvantages that cannot be overlooked. ABAC is able to support more complex policies, but it is also inherently more complex to operate. There are many more attributes than roles, which increases management complexity. Simple changes to attributes or policy can have unintended effects that do not readily manifest. Simple audit queries, such as the list of resources accessible to a user, are difficult to answer.

According to the discussion on National Cybersecurity Center of Excellence (NCCoE) on the topic of the slow adoption of ABAC [3], *"one obstacle is lack of detailed guidance on how to integrate and configure ABAC components; hence the Practice Guide"*, i.e., there is a lack of well-documented (preferably with some before-and-after metrics) "case studies" of how ABAC has delivered one or more business benefits. Innovative access control solution must not only be technically sound but also address real business problems and provide tangible benefits. Infrastructure-centric ABAC will only provide evolutionary and marginal benefits to most organizations and their current use-cases. The benefits are outweighed by the cost and risk in making drastic change to their mission-critical infrastructures.

ABE-based ABAC, on the other hand, provides revolutionary capabilities. The problem here is that those capabilities only make sense in specific data-centric use-cases, where data and its attributes play a significant role for the application on how to use and access the data. However, there are not many use cases in current business practices. For a large part, there is a lack of support from traditional access control and security technologies.

In reality, the adoption of ABAC must be driven by the application of ABE-based ABAC to new data-centric information sharing paradigms. Only after ABE-based ABAC is widely in use will we see adoption of infrastructure-centric ABAC. In Part II, we present several ABE use-case studies to demonstrate how to use ABE-based ABAC for specific applications. Chapter 8 presents how to use ABE to realize efficient secure group communication; Chapter 9 describes how to use ABE to protect users' identities in a privacy-preserving communication environment; Chapter 10 presents how to use ABE to support mobile and IoT devices to sense data and then store data in a mobile cloud, where ABE-based computational offloading plays a key role to support light-weight IoT devices; Chapter 11 shows an ABE-based naming scheme to support name-based content routing and accessing in an Information-Centric Networking (ICN) setup; Chapter 12 shows how to use ABE to enforce policy-based data access control in a highly dynamic and mobile networking environment; finally, Chapter 13 presents an ABE-based data access control to create channels and protect blockchain data.

8 Efficient Key Management for Secure Multicast Communication

Existing rooted-tree based group key distribution schemes [46, 171, 197, 224] attain information theoretical storage optimality. In this chapter, the focus is on how to address the communication overhead, which is denoted as storage-communication-optimality condition. Basically, the ciphertext size is proportional to the number of used attributes in the encryption based on existing CP-ABE solutions. Thus, this solution is to achieve the constant-size of ciphertext size regardless the number of attributes to be used for ABE. In literature, a flat table scheme [49] claims the storage-communication optimality, however it is vulnerable to collusion attacks, and thus, none of existing rooted-tree based schemes is optimal under the new storage-communication-optimality condition.

In this chapter, we presented how to use ABE schemes to build a collusion resistant Optimal Group Key (OGK) management scheme, and thus to address both the collusion issue and ciphertext size issue. OGK is to secure multicast group communication, and it is the first scheme that achieves storage-communication optimality. Moreover, OGK is very flexible in that it does not rely on a centralized party to initiate and establish a secure communication group, and it allows dynamic subgroup communication initialized by each group member. OGK requires $O(\log N)$ storage overhead for each group member. Thus, it is suitable for applications requiring large-scale secure group communication. The performance evaluations demonstrate that OGK out-performs all existing secure group communication schemes under the storage-communication optimal condition. Finally, OGK can achieve several security properties such as forward/backward secrecy, collusion resistant, and IND-CPA security.

8.1 INTRODUCTION

IP multicast can be used to distribute data to a group of receivers efficiently. Existing multicast group key distribution schemes [209, 224] secure the one-to-many communication by encrypting the data using a Group Key (GK). Thus, only legitimated group members should have access to update-to-date GKs when group members join or leave the group dynamically. This requirement is usually achieved through a group rekeying procedure, in which a centralized group controller updates key material for all legitimate group members.

Group membership removal is usually more challenging than group membership addition because the number of impacted group members is exponentially large. To facilitate membership revocation, rooted-tree based key distribution schemes have been proposed, such as [224, 197, 46, 171]. In these schemes (illustrated in Figure 8.1 and 8.2), each member is distributed $\log N$ auxiliary secrets for group management (i.e., revocation and addition). In [175], the authors proved that assigning $\log N$ secrets to each member is information-theoretical optimal in terms of minimizing storage overhead when group size is N. However, as we will present in the latter part of this chapter, existing solutions do not fully utilize the pre-installed secrets to minimize the communication overheads.

The presented OGK scheme can achieve theoretical optimal performance considering both storage and communication overheads. In OGK, a group controller (GC) is responsible for key generation and distribution for each group member only at the beginning when the member joins the overall group. Each group member can initiate a secure group/subgroup communication without relying on the GC. Data targeting to a group/subgroup is encrypted by a GK that can only be decrypted by legitimated group members. When joining the overall group, each group member (GM) is assigned a unique n-bit ID and a set of secrets, in which each bit of the ID is one-to-one mapped to a unique secret. We must note that GMs may have the same bit-assignment at particular bit positions (i.e., value "0" or "1") in some positions of their IDs, however the corresponding secrets are different and they are masked by using distinct random numbers to prevent collusion problems. In this way, GMs cannot share their secrets to derive others' predistributed secrets. In this chapter, we denote the set of pre-distributed secrets as the GM's private key.

Whenever GMs are revoked from the group, the GC broadcasts an encrypted key-update message. Only the remaining GMs are able to recover the message and then update the GK as well as their private keys. To achieve *storage-communication optimality*, which is formally defined in Definition 8.3, it uses tree-based construction that is based on Flat Table (FT) [49] approach. In order to minimize the number of encrypted key-update messages, the presented solution utilizes the minimized Boolean function in the form of Sum-of-Product Expression (SOPE) that is calculated based on the IDs of remaining GMs. As a result, remaining GMs can combine the pre-distributed secrets to decrypt the updated GK.

OGK can achieve the storage-communication optimality with constant ciphertext size. Moreover, OGK is immune to collusion attack. It outperforms existing group key management schemes in terms of communication and storage efficiency. In [55], the authors utilized the Ciphertext-Policy Attributed-Based-Encryption (CP-ABE [25]) scheme to implement FT so that it is secure against collusion attack, which looks similar to the presented solution. OGK adopts a different approach for key distribution that significantly improves the communication efficiency compared to using CP-ABE directly. For example, as presented in [55], the size of each key update message is linearly proportional to the numbers of involved attributes [55, 25]. However, in OGK, the message size is substantially reduced to a constant size.

Based on the storage-communication optimality, OGK also supports dynamic subgroup communication efficiently. OGK allows each GM to initialize a secure subgroup communication with any subset of GMs. Moreover, the numbers of required messages for subgroup setup is minimized. Later in this chapter the storage-communication optimality is proved based on information theory. Moreover, we show that the construction of OGK is secure under Chosen-Plaintext-Attack (CPA) based on Decisional Bilinear Diffie-Hellman (DBDH) assumption, and OGK also has the proved security features such as collusion resistance, forward group secrecy and backward group secrecy.

In summary, OGK achieves all of following properties:

- Given any number of revoked GMs, the number of encrypted key-update messages is information theoretically minimized to $\approx O(\log N)$.
- The size of each encrypted key-update message is constant.
- The communication overhead of GM addition is $O(1)$, i.e., only one multicast message is required.
- The storage overhead of the GC and each GM is $O(\log N)$ if the GC does not store IDs of GMs.
- OGK supports dynamic subgroup communication efficiently.
- OGK is proved to be collusion resistant.
- OGK is proved to provide forward and backward group key secrecy.
- OGK is proved to be IND-CPA secure.

8.2 RELATED WORKS

Multicast key distribution schemes have been investigated intensively in past two decades. Some of the works include but are not limited to [47, 46, 172, 224, 159, 228]. Due to the richness of related research, we cannot list all the related work in this area. We refer readers to [161] as two excellent surveys.

The rooted-tree structure (see Figure 8.1 and Figure 8.2) is constructed such that each group member is assigned a unique leaf node in the tree. Every node in the tree, including leaf and non-leaf nodes, is assigned a unique auxiliary secret. Each group member is pre-distributed a set of auxiliary symmetric secrets (or keys) that are along the path from the leaf to the root, in which the root secret is GK for the entire group. Using rooted-tree based solutions, an auxiliary secret can be shared among a partition of members, and a member can be involved in multiple partitions. Typically, a rooted-tree based solution requires $O(\log_a N)$ storage overhead for each member [46], where N is the group size. The rooted-tree based multicast group key distribution scheme can be divided into two categories: Non-flat-table schemes (Figure 8.1) and flat-table schemes (Figure 8.2).

Non-flat-table include most famous rooted-tree based schemes, such as OFT [197], LKH [224], and ELK [171]). One important feature of these schemes is there are a^d distinct secrets at level d in the key distribution tree as illustrated in Figure 8.1. In other words, each node is associated with a unique secret. We note that the secrets are not necessarily just pre-distributed random symmetric keys [224]. They may be generated using one-way hash function [197] or pseudo random number generator [171]. Non-flat-table schemes only improve the efficiency marginally. This is because, in these solutions, based on the $\log_a N$ pre-distributed auxiliary keys, each

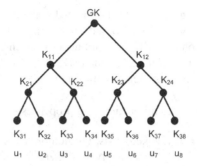

Figure 8.1: A tree example of non-flat table scheme.

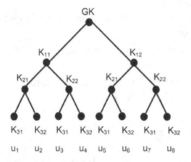

Figure 8.2: A tree example of flat table scheme.

group member can merely decrypt $\log N$ encrypted streams, as illustrated in Figure 8.1. In this example, three auxiliary non-root keys are assigned to the group member u_2: K_{11}, K_{21}, and K_{32}. Note that combining multiple keys cannot generate new valid keys. This is due to the fact that members holding K_{21} are a subset of members holding K_{11}, and members holding K_{32} are a subset of members holding K_{21}. Using these auxiliary keys, u_2 can decrypt three distinct encrypted streams:

Encrypted Streams	Accessible Members
K_{11}	$\{u_1, u_2, u_3, u_4\}$
K_{21}	$\{u_1, u_2\}$
K_{32}	$\{u_2\}$

Flat-table schemes [49, 47] adopt a slightly different construction, as illustrated in Figure 8.2. In flat-table schemes, each group member is issued a unique binary ID with n bits: $b_0 b_1 \dots b_{n-2} b_{n-1}$. In addition to the GK, group controller generates $2n$ auxiliary key encryption keys (KEK) $\{K_{i,b} | i \in \mathbb{Z}_n, b \in \{0,1\}\}$. A group member with ID $b_0 b_1 \dots b_{n-2} b_{n-1}$ holds KEKs $\{K_{i,b_i} | i \in \mathbb{Z}_n\}$. The KEKs are organized in the key distribution tree in Figure 8.2, where each level corresponding to one bit position in a user's ID. Thus, at each level in the flat-table key distribution tree, there are exact 2 distinct KEKs, which map to a bit position in ID. For example, in the Figure 8.2, member with ID 011 is predistributed keys $\{K_{11}, K_{22}, K_{32}\}$. In flat-table, the number of partitions each group member can participate in is maximized to $2^{\log N} - 1 = N - 1$. As illustrated in Figure 8.2, 3 non-root keys are distributed to the group member u_2: K_{11}, K_{21}, and K_{32}. Using these auxiliary keys, u_2 can decrypt 7 encrypted streams:

Encrypted Streams	Accessible Members
K_{11}	$\{u_1, u_2, u_3, u_4\}$
K_{21}	$\{u_1, u_2, u_5, u_6\}$
K_{32}	$\{u_2, u_4, u_6, u_8\}$
K_{11} and K_{21}	$\{u_1, u_2\}$
K_{11} and K_{32}	$\{u_2, u_4\}$
K_{21} and K_{32}	$\{u_2, u_6\}$
K_{11} and K_{21} and K_{32}	$\{u_2\}$

Despite its efficiency, flat-table schemes are vulnerable to collusion attacks since FT solutions simply adopt the symmetric KEKs. For example, GMs u_2 (001) and u_3 (010) can decrypt ciphertexts destined to other GMs, e.g., u_4 (011), u_1 (000), by combining their symmetric KEKs. To prevent the collusion attacks, Cheung et al. [55] proposed CP-ABE-FT to implement the FT using CP-ABE. However, message size of CP-ABE-FT is linearly growing [55] and, thus, the communication overhead is actually $\log^2 N$. As a contrast, OGK features collusion resistance and a constant message size. Thus, the OGK communication overhead is $\log N$. Also, CP-ABE-FT utilizes a periodic refreshment mechanism to ensure forward secrecy. If the ID of a revoked GM is re-assigned to another GM before the refreshment, the revoked GM can regain the access to group data and then the group forward secrecy is compromised.

Broadcast Encryption (BE) was introduced by Fiat and Naor et al. in [81] and then followed by [35, 85, 93, 164, 240]. In BE a broadcaster encrypts a message for some set of users who are listening to a broadcasting channel and use their private keys to decrypt the message. Compared with traditional one-to-one encryption schemes, BE features superior efficiency. Instead of sending messages encrypted with each individual recipient's public key, the broadcast encryptor broadcast one encrypted message to be decrypted by multiple recipients with their own private keys.

Although existing BE schemes always feature small or constant ciphertext, the number of public keys or private keys is linear in the max number of non-colluding users in the system. In the case of the BE scheme is fully collusion-resistant, the number of public/private keys each user needs to store equals to the number of users in the system. For example, in the existing BE system with N users, each user $u_{i \in \{1,...,N\}}$ is generated a public key PK_i and a private key SK_i. To encrypt a message to a set of users S, the encrypting algorithm takes input of the set of public keys for all recipients $\{PK_i | \forall u_i \in S\}$ and outputs the ciphertext. To decrypt a message, the decrypting algorithm takes input of the private key SK_i of user u_i and the set of all public keys $\{PK_i | \forall u_i \in S\}$ to recover original message. OGK supports many-to-many subgroup communication with $O(\log N)$ storage overhead on GMs, as contrast to $O(N)$ storage overhead in BE [35].

8.3 SYSTEM MODELS AND BACKGROUND

8.3.1 NOTATIONS

The notations used in this solution is listed in Table 8.1.

Table 8.1

Notations used in the presented scheme

Symbols	Descriptions
G	the Broadcasting Group Includes All GMs
L	a Subset of GMs
u	a GM
B	Bit-Assignment
S	Set of Bit-Assignments
GC	Group Controller
GM	Group Member

8.3.2 COMMUNICATION MODEL

The communication model of OGK is based on multicast. All Group Members (GM) belong to a multicast group $G = \{u_1, u_2, \ldots, u_{|G|}\}$. Each GM u can send or receive diagrams. The multicast group is associated with a trusted server, referred to as Group Controller (GC), responsible for managing the membership. Each GM can initialize a secure subgroup communication with any subset of GMs. The subgroup traffic is multicasted to whole group while only a designated subset of GMs can decrypt the data.

8.3.3 BILINEAR PAIRING

Pairing is a *bilinear* map function $e : \mathbb{G}_1 \times \mathbb{G}_2 \to \mathbb{G}_T$, where $\mathbb{G}_1$, $\mathbb{G}_2$ and $\mathbb{G}_T$ are three cyclic groups with large prime order p. The $\mathbb{G}_1$ and $\mathbb{G}_2$ are additive groups and $\mathbb{G}_T$ is a multiplicative group. The discrete logarithm problem on $\mathbb{G}_1$, $\mathbb{G}_2$ and $\mathbb{G}_T$ are hard. Pairing has the bilinearity property:

$$e([a]g, [b]h) = e(g, h)^{ab}, \ \forall g \in \mathbb{G}_1, h \in \mathbb{G}_2, a, b \in \mathbb{Z}_p^*.$$

8.3.4   ATTACK MODELS

We assume that the symmetric encryption algorithm E and one-way hash function H used in this solution is a random oracle. Additionally, we assume that the Discrete Logarithm Problem (DLP) on groups G_1, G_2, and G_T is intractable. In addition, the GC is well guarded and trustable. Finally, we assume that there is no collusion between legitimate GMs and revoked GMs.

The security analysis will focus on collusion resistance, forward secrecy, and backward secrecy. The attackers' goal is to reveal multicasted data. In particular, we can consider the attacking scenarios in the following cases:

1. *Breaking the Group Secrecy*: Non-GMs try to reveal the multicasted group data with more than negligible probability.
2. *Breaking Backward Secrecy*: GMs try to reveal any group data that were transmitted before they joined the group with more than negligible probability.
3. *Breaking Forward Secrecy*: GMs try to continue to reveal the group data that are transmitted after they left the group with more than negligible probability.
4. *Collusion Attacks*: Multiple revoked GMs combine their pre-distributed secrets to decrypt the ciphertext not intended to them. One example of this attack is that when multiple GMs are revoked from the group, they try to collude to continue decrypting group data. Another example is that when a secure conference is held among a subgroup of GMs, some excluded GMs try to listen to the conference.

In all of these scenarios, we assume that attackers can receive and stores all transmitted messages. However, there is no such a compromised insider GM that works as a decryption proxy for attackers.

8.4   CONSTRUCTIONS OF OGK

8.4.1   ID AND BIT-ASSIGNMENT

In OGK, each GM is associated with a unique n-bit binary ID: $b_0 b_1 \ldots b_{n-2} b_{n-1}$, where $n = \log N$. The ID is issued by the GC when a GM joins the group. Once the GM left the group, his/her ID can be re-assigned to other joining GMs.

We can use a logic literal, which is called *bit-assignment*, B_i or $\overline{B_i}$ to indicate the binary value at position i in a particular ID. B_i indicates the $b_i = 1$; $\overline{B_i}$ indicates the $b_i = 0$. For a group with N GMs, the length of an ID is $n = \log N$ and the total number of bit-assignments is $2n$; that is, two binary values are mapped to one-bit position. We call the set of all possible bit-assignments to be *Universe U*, which contains $2n$ bit-assignments.

A GM u is uniquely identified by the set of bit-assignments S_u associated with u's ID. Also, multiple GMs may have a common subset of bit-assignments. For example, in Figure 8.3, a GM u_1's ID is 000 and a GM u_2's ID is 001, $S_{u_1} = \{\overline{B}_0, \overline{B}_1, \overline{B}_2\}$ and $S_{u_2} = \{\overline{B}_0, \overline{B}_1, B_2\}$ and $S_{u_1} \cap S_{u_2} = \{\overline{B}_0, \overline{B}_1\}$.

In OGK, the GMs can be organized as leafs in a binary tree with each non-root node marked with a bit-assignment (Figure 8.3). Note that there are only $2n$ distinct non-root nodes in the tree and each level contains 2 distinct nodes. This is fundamentally different from existing tree-based schemes in [197, 224, 171], where there are 2^d distinct nodes at level d. The ID of a GM can be represented by the bit-assignment nodes from the root down to leaves. Thus, any two GMs will have at least one different bit-assignment.

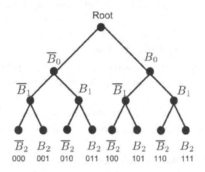

Figure 8.3: An example of bit-assignments for a 3-bit ID space presented in a binary tree structure.

8.4.2 GROUP SETUP

Here, we describe how the GC sets up the multicast group. First, GC chooses *bilinear* map over group G_1, G_2 and G_T of prime order p. Assume the generator of G_1 is g and generator of G_2 is h. Also, GC chooses a publicly known one-way function H. Then, it chooses two non-trivial random numbers $\alpha, \beta \in \mathbb{Z}_p^*$. For simplicity, we can map the universe of bit-assignments U to the first $|U|$ members of $\mathbb{Z}_p^*$, i.e., the integers $\{1, 2, \ldots, |U|\}$. For each bit-assignment $B \in U$, GC chooses a non-trivial random number $y_B \in \mathbb{Z}_p$. We denote this set of $2n$ random numbers as:

$$Y_B = \{y_{B_0}, y_{\overline{B}_0}, \ldots, y_{B_{n-1}}, y_{\overline{B}_{n-1}}\}.$$

For each $y_B \in Y_B$, GC generates a tuple $< e(g, h)^{\alpha y_B}, g^{\beta y_B} >$. We denote the set of $2n$ tuples as:

$$E_B = \{< e(g, h)^{\alpha y_B}, g^{\beta y_B} > | \forall y_B \in Y_B\}.$$

GC publishes the group public parameter:

$$GP = \{e, g \in G_1, h \in G_2, H, E_B\}.$$

On the other hand, GC protects the group master key:

$$MK = \{\alpha, \beta, Y_B\}.$$

8.4.3 GM JOINING AND KEY GENERATION

When a new GM u joins the group, u needs to setup a secure channel with the GC using either a pre-shared key or public key certificates. GC then checks whether the GM is authorized to join. Once the checking is passed, GC assigns a unique ID $b_{n-1}^u b_{n-2}^u ... b_0^u$ and a set of bit assignments S_u to u.

Once u is admitted to the group, GC runs key generation algorithm **KeyGen**(MK, S_u) to generate private key SK_u for u, where MK is the group master key and S_u is the set of bit-assignments in u' ID. The algorithm first chooses a non-trivial random number $r \in \mathbb{Z}_p^*$. Then, it computes $h^{\frac{\alpha+r}{\beta}} \in \mathbb{G}_2$. Finally, for each bit-assignment $B \in S_u$, the **KeyGen** algorithm calculates a blinded secret share $h^{r\gamma_B} \in \mathbb{G}_2$. The outputted private key:

$$SK_u : \{D = h^{\frac{\alpha+r}{\beta}}, \forall B \in S_u : D_B = h^{r\gamma_B}\}.$$

If u is the first GM in the group, GC will generate an initial GK and sends the private key $\{SK_u, GK\}$ to the new GM u through a secure channel. If u is not the first joining GM, to preserve backward secrecy, GC generates another random key GK' and multicast $\{GK'\}_{GK}$. Each GM other than u can decrypt the message and replace GK with GK'. Finally, GC sends $\{SK_u, GK'\}$ to the new GM u through a secure unicast channel. In the *join* process, besides the unicast communication, GC only needs to multicast one message, i.e., $\{GK'\}_{GK}$. Thus, the communication overhead for GMs join is $O(1)$.

One important observation is that GC does not need to store the ID or private keys of any GMs. Thus, the storage overhead of GC can be significantly reduced to $O(\log N)$, since GC is only required to store the system parameters and master key.

8.4.4 ENCRYPTION AND DECRYPTION

As we have mentioned, OGK allows GC and GMs to securely communicate with any subset of GMs. Whenever, GMs are revoked from the group, GC needs to multicast a key update message to all remaining GMs, who will update their GK as well as private keys. On the other hand, GMs can initialize a secure subgroup communication with any subset of GMs.

In this section, we present how a GC or GM can encrypt a message with a set of bit-assignment S, so that only GMs whose IDs satisfy S can decrypt the message. For example, in a three-bit-ID group, if a ciphertext is encrypted by using bit-assignment $S = \{\overline{B}_0, B_1\}$, GMs with IDs 010 and 011 can decrypt the ciphertext.

Encryption

Encrypt(GP, S, M) encryption algorithm takes inputs of the group parameter GP, a set of bit-assignment S, the message M, and returns the ciphertext CT. Given the set of bit-assignment S, it is easy to calculate the following terms:

$$e(g,h)^{\alpha Y_S} = e(g,h)^{\alpha \sum_{B \in S} y_B} = \prod_{B \in S} e(g,h)^{\alpha y_B},$$

$$g^{\beta Y_S} = g^{\beta \sum_{B \in S} y_B} = \prod_{B \in S} g^{\beta y_B}.$$

For example, if $S = \{\overline{B}_0, B_1, B_2\}$, $e(g,h)^{\alpha Y_S} = e(g,h)^{\alpha(y_{\overline{B}_0} + y_{B_1} + y_{B_2})}$.

After calculating $e(g,h)^{\alpha Y_S}$ and $g^{\beta Y_S}$, the **Encrypt** Algorithm 8.1 generates a non-trivial random number $t \in \mathbb{Z}_p^*$. Then, the algorithm computes $C_0 = M \oplus e(g,h)^{\alpha t Y_S}$, $C_1 = g^{\beta t Y_S}$, $C_2 = g^t$, where $\oplus$ is bitwise XOR operation. Thus, the ciphertext is as:

$$CT : \{S, C_0 = M \oplus e(g,h)^{\alpha t Y_S}, C_1 = g^{\beta t Y_S}, C_2 = g^t\}.$$

Algorithm 8.1 Encrypt(MK, S, M)

Compute $e(g,h)^{\alpha Y_S} = \prod_{B \in S} e(g,h)^{\alpha y_B}$;
Compute $g^{\beta Y_S} = \prod_{B \in S} g^{\beta y_B}$;
Randomly select $t \in \mathbb{Z}_p$;
Compute $C_0 = M \oplus e(g,h)^{\alpha t Y_S}$;
Compute $C_1 = g^{\beta t Y_S}$;
Compute $C_2 = g^t$;
return
$CT : \{S, C_0 = M \oplus e(g,h)^{\alpha t Y_S}, C_1 = g^{\beta t Y_S}, C_2 = g^t\}$;

Decryption

On receiving the CT, those GMs who satisfy the bit-assignment $CT.S$ can decrypt the CT by performing decryption algorithm **Decrypt(GP, SK, CT)**.

The **Decrypt** Algorithm 8.2 first checks whether the GM u is eligible to decrypt the message by testing whether $CT.S \subseteq S_u$, where $CT.S$ represents the bit assignments associated with the ciphertext CT. Then, for each bit assignment $B \in CT.S$, the algorithm uses u's pre-distributed secret shares $D_B = h^{r y_B}$ to compute:

$$F = \prod_{B \in CT.S} D_B = \prod_{B \in CT.S} h^{r y_B} = h^{r \sum_{B \in CT.S} y_B}$$
$$= h^{r Y_{CT.S}}.$$

Next, the algorithm computes:

$$A_1 = e(C_1, D) = e(g,h)^{(\alpha + r) t Y_{CT.S}},$$

and

$$A_2 = e(C_2, F) = e(g,h)^{r t Y_{CT.S}}.$$

Algorithm 8.2 Decrypt(GP, SK, CT)

---

if $CT.S \nsubseteq S_u$ **then**
  return $\perp$;
end if
Compute $F = \prod_{B \in CT.S} h^{r_{yB}} = g^{rY_{CT.S}}$;
Compute $A_1 = e(C_1, D) = e(g, h)^{(\alpha+r)tY_{CT.S}}$
Compute $A_2 = e(C_2, F) = e(g, h)^{rtY_{CT.S}}$
Compute $A_1/A_2 = A_3 = e(g, h)^{\alpha t Y_{CT.S}}$
Compute $C_0 \oplus A_3 = M$
return M;

Then the algorithm divides A_1 by A_2 and gets:

$$A_3 = A_1/A_2 = e(g, h)^{\alpha t Y_{CT.S}},$$

which blinds the plaintext in ciphertext. Finally, the algorithm unblinds the ciphertext by calculating $C_0 \oplus A_3 = M$.

8.4.5   ENCRYPTION FOR SUBGROUPS OF GMS

In this subsection, we present how GC or GMs can securely communicate with arbitrary subgroup members optimally. We first define some of the terms used in the following presentations:

- *Literal*: A variable or its complement, e.g., B_1, $\overline{B}_1$, etc.
- *Product Term*: Literals connected by AND gate, e.g., $\overline{B}_2 B_1 \overline{B}_0$.
- *Sum-of-Product Expression (SOPE)*: Product terms connected by OR, e.g., $\overline{B}_2 B_1 B_0 + B_2$.

Given a subgroup of GMs L, a Boolean membership function $M_L(B_0, B_1, \ldots, B_{n-2}, B_{n-1})$, which is in the form of SOPE, is used to determine the membership of this subgroup. Formally, the following properties of membership functions hold:

$$M_L(b_0^u, b_1^u, \ldots, b_{n-2}^u, b_{n-1}^u) = \begin{cases} 0 & \text{iff } u \in G \setminus L, \\ 1 & \text{iff } u \in L. \end{cases}$$

For example, if the subgroup $L = \{000, 001, 011, 111\}$, then $M = \overline{B}_0 \overline{B}_1 \overline{B}_2 + \overline{B}_0 \overline{B}_1 B_2 + \overline{B}_0 B_1 B_2 + B_0 B_1 B_2$.

The GC or a GM runs the Quine-McCluskey algorithm [157] to reduce M_L to minimal SOPE M_L^{min}. The reduction can consider *do not care* values on those IDs that are not currently assigned to any GM to further reduce the size of M_L^{min}. Since M_L^{min} is in the form of SOPE, encryption is performed on each product term. That is, for each product term E in M_L^{min}, the **Encrypt** algorithm encrypts the message with the set of bit-assignment S that contains all literals in E. The total number of encrypted messages equals the number of product terms in M_{min}.

For example, if $L = \{000, 001, 011, 111\}$, $M_L^{min} = \overline{B}_0\overline{B}_1 + B_1B_2$. We can find that M_L^{min} contains 2 product terms. the message M for the subgroup L can be encrypted as $M_{\{\overline{B}_0, \overline{B}_1\}}$ and $M_{\{B_1, B_2\}}$ respectively.

8.4.6 GM LEAVING

Key Update

When several GMs (denoted by set L) are revoked from the group, GC needs to update the $\{MK, GP, GK\}$ as well as the private key of each remaining GM $u \in G \setminus L$. We present how this process can be done efficiently.

GC first changes MK to $MK' = \{\alpha', \beta, Y_B\}$, where α' is randomly selected in $\mathbb{Z}_p$. Also, group public parameter GP is updated accordingly. Then, GC multicasts an encrypted key-update factor $kuf = h^{\frac{\alpha'-\alpha}{\beta}}$. Note that kuf is encrypted, and it cannot be decrypted by any $u \in L$.

Each GM $u \in G \setminus L$ updates the component D in its private key SK_u using the kuf. The new D can be updated by the following method: $D \cdot h^{\frac{\alpha'-\alpha}{\beta}} = h^{\frac{\alpha+r}{\beta}} \cdot h^{\frac{\alpha'-\alpha}{\beta}} = h^{\frac{\alpha+r+\alpha'-\alpha}{\beta}} = h^{\frac{\alpha'+r}{\beta}}$. Also, each $u \in G \setminus L$ updates their GK simply by computing $GK' = H(h^{\frac{\alpha'-\alpha}{\beta}})$.

Single or Multiple Leave

We first consider that only one GM leaves the group. For example, if the leaving GM u's ID is 101 with bit-assignment $S_u = \{B_0, \overline{B}_1, B_2\}$. The key updating message is encrypted as $\{kuf\}_{\{\overline{B}_0\}}$, $\{kuf\}_{\{B_1\}}$, $\{kuf\}_{\{\overline{B}_2\}}$ and is multicasted to the entire group. If ID 100 is not assigned, $\{kuf\}_{\{\overline{B}_2\}}$ is not needed. Although the leaving member may intercept the transmitted messages, it cannot decrypt them since every message is encrypted with a bit assignment that the leaving member does not possess. On the other hand, each of remaining GMs can decrypt at least one of the multicasted messages.

We now focus on the case when multiple GMs leave the multicast group. Given the set of leaving GMs L, GC can easily derive the set of remaining GMs, i.e. $G \setminus L$, as well as the set of unassigned IDs if GC stores all assigned IDs. If GC does not store assigned ID, GC can assume all IDs are assigned. Then, the GC runs the Quine-McCluskey algorithm [157] to reduce the membership function $M_{G \setminus L}$ to minimal SOPE. Then, GC can encrypt the key updating factor for each product term. The total number of encrypted key updating factors equals to the number of product terms in M_{min}. For example, we assume that two GMs $\{000, 010\}$ leave, five GMS $\{001, 011, 100, 101, 110\}$ remain, and $\{111\}$ is not assigned to any GM (i.e., the ID bit assignments are *do not care*). With the considerations *do-not-care* values, M can be reduced to $M_{G \setminus L}^{min} = B_0 + B_2$. GC needs to multicast two messages $\{kuf\}_{\{B_0\}}$ and $\{kuf\}_{\{B_2\}}$.

8.5 INFORMATION THEORETICAL STORAGE-COMMUNICATION-OPTIMALITY

In this section, we investigate the optimality of OGK through an information theoretical approach similar to the models in [175]. We first proved that $O(\log N)$ attains information-theoretical lower bound of storage requirements. Then, we define the Storage-Communication-Optimality condition.

8.5.1 OPTIMAL STORAGE

To be uniquely identified, each user's ID should not be a prefix of any other user's, i.e. the bit-assignments should be *prefix-free*. For example, suppose a user u' is issued an ID 00, which is prefix of u_1 with ID 000 and u_2 with ID 001. When an encryptor tries to reach u_1 and u_2, the minimized membership function is $M = \overline{B}_0 \overline{B}_1$, which is also satisfied by u'. Similarly, it is also imperative that a user's bit-assignments should not be a subset of any other users'.

Theorem 8.1. *We denote the number of bit-assignments (or number of bits in the ID) for a user u_i as l_i. For an multicast communication group with N users and the IDs of users satisfy the prefix-free condition, the set $\{l_1, l_2, \ldots, l_N\}$ satisfies the Kraft inequality:*

$$\sum_{i=1}^{N} 2^{-l_i} \leq 1.$$

□

Proof of Theorem 8.1. *Refer to proof of Theorem 3.2 and let $D = 2$.*

The prefix-free condition is necessary and sufficient condition for addressing any user with their bit-assignments.

Assuming l_i bit-assignments are required to identify u_i and the probability to send a message to u_i is p_i, we can model the storage overhead as:

$$\sum_{i=1}^{N} p_i l_i. \tag{8.1}$$

Intuitively, this formation argues that the storage overhead from a sender's perspective is the average number of bit-assignments required to address to any particular receiver. Thus, an optimization problem is formulated to minimize the storage overhead for a broadcast encryption system:

$$\min_{l_i} \sum_{i=1}^{N} p_i l_i$$

s.t.

$$\sum_{i=1}^{N} 2^{-l_i} \leq 1.$$

This problem can be further rewritten as a Lagrangian optimization problem as:

$$\min_{l_i}\{\sum_{i=1}^{N} p_i l_i + \lambda(\sum_{i=1}^{N} d^{-l_i} - 1)\}, \tag{8.2}$$

where λ is the Lagrangian multiplier. The optimization problem is identical to the optimal codeword-length selection problem [62] in information theory. Before giving the solution to this optimization problem, we define the entropy of targeting one user in the system:

Definition 8.1. *The entropy H of targeting a user is*

$$H = -\sum_{i=1}^{N} p_i \log p_i.$$

$\square$

Theorem 8.2. *For a system of N users with prefix free distribution of bit-assignments, the optimal (i.e., minimal) average number of storage overhead required for a sender to address a receiver, written as $\sum_{i=1}^{N} p_i l_i$ can be given by the binary entropy*

$$H = -\sum_{i=1}^{N} p_i \log p_i.$$

$\square$

Proof of Theorem 8.2. *The theorem is equivalent to optimal codeword-length selection problem, and proof is available in [62].* $\square$

Since the average number of bit-assignments required for addressing one particular receiver is given by the entropy of targeting a user, we now try to derive the upper and lower bounds of the entropy:

$$\max_{p_i}(-\sum_{i=1}^{N} p_i \log p_i)$$

and

$$\min_{p_i}(-\sum_{i=1}^{N} p_i \log p_i)$$

s.t.

$$\sum_{i=1}^{N} p_i = 1.$$

The upper bound $H_{max} = -\sum_{i=1}^{N} \frac{1}{N} \log N^{-1} = \log N$ is yielded when $p_i = 1/N$, $\forall i \in \{1, 2, \ldots, N\}$, when each user has equal possibility to be addressed as the receiver. When there is no a priori information about the probability distribution of targeting one of the users, $l = H_{max} = \log_d N$ correspond to the optimal strategy to minimize the average number of storage overhead required for each user. On the other hand, the lower bound $H_{min} = 0$ is achieved when $p_i = 1$ for $\exists i \in \{1, 2, \ldots, N\}$, which is an extreme case where there is no randomness and only one user is reachable.

8.5.2 STORAGE-COMMUNICATION-OPTIMALITY

Now that we have proved that $O(\log N)$ is the optimal storage strategy, we move to the optimal condition considering both storage and communication overhead. The authors in [175] showed that the assignment of $O(\log N)$ secrets per group member is the best strategy for group communication schemes. Thus, we can further claim that, given the $\log N$ pre-distributed secrets, the optimality is attained only if the number of encrypted streams that each group member can participate is maximized.

Formally, the Storage-Communication-Optimality is defined as follows:

Definition 8.2. Storage-Communication-Optimality condition: *for a group of N members, each group member can combine any of the pre-distributed* $\log N$ *secrets to decrypt* $2^{\log N} - 1 = N - 1$ *distinct encrypted streams.*

Given the formal definition of Storage-Communication-Optimality, we are going to prove that OGK achieves the Storage-Communication-Optimality in the rest part of this section. Next, we define "encryption stream" as follows.

Definition 8.3. Encrypted Stream: *An encrypted stream* ES_S *includes all ciphertexts encrypted by the* **Encrypt**(GP, S, M), *where S is the set of bit-assignments.* ES_S *can only be decrypted by the set of users whose IDs satisfy the S.*

Then, we investigate the number of users that can decrypt a particular encrypted stream.

Lemma 8.1. *For a set of bit-assignments S with x bit-assignments, i.e.,* $|S| = x$, *the number of IDs that satisfy S is* 2^{n-x}, *where n is the number of bits in a binary ID.*

Proof of Lemma 8.1. *Assume each user is identified by an n-bit binary ID. For an ID that satisfies the S, there are* $n - x$ *unfixed bits that can be either 1 or 0. Thus, the number of combinations of the* $n - x$ *unfixed bits is* 2^{n-x}. $\quad\square$

Lemma 8.2. *Two encrypted streams* ES_{S_1} *and* ES_{S_2} *can be decrypted by different sets of users if and only if* $S_1 \neq S_2$.

Proof of Lemma 8.2. *We consider two conditions: (1)* $|S_1| \neq |S_2|$ *and (2)* $|S_1| = |S_2|$ *and* $S_1 \neq S_2$. *In the condition (1), it easy to prove that if* $|S_1| \neq |S_2|$, *the number of IDs that satisfy* S_1 *is different from the number of IDs that satisfy the* S_2. *Thus,* ES_{S_1} *and* ES_{S_2} *can be decrypted by different set of users if* $|S_1| \neq |S_2|$.

In the condition (2), $|S_1| = |S_2|$ *and* $S_1 \neq S_2$. *There must exist at least one bit-assignment* $B' \in S_1$ *and* $B' \notin S_2$. *Since* $B' \notin S_2$, *There exists at least one user with the bit-assignment* B' *satisfying* S_2 *but not satisfying* S_1. *Thus* ES_{S_1} *and* ES_{S_2} *can be decrypted by a different set of users if* $|S_1| = |S_2|$ *and* $S_1 \neq S_2$. $\quad\square$

Theorem 8.3. Storage-Communication-Optimality of OGK: *OGK achieves storage-communication optimality.*

Proof of Theorem 8.3. *It is easy to show that the storage overhead incurred for each group member is $O(\log N)$. Each group member (e.g., u) needs to store group public parameter $GP = \{e, g \in \mathbb{G}_1, h \in \mathbb{G}_2, H, E_B\}$ and private key $SK_u : \{D = h^{\frac{\alpha+r}{\beta}}, \forall B \in S_u : D_B = h^{r y_B}\}$.*

Now, we prove that each user can decrypt $2^{\log N} - 1 = N - 1$ encrypted streams. Since there are $\log N$ bit-assignments for group member u, the total combinations of bit-assignments of non-empty sets are $2^{\log N} - 1 = N - 1$. According to Lemma 8.2, each of the $N - 1$ sets of bit-assignments is corresponding to a distinct encrypted stream. Thus, each group member can decrypt $N - 1$ distinct encrypted streams. Thus, the theorem is proved. □

8.6 IMPLEMENTATION OF OGK

In this section, we discuss the practical issues in implementing OGK, including choice of parameters and optimization methods on further reducing ciphertext size. The implementation uses the Pairing Based Cryptography (PBC) library. The theoretical and experimental performance assessment will be given in Section 8.7.

8.6.1 PARAMETERS

OGK is implemented over two parameter sets, each of which is specially optimized for different purposes. The Type $-$ A curve [152] is a supersingular curve $y^2 = x^3 + x$ over 512-bit finite field, which defines a160-bit elliptic curve group and features fastest pairing computation. On the other hand, the Type $-$ D curve [152] is chosen using MNT method [160] and has shortest group elements. Note that each element can be compressed to reduce size. In the actual implementation and performance evaluation presented in Section 8.7, we adopt the compressed Type $-$ D element to minimize storage and communication overhead. The benchmark was performed on a modern workstation that has a 3.0GHz Pentium 4-core CPU with 2 MB cache and 1.5 GB memory and runs Linux 2.6.32 kernel. In the Table 8.2, we compared the Type $-$ A and Type $-$ D parameters.

8.6.2 FURTHER REDUCING CIPHERTEXT SIZE

If we further investigate into the ciphertext, we can reduce the total multicast data size by combining common C_2 components for different product terms in the same membership function. For example, if $L = \{000, 001, 011, 111\}$, $M_L^{min} = \overline{B}_0 \overline{B}_1 + B_1 B_2$. We can find that M_L^{min} contains 2 product terms. the message M for L can be encrypted as $M_{\{\overline{B}_0, \overline{B}_1\}}$ and $M_{\{B_1, B_2\}}$. As presented in Section 8.4.4, the two encrypted messages are constructed as:

$$\{S_1 = \{\overline{B}_0, \overline{B}_1\}, C_0 = Me(g,h)^{\alpha t y_{S_1}}, C_1 = g^{\beta t y_{S_1}}, C_2 = g^t\}$$

Table 8.2

Comparison between Type − A **and** Type − D **curves**

	Type − A	Type − D
Base Field Size(bits)	512	159
Embedded Degree(k)	2	6
DLP Security(bits)	1024	954
Pairing (ms)	6.4	15.4
G_1 Element Size(bytes)	65	21
G_2 Element Size(bytes)	65	61
G_T Element Size(bytes)	128	120
Exp G_1(ms)	7.2	3.6
Exp G_2(ms)	7.3	21.1
Random G_1(ms)	8.4	3.5
Random G_2(ms)	8.5	20.9
Random G_T(ms)	3.0	8.0

and

$$\{S_2 = \{B_1, B_2\}, C_0 = Me(g,h)^{\alpha t Y_{S_2}}, C_1 = g^{\beta t Y_{S_2}}, C_2 = g^t\}.$$

Note that the C_2 component in these two messages are identical for the same random t.

8.7 PERFORMANCE ANALYSIS

In this section, a comparative study is performed based on the following related schemes: Flat-table scheme (FT) [49], FT implemented using CP-ABE (FT-ABE) [55], Subset-Diff broadcast encryption scheme [81], BGW broadcasting encryption [34], and non-flat-table tree-based schemes (e.g., OFT [197], LKH [224], ELK [171], etc.). The performance is assessed in terms of communication overhead (number and size of messages incurred by user revocation operations), storage overhead (group data stored on the GC and GM), and computation overhead (number of cryptographic operations needed in encryption and decryption operations). We denote the group size be N, the number of leaving GMs to be l. Also, for the Subset-Diff scheme, t denotes the maximum number of colluding users to compromise the ciphertext. The summary of comparative results is presented in Table 8.3.

8.7.1 COMMUNICATION OVERHEAD

The comparison of communication overheads is based on two metrics: 1) the number of messages and 2) the total size of messages. Key update messages are required for user revocation, and each message contains a key or key update materials. In

Table 8.3

Comparison of communication overhead and storage overhead in different group key management schemes

Scheme	Communication Overhead			Storage Overhead	
	join	single *leave*	multiple *leaves*	GC	GM
OGK	$O(1)$	$O(\log N)$	$\approx O(\log N)$	$O(\log N)/O(N)$	$O(\log N)$
Flat-Table [49]	$O(\log N)$	$O(\log N)$	$\approx O(\log N)$	$O(\log N)/O(N)$	$O(\log N)$
Flat-Table-ABE [55]	$O(1)$	$O(\log N)$	$\approx O(\log^2 N)$	$O(\log N)/O(N)$	$O(\log N)$
Subset-Diff [81]	N/A	$O(t^2 \cdot log^2(t) \cdot \log N)$	$O(t^2 \cdot log^2(t) \cdot \log N)$	$O(N)$	$O(log^2(N))$
BGW [34]	N/A	$O(N^{\frac{1}{2}})$	$O(N^{\frac{1}{2}})$	$O(N^{\frac{1}{2}})$	$O(N^{\frac{1}{2}})$
Non-Flat-Table-Tree [197, 224, 171]	$O(1)$	$O(\log N)$	$O(l \cdot \log N)$	$O(N)$	$O(\log N)$

N: the number of group members; l: the number of leaving members; t: maximum number of colluding users to compromise the ciphertext.

broadcast encryption schemes, each message contains an encrypted data encrypting key.

In BGW scheme, there are two constructions. In the first construction, the message size is constant $O(1)$ while the storage overhead is $O(N)$; in the second construction, the message size is $O(N^{\frac{1}{2}})$ as reported in [34]. It should be noted that the communication overhead are measured by the number of messages and the total size of messages broadcasted in the system.

For tree-based multicast key distribution schemes such as OFT [197], LKH [224], ELK [171], etc., the communication overhead for a GM leaving depends on the number of keys in the tree that need to be updated [202, 171]. Some tree-based schemes tried to optimize the number of messages to update all the affected keys in the case of multiple *leaves*. In ELK [171], which is known to be one of the most efficient tree-based schemes, the communication overhead for multiple *leaves* is $O(a - l)$, where a is the number of affected keys and l is the number of leaving GMs. Since there are $\log N$ nodes on the path from root to leaf in the tree structure, the total number of affected keys when l GMs leave the group is bounded by $O(l \cdot \log N)$.

When revoking multiple GMs from the OGK group, the number of messages depends on the number of product terms in the M_{min}. In [189], the authors derived an upper bound and lower bound on the average number of products in a minimized SOPE. For example, $\{000, 010\}$ are leaving GMs, and $\{001, 011, 100, 101, 110\}$ are remaining GMs, and $\{111\}$ is not assigned (i.e., *do not care*). In this example, the minimized SOPE is $M_{min} = B_0 + B_2$ and OGK requires 2 messages while tree-based schemes needs at least three messages. Now, we prove that OGK achieves storage-communication-optimal:

Number of Messages: Worst Cases

Firstly, we analyze the OGK performance in worst cases. In the worst cases, OGK out-performs all the tree-based schemes except flat-table. Since OGK requires same number of messages as flat-table when revoking a set of GMs, we utilize some of the performance results from [49].

Lemma 8.3 (Worst case of revoking 2 GMs). *When revoking two GMs from a group with $N = 2^n$ GMs, the number of key updating messages is at most n. The worst case is achieved when the Hamming distance between 2 GMs is n.*

Proof of Lemma 8.3. *Please refer to Theorem 3.5.*

As a comparison, in the same scenario, the number of keys to be updated is $2n - 1$, thus ELK requires $2n - 3$ messages while OGK requires n messages.

Lemma 8.4 (worst case of revoking multiple GMs). *The worst case of revoking multiple GMs happens when both of following conditions hold: 1) there are $N/2$ GMs to be revoked; 2) the Hamming distance between IDs of any two remaining GMs is at least 2. In the worst case, the number of key updating messages is $N/2$.*

Proof of Lemma 8.4. *Please refer to Lemma 3.3.*

In this case, the number of keys to be updated is $N - N/2 = N/2$ for ELK, since there are N non-leaf keys to be updated and the number of leaving GMs is $N/2$. We can see that, in this particular worst case, OGK's performance is same as the ELK approach. We argue that the worst cases happens in very low probability.

Lemma 8.5 (Worst case possibility). *When GMs are revoked in uniform probability, the worst case scenario happens with probability $\frac{1}{2^{N-1}}$.* □

Proof of Lemma 8.5. *In the worst case, the Hamming distance of IDs of $N/2$ revoked GMs should be at least 2. As shown in the Karnaugh table in Figure 8.4, each cell represents an ID. For any cell marked 0 and any cell marked 1, the Hamming distance is at least 2. Thus, the worst cases happens in two cases: (1) the encryptor wants to reach $N/2$ receivers marked 1 in Figure 8.4; (2) the encryptor wants to reach $N/2$ receivers marked 0 in Figure 8.4.*

□

Number of Messages: Average Cases

To evaluate the number of multicast messages in an average case, we simulated OGK along with ELK [171] that is considered as one of efficient group key management

b_0b_1 \ b_2b_3	00	01	11	10
00	1	0	1	0
01	0	1	0	1
11	1	0	1	0
10	0	1	0	1

Figure 8.4: Worst cases of broadcast encryption to $N/2$ receivers.

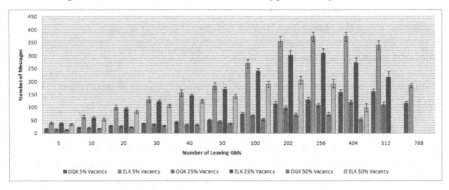

Figure 8.5: Number of messages of multiple leave for a group with 1024 GMs.

schemes. In the simulation, we consider the group size to be 1024 and each case is repeated 100 times to calculate the confidence intervals of 95% in Table 8.4. The number of messages required are shown in Figure 8.5, where we consider the three cases: 5%, 25%, 50% IDs are not assigned (i.e., *do not care* value). For each case the comparison between OGK and ELK is plotted side-by-side in the figure. Additionally, we consider different numbers of leaving members: 5, 10, 20, 30, 40, 50, 100, 202, 256, 404, 512, and 768 that are also plotted horizontally in the figure, to cover the various circumferences such as a small number of leaving members, half of the leaving members, and majority of leaving members. From the simulation results, we can derive that OGK performs better than ELK and the average complexity is $O(\log N)$. From the figure, we can see that OGK performs significantly better than ELK in all cases.

Total Message Size

Finally, we look into the message size of OGK, FT-CP-ABE [55], and symmetric key tree-based schemes. In Table 8.5, we compare the total ciphertext size using OGK, ELK, and FT-CP-ABE in a 4096 group.

As mentioned in [55], in FT-CP-ABE, the size of ciphertext grows linearly based on the increase of the number of attributes in the access policy [55, 25].

Table 8.4

Confidence intervals ranges for number of messages of multiple leave for a group with 1024 GMs

Scheme(Vacancy)	OGK 5%	ELK 5%	OGK 25%	ELK 25%	OGK 50%	ELK 50%
Max	13.2%	12.8%	14.3%	11.4%	16.7%	16.3%
Min	6.7%	4.2%	6.8%	4.2%	8.6%	5.5%

Table 8.5

Total message size for 4096 group in bytes

Percentage of Vacancy	OGK			ELK			FT-CP-ABE		
	5%	25%	50%	5%	25%	50%	5%	25%	50%
41 GMs are leaving	7280	5824	4550	15872	14848	13248	133120	96896	68200
205 GMs are leaving	16926	15834	11466	46272	42496	35904	309504	263436	171864
410 GMs are leaving	42224	39494	28938	67904	61568	49344	772096	657076	433752
820 GMs are leaving	65338	55510	43862	89984	76096	52864	1194752	923540	657448
1024 GMs are leaving	78078	66976	47138	96064	77888	48640	1427712	1114304	706552

Experimentally, the message size in FT-CP-ABE starts at about 650 bytes, and each additional attribute adds about 300 bytes. In a system with 10-bit ID or 1024 GMs, the number of attributes using FT-CP-ABE ciphertext is at most 10 and the message size may be as large as 650+9*300=3350 bytes. Since the number of attributes in the access policy is bounded by $\log N$, we can conclude that the communication overhead of FT-CP-ABE is in the order of $O(\log^2 N)$. In OGK, every ciphertext contains exactly two group members on G_1, i.e., $\{C_0, C_1\}$ and one group member on G_T, i.e., G_0. Moreover, as we have mentioned in Section 8.6.2, the elements C_2 of all messages can be combined to further reduce the total ciphertext size. Using Type $-$ D with the element combination, the size of C_1 or C_2 is about 21 bytes and the size of G_0 is about 120 bytes. The encoded S takes n bytes, where n is the number of bits in ID. Thus, the first message takes $n + 170$ bytes, e.g., if $n = 10$, the first message takes 180 bytes; if $n = 12$, the first message takes 182 bytes. Each additional message will add $n + 145$ bytes. In summary, OGK's ciphertext size is smaller than the ciphertext size reported in FT-CP-ABE [55].

Existing tree-based schemes using symmetric encryption algorithms, such as AES, enjoys the advantage of small ciphertext. Each message contains a identifier to denote which key-encryption key (KEK) in the key tree is used to encrypt the message. In a system with N users, the number of KEKs in the binary key tree is $2N - 1$. Thus, it takes $\log(2N - 1)/8$ bytes to encode the identifier. In the simulation, we assume each message is 64 bytes, which includes: 1) a KEK identifier ($\log(2N - 1)/8$ bytes); 2) a symmetric key, e.g. AES key 32 bytes; and 3) an error-detecting code. However, based on the evaluation results in Table 8.5, the total message size of OGK will be smaller than symmetric key-based schemes when the size of a group is large and the number of leaving GMs is relatively small, e.g. $< \%25$, thanks to significantly reduced number numbers of transmitted messages. It can be expected in large scale systems, where the size of a multicast group is larger than 4096, OGK will be more efficient than other schemes.

8.7.2 STORAGE OVERHEAD

In OGK, the storage overhead for GC can be either $O(N)$ or $O(\log N)$. In the $O(N)$ case, GC stores all IDs and marks those IDs that are assigned (or not assigned). In this case, it is trivial to know the unassigned IDs and use them as do-not-care values when performing Boolean function minimizations. In the $O(\log N)$ case, GC stores no IDs but only maintains a SOPE of unassigned IDs. Initially, all IDs are not assigned. Then, if 000 is assigned, the SOPE of unassigned IDs is $B_0 + B_1 + B_2$. Whenever IDs are assigned to new users or reclaimed from revoked users, GC can always update this SOPE. When performing the Boolean function minimizations, GC can easily enumerate unassigned IDs from the SOPE. In both cases, the GC also needs to store a constant-size group master key MK and group public parameter GP, which is $O(\log N)$ in length.

The storage overhead for a GM is $O(\log N)$ since GM stores a private key component for each bit in its ID. Although the GC or GMs may need the list of GMs' IDs along with the list of *do not care* IDs to perform Boolean function minimization, we can argue that this does not incur extra storage overhead based on the following facts.

- An encryptor does not need to store the GMs' IDs after the multicast the data; thus, the storage space can be released.
- The GC can periodically publish the minimized SOPE of all unassigned IDs, which can be used by the encryptor to further reduce the number of messages.

In Table 8.6, we summarize the key sizes for 1024 and 4096 group based on the implementation. We note that adding 1 bit to a user's ID will add 296 bytes to each public key, 24 bytes to the master key, and 66 bytes to each private key using $Type - D$ curve with element compression.

Table 8.6

Key sizes for 1024 and 4096 group

	1024 group	4096 group
Public Key (bytes)	4054	4650
Master Key (bytes)	528	624
Private Key (bytes)	725	857

8.7.3   COMPUTATION OVERHEAD

In this section, we compare the computation overhead of those asymmetric key-based schemes. In an ACP scheme, the author reports that the encryption needs $O(N^2)$ finite field operations when the sub-group size if N; in the BGW scheme, the encryption and decryption require $O(N)$ operations on the *bilinear* group, which are heavier than finite field operations [178]. In OGK, each encryption requires $\log N$ operations on the *bilinear* groups, and the decryption requires 2 pairings. Thus, the complexities of encryption and decryption are bounded by $O(\log N)$ and $O(1)$ respectively. Although the problem of minimizing SOPE is NP-hard, efficient approximations are widely known. Thus, OGK is much more efficient than ACP and BGW when group size is large.

Experimentally, we summarized the benchmark results for OGK operations. The benchmark was performed on a modern workstation, which has a 3.0 GHz Pentium 4 CPU with 2 MB cache and 1.5 GB memory and runs Linux 2.6.32 kernel. Also, the Type − D curve is used. The computation overhead for 1024 and 4096 groups is presented in Table 8.7.

Table 8.7
Computation overhead for 1024 and 4096 groups

	1024 group	4096 group
Setup (ms)	398	405
Kengen (ms)	293	307
Encrypt (ms)	12	13
Decrypt (ms)	36	38

8.8   SECURITY PROOF OF OGK SCHEME

In this section, we formally reduce the Chosen Plaintext Security (CPA) security of the scheme to the standard Decisional Bilinear Diffie-Hellman (DBDH) assumption. This implies that breaking CPA security under the scheme is at least as difficult as breaking the DBDH assumption. The DBDH is generally considered a hard problem. In the rest of this section, we assume that a symmetric model of pairing, namely $e : \mathbb{G} \times \mathbb{G} \to \mathbb{G}_T$, to replace $e : \mathbb{G}_1 \times \mathbb{G}_2 \to \mathbb{G}_T$ in the scheme.

Definition 8.4. Decisional Bilinear Diffie-Hellman (DBDH): *Let $e : \mathbb{G} \times \mathbb{G} \to \mathbb{G}_1$ be an efficiently computable Bilinear map, where G has prime order p. The DBDH assumption is said to hold in G if no probabilistic polynomial-time adversary is able to distinguish the tuples $< g, g^a, g^b, g^c, e(g,g)^{abc} >$ and $< g, g^a, g^b, g^c, e(g,g)^z >$ with non-negligible advantage, where $a, b, c, z \in \mathbb{Z}_p$ and generator $g \in G$ are chosen independently and uniformly at random.*

The OGK is said to be secure against CPA if no probabilistic polynomial-time adversaries have non-negligible advantage in this game (noted as OGK-CPA game).

1. **Init**: The adversary chooses the challenge set of bit-assignments $\bar{S}$ and gives it to the challenger.
2. **Setup**: The challenger runs the Setup algorithm and gives the adversary GP.
3. **Phase 1**: The adversary submits S for a **KeyGen** query. Provided $\bar{S} \nsubseteq S$, the challenger answers a secret key SK for S. This can be repeated adaptively.
4. **Challenge**: The adversary submits two messages M_0 and M_1 of equal length. The challenger chooses $\mu \in \{0,1\}$ randomly and encrypts M_μ to $\bar{S}$. The resulting ciphertext CT is given to the adversary.
5. **Phase 2**: Same as Phase 1.
6. **Guess**: The adversary outputs a guess μ' of μ.

Theorem 8.4. *If a probabilistic polynomial-time adversary wins the OGK-CPA game with non-negligible advantage, then we can construct a simulator that distinguishes a DBDH tuple from a random tuple with non-negligible advantage.*

Proof of Theorem 8.4. *Suppose that an adversary $\mathscr{A}$ wins the CPA game for OGK with the advantage ε. Then, we can construct a Simulator $\mathscr{B}$ that breaks BDHE assumption with the advantage $\varepsilon/2$. The simulator $\mathscr{B}$ takes an input of a random BDHE challenge $< g, g^a, g^b, g^c, Z > = < g, A, B, C, Z >$, where Z is either $e(g,g)^{abc}$ or a random element $\mathbf{G}_1$. $\mathscr{B}$ now plays the role of challenger in the pre-defined CPA game:*
Init: *$\mathscr{A}$ sends to $\mathscr{B}$ the set of bit-assignments $\bar{S}$ that $\mathscr{A}$ wants to be challenged.*
Setup: *$\mathscr{B}$ sets up the system to generate GP. First, $\mathscr{B}$ selects $2n$ random numbers $Y_B' = \{y_{B_0}', y_{\bar{B}_0}', \dots, y_{B_{n-1}}', y_{\bar{B}_{n-1}}'\}$ and then re-computes the random numbers $Y_B = \{y_{B_0}, y_{\bar{B}_0}, \dots, y_{B_{n-1}}, y_{\bar{B}_{n-1}}\}$ by using a random $\gamma \in \mathbb{Z}_p$:*

$$y_B = \begin{cases} \gamma \cdot y_B' & B \notin \bar{S}, \\ y_B' & B \in \bar{S}. \end{cases}$$

Then, $\mathscr{B}$ sets $Y_{\bar{S}} = \sum_{B \in \bar{S}} y_B$, $\alpha = a$, and $h = g^c = C$ and randomly chooses $\beta \in \mathbb{Z}_p$. Also, $\mathscr{B}$ generates the group public parameter $GP = \{e, g, h, E_B\}$. Especially, for $\forall y_B \in Y_B$, $\mathscr{B}$ adds a tuple to E_B:

$$
< e(g,h)^{\alpha y_B}, g^{\beta y_B} >
$$
$$
= \begin{cases} < e(A,C)^{\gamma y_B'}, g^{\beta \gamma y_B'} > & B \notin \bar{S}, \\ < e(A,C)^{y_B'}, g^{\beta y_B'} > & B \in \bar{S}. \end{cases}
$$

Phase 1: *The adversary $\mathscr{A}$ submits a set of bit-assignments S for a private key query, where $\bar{S} \nsubseteq S$. The $\mathscr{B}$ randomly picks $r \in \mathbb{Z}_p$ set $D = g^{\frac{a+r}{\beta}} = (A \cdot g^r)^{\frac{1}{\beta}}$ and*

$$
D_B = \begin{cases} g^{r\gamma y_B'} & B \notin \bar{S}, \\ g^{r y_B'} & B \in \bar{S}. \end{cases}
$$

It is easy to verify that the private key SK is valid.

Challenge: *The adversary $\mathscr{A}$ submits two messages M_0 and M_1 of equal length. The $\mathscr{B}$ sets $t = b$ and chooses $\mu \in \{1,0\}$ at random and sets $C_0 = M_\mu \oplus Z^{Y_{\overline{S}}}$ and $C_1 = g^{\beta t Y_{\overline{S}}} = B^{\beta Y_{\overline{S}}}$ and $C_2 = g^t = B$. The $\mathscr{B}$ gives the $\mathscr{A}$ the following ciphertext $CT = \{\overline{S}, C_0, C_1, C_2\}$. It is easy to verify that the ciphertexts are valid.*

Phase 2: *Repeat as **Phase 1**.*

Guess: *$\mathscr{A}$ produces a guess μ' of μ. If $\mu' = \mu$, $\mathscr{B}$ wins the DBDH game. Otherwise, $\mathscr{B}$ fails DBDH game.*

If $Z = e(g,g)^{abc}$, then CT is a valid ciphertext due to

$$C_0 = M_\mu \oplus Z^{Y_{\overline{S}}} = M_\mu \oplus e(g,g)^{abc Y_{\overline{S}}} = M_\mu \oplus e(g,h)^{\alpha t Y_{\overline{S}}},$$

in which the advantage of $\mathscr{A}$ is ε. Hence, $P[\mu' = \mu | Z = e(g,g)^{abc}] > 1/2 + \varepsilon$.

If $Z = e(g,g)^z$, then C_0 is completely random from the view of $\mathscr{A}$. Therefore $\mu' = \mu$ holds with probability exactly $1/2$, regardless of the distribution on μ'. Hence, $P[\mu' = \mu | Z = e(g,g)^z] = 1/2$.

Hence, we have the equation

$$
\begin{aligned}
\Pr[\mu' = \mu] \ &= \ P[\mu' = \mu | Z = e(g,g)^{abc}] \Pr[Z = e(g,g)^{abc}] \\
&+ P[\mu' = \mu | Z = e(g,g)^z] \Pr[Z = e(g,g)^z] \\
&> \ \frac{1}{2}(\frac{1}{2} + \varepsilon) + \frac{1}{2} \cdot \frac{1}{2} = \frac{1}{2} + \frac{\varepsilon}{2}.
\end{aligned}
$$

It follows that $\mathscr{B}$'s advantage in the DBDH game is $\varepsilon/2$.

$\square$

Proven that the scheme is CPA secure, we are ready to prove collusion resistance and forward/backward secrecy of OGK.

Lemma 8.6 (Probability of Collision Attacks). *For any $S \neq S'$, the probability that $Y_S \neq Y_{S'}$ is overwhelming.*

Proof of Lemma 8.6. *If there exist S and S' ($S \neq S'$) such that $Y_S = Y_{S'}$, a GM with bit-assignments S' will be able to decrypt the ciphertext encrypted with bit-assignments S. Note that that the assumption holds with overwhelming probability $\frac{p(p-1)\cdots(p-N-1)}{p^N} > \frac{(p-N-1)^N}{p^N} = (1 - \frac{N-1}{p}) > 1 - \frac{N(N-1)}{p} > 1 - \frac{N^2}{p}$, where $N = 2^n$ and n is the number of bits in the ID.*

Lemma 8.7 (Collusion Resistance). *Leaving GMs cannot collude to decrypt multi-casted messages targeted to other GMs.*

Proof of Lemma 8.7. *We refer to the collusion attack as any combinations of GMs attempting to derive other GM's private keys by sharing their private keys. We first show that any two GMs cannot collude using their private keys since their private keys are embedded with different random numbers r. Given the private keys of two attackers a_1 and a_2, $SK_{a_1} = \{D = g^{\frac{\alpha + r a_1}{\beta}}, \forall B \in S_{a_1} : D_B = h^{r a_1 y_B}\},$*

$SK_{a_2} = \{D = g^{\frac{\alpha + ra_2}{\beta}}, \forall B \in S_{a_2} : D_B = h^{ra_2 y_B}\}$, the probability of deriving $SK_v = \{D = g^{\frac{\alpha + rv}{\beta}}, \forall k \in S_v : D_A = h^{rv y_k}\}$, where $S_v \subseteq S_{a_1} \bigcup S_{a_2}$ and $S_v \neq S_{a_1}$, $S_v \neq S_{a_2}$, is negligible, i.e., $\frac{1}{p}$, and the collusion resistance can be reduced to the DLP on G_1. Furthermore, adding more colluding attackers will not help due to the hardness of DLP.

Lemma 8.8 (Backward Group Secrecy). *OGK provides backward group secrecy.*

Proof of Lemma 8.8. *When new GMs join the group, a new random GK' is encrypted ($\{GK'\}_{GK}$) and multicasted. Furthermore, the private key of joining GMs are generated under a random system master key α_1. We note that all the previous key updating factors are encrypted using different random α's. Assuming one of the key-update factors kuf are encrypted by α_2, we have $kuf e(g, h)^{\alpha_2 t Y_S}$. Now, if a GM with bit-assignments S_u tries to decrypt the message and his/her bit-assignments $S \subseteq S_u$, the result he/she derived is $M \oplus e(g, h)^{(\alpha_2 - \alpha_1) t Y_S}$. In other words, although the GM's bit-assignments S_u satisfy $CT.S$, he/she cannot decrypt a message with outdated system parameters.*

It is easy to check that given facts: (1) randomness of GK', (2) the randomness of $\alpha's$, (3) security of symmetric encryption, and (4) DLP on G_1, G_2 and G_T, the new joining group member cannot derive any previous GKs or key-update factors. Thus, the group backward secrecy is guaranteed.

Lemma 8.9 (Forward Group Secrecy). *OGK provides forward group secrecy.*

Proof of Lemma 8.9. *When GMs leave the group, the GC updates the system parameters to MK' using a new random α' and multicasts the encrypted $g^{\frac{\alpha' - \alpha}{\beta}}$ to all the GMs in $G \setminus L$. The remaining GMs will update their private keys and the GK using the key updating factor $g^{\frac{\alpha' - \alpha}{\beta}}$. Based on Lemma 8.7 (Collusion Resistance), the leaving GMs cannot decrypt the key update factor. Thus, the leaving GMs cannot decrypt future encrypted messages since GK has been changed.*

Even if a revoked GM stores all encrypted key-updating messages after he/she left and re-join the group, he or she cannot decrypt a previous key updating message, since these messages are encrypted under different master keys, which is intractable to the joining GM based on Lemma 8.8 (Backward Group Secrecy). Moreover, using the key updating factor $g^{\frac{\alpha' - \alpha}{\beta}}$ to derive g^α and β is hard due to the DLP.

8.9 SUMMARY

In this chapter, we presented a key distribution scheme – OGK that attains storage-communication optimality without collusion vulnerability. OGK also supports dynamic subgroup communication initialized by each group member. Additionally, OGK requires $O(\log N)$ storage overhead at the group controller, which makes OGK suitable for applications containing a large number of multicasting group members. Moreover, adding members in OGK requires just one multicasting message. OGK is the first work with such features.

According to the theoretical and experimental analysis, OGK out-performs all existing solutions in the multicast and broadcast application domain. Moreover, we defined the storage-communication optimality from an information theory perspective. Finally, we discussed expanding scalability of OGK by clustering the multicast group and further reducing the communication overhead by combining common components in the ciphertext. To improve the security of OGK scheme from IND-CPA to IND-CCA scheme, it is possible to convert the presented scheme by using the Optimal Asymmetric Encryption Padding (OAEP) model, which is left for future investigation.

9 Gradual Identity Exposure Using Attribute-Based Encryption

Many Attribute-Based Encryption (ABE) schemes do not protect receivers' privacy, such that all the attributes to describe the eligible receivers are transmitted in plaintexts. Hidden policy-based ABE schemes have been proposed to protect receivers' privacy by using a construction that requires every user in the system to decrypt the ciphertext using all the attributes they possess, which incurs great computation and communication overhead. To address this issue, in this chapter, we propose a new concept – Gradual Identity Exposure (GIE) – to protect data receivers' identity. Our approach is to reveal the receivers' information gradually by allowing ciphertext recipients for decrypting the message using their possessed attributes one-by-one (but not all). If the receiver does not possess one attribute in this procedure, the rest of attributes are still hidden. Compared to hidden-policy based solutions, GIE provides significant performance improvement in terms of reducing both computation and communication overhead. We also present a theoretical framework to model the GIE with several new proposed concepts.

9.1 INTRODUCTION

The introduction of Identity Based Encryption (IBE) schemes [32] significantly enriched the identity management research by combining identity management with key management and encryption/decryption procedures. In IBE, the identity is not only the identifiable information, e.g., email address, ID number, etc., but also the public key of the identity carrier. IBE allows a sender to encrypt a message with the receiver's identity as the public key. Attribute-Based Encryption (ABE) [25] scheme extended the basic construction of IBE. In ABE, the identity is extended to a set of descriptive attributes that define, classify or annotate the user to which they are assigned. The encryptor can enforce an access policy, defined as a set of attributes, with encryption. Only those receivers whose attributes satisfy the access policy can decrypt the ciphertext.

Original ABE schemes do not consider the anonymity of data recipients. The data access policy is attached to the ciphertext in plaintext form. Thus, passive attackers can locate and track a user or infer the sensitivity of ciphertext by eavesdropping the user's identity in the networks. For example, the access policy *"ComputerScience"*, *"DepartmentChair" AND* – *"ComputerScience"*, *"Student"* implies the recipient's roles or positions.

Hidden policy [165, 127, 230] schemes have been proposed to protect the ciphertext recipients' privacy. In the hidden policy solutions, the data access policy is not attached with the ciphertext. Here, we consider the set of attributes as policies enforced for data access control using ABE schemes. Since policies also describe who are eligible data receivers (i.e., who can decrypt the ciphertexts), they can be used as an identity to represent a group or one user. Although hidden policy schemes ensure the perfect anonymity, they incur significant computation overhead for each user in the system. For example, in [165], the access policies must be pre-defined to avoid ambiguity. An access policy must contain all attributes predefined in the entire system. As a result, the number of attributes may proliferate into thousands in a large system even though many of the attributes were not actually used to encrypt the data, and thus the encryption/decryption process can be very expensive. Another critical drawback of hidden policy approach is that each receiver is required to "try" to decrypt all the ciphertexts they received. Only after the receivers finished the decryption process can they know whether they satisfy the associated policies. While the decryption can be comparatively efficient for hidden policy schemes, where receivers only need to check his own identity, hidden policy schemes require much more computation because the policy may contain a large number of attributes, and the receivers need to check all attributes to perform decryption.

Using pseudonyms to protect users' identities is another main technique to provide anonymity. The major restriction of using pseudonyms is that the communication peers must negotiate pseudonyms in advance, which greatly reduces the flexibility of applications. For examples, in delay-tolerant types of applications, a message sender may not know the pseudonyms of receivers. Using pseudonyms is even difficult when the receivers are in groups and the groups are highly dynamic. Another issue of using pseudonyms is that pseudonyms usually significantly complicate data access control. For example, in role-based access control, an identity is actually a set of descriptive attributes to describe the capability of a user. Anonymizing each attribute will make the system extremely difficult to deploy data access control rules. This is especially true when access control rules are generated dynamically. Even if a data access control model only requires the user's real identity, using pseudonym requires a pre-established translation mechanism to deploy the user's pseudonyms in advance. Tracking is another critical issue of using pseudonyms. If only one pseudonym is used for each entity, attackers can easily exploit vulnerabilities of a security system. For example, in a shared database system, data sets belonging to different parties can be encrypted and labeled with different IDs to identify their owners or parties who can decrypt the data. If the same pseudonym is used as the ID for multiple data sets, attackers can easily identify them and deploy attacks more effectively (i.e., destroy data sets with same IDs). To prevent a pseudonym from being tracked to identify the source of actions or locations, a user requires using different pseudonyms with respect to different actions and locations, which make the pseudonym-based identity management schemes even difficult and complicated, i.e., a synchronization mechanism is required to track the pseudonyms.

To address the above identified drawbacks, we develop a new identity management scheme to protect recipients' anonymity and reduce incurred computation complexity. We present a new concept – Gradual Identity Exposure (GIE) – for identity management. GIE has the following capabilities:

- The user's identity is exposed gradually based on receivers' authorized capabilities. At each step, the decryptor needs to satisfy certain attributes to expose next step attributes. Otherwise, decryption fails immediately and the decryptor learns nothing more than the attributes he/she is entitled. This is fundamentally different to the rigid concept of using hidden policy: "Try to decrypt the entire ciphertext, if it is decrypted, the policy will be revealed; if it cannot be decrypted, no policy will be revealed".
- GIE is flexible in that it does not require a pre-established policy agreement. Each user can specify a GIE scheduler, i.e., a procedure to expose an identity gradually, based on his/her security requirements without negotiating with the message receivers. This property makes the identity management adaptable in various unpredictable application scenarios.

GIE does need to expose some attributes information to the receivers at beginning. We find that this property is quite useful in some environments, such as Enterprise networks, since receivers can learn some information they are authorized to know. For example, in a large company's intra-networks, messages are encrypted and broadcasted to all departments. Each department has a server to backup messages targeted to its department, and this backup server only has the attributes assigned to department, e.g. "R&D". One sensitive message may be encrypted by policy "R&D", "StaffEngineer", "Female" in sequence, where attribute "R&D" is revealed first. Although the server cannot learn all attributes in the policy, it can identify attribute "R&D" and backup the message.

To measure the amount of revealed information during the identity exposure procedure, we present an information theoretical approach to measure the uncertainty reduction[1] for each exposure step using set theory. Based on the GIE model, we further propose a new concept: Optimal Identity Exposure (OIE), which requires the identity exposure procedure to reveal the minimal amount of information when a receiver fails the decryption process. Based on our investigations, we found that the OIE cannot be guaranteed for an arbitrarily selected composition attributes. However, we prove that if OIE exists in a given set of attributes, we can always find the optimality by using a deterministic polynomial searching algorithm. To handle the scenarios that OIE does not exist, we propose two algorithms: Pre-k-optimal Identity Exposure and Post-k-Optimal Identity Exposure to reveal an identity at the k far-end and close-end optimal steps, respectively, in the identity exposure procedure. Intuitively, the Pre-k-optimal ensures the maximal anonymity in beginning steps, while Post-k-optimal ensures in ending steps.

---

[1]We use reductions of sets to represent the uncertainty reduction. In this chapter, without special notice, we use uncertainty reduction and set reduction interchangeably, although uncertainty is usually represented using Shannon information theory.

The construction of GIE is based on Ciphertext-Policy Attribute-Based Encryption (CP-ABE) [25] by disclosing attributes and corresponding tree structure gradually in the process of decryption. In this way, attackers can only learn partial information of receivers' identities, since they will stop at a certain step during the decryption procedure. Our presented techniques do not increase the complexity of encrypting and decrypting messages with the original CP-ABE scheme. GIE also inherits the security features, such as collusion resistance, provided by the CP-ABE scheme presented in [25].

In summary, GIE focuses on the following salient features:

* We present a new identity management concept by gradually exposing users' information to protect users' privacy. This approach is flexible, and it does not need pre-establishment of secrets or agreements between communication peers.
* We present a new construction for ABE scheme to achieve the gradual identity exposure. Our solution inherits all security properties of original CP-ABE. Moreover, the new solutions hide the attribute tree through gradual identity exposure procedures, which is robust to counter colluding attackers sharing information to derive the hidden identity. Furthermore, with the added new privacy-preservation features, our solution is efficient in that: it does not increase the computation overhead compared to the original CP-ABE solution; and it does not introduce heavy communication overhead compared to existing attributes anonymizing solutions.
* We present a new anonymity measurement model based on set theory to capture the gradual uncertainty reduction of the proposed solution. Based on the proposed anonymity measurement model, we can evaluate several newly introduced concepts such as Optimal Identity Exposure, Pre-k-optimal Identity Exposure, and Post-k-Optimal Identity Exposure.

The rest of this chapter is arranged as follows: related work is presented in Section 9.2; in Section 9.3, the system and models required to build gradual identity exposure are presented; the detailed system construction is presented in Section 9.4; the anonymity measurement model of our solutions is presented in Section 9.5; in Section 9.6 the security analysis of GIE encryption scheme and performance evaluations are presented; finally, we conclude our research and present several ongoing research challenges in Section 9.7.

9.2 RELATED WORKS AND BACKGROUND

The Identity-Based Encryption (IBE) scheme was first introduced in [195] with several works [69, 206, 207, 155] followed up. However, the first fully functional IBE was proposed in [32]. In Identity-Based Encryption (IBE), an identity or ID is a string one-to-one mapped to each user. A user can acquire a private key corresponding to his ID in an off-line manner from trusted authority, and the ID is used as his public key. The ciphertext encrypted by a particular ID can only be decrypted by the user with corresponding private key, i.e., the encryption is one-to-one. Attribute-based encryption is a novel extension from IBE by enabling expressive access policy to control the decryption process as presented in [25, 186, 173, 50, 95, 56, 168, 127]. Particularly, BSW CP-ABE scheme [25] uses an access policy to encrypt the message. The ciphertext can be successfully decrypted only if the decryptor's set of attributes matches the access policy. The computation and communication overhead of BSW scheme is linear depending on the number of attributes in the ciphertext. Authors in [56] proposed a provable secure CP-ABE scheme, namely CN CP-ABE. In CN CP-ABE, all the attributes need to be pre-defined and the computation and communication overheads linearly depend on the number of attributes pre-defined by the system. However, both BSW and CN schemes do not consider hiding the access policy, and thus the access policy is public known to all receivers.

The anonymous IBE schemes [31, 42, 87] preserve recipient anonymity by hiding the identity in the ciphertext. To protect the privacy of access policy, a KSW scheme [127] and NYO scheme [165] was proposed, where the encryptor specified access policy is hidden. Also, a YRL scheme was proposed in [230] based on a BSW scheme as a group key management scheme providing group membership anonymity. This main difference between our scheme and existing hidden policy attribute-based encryption scheme is those schemes require all receivers to "try" decrypting all ciphertexts they received. Only after the receiver finished the decryption process can he know whether he is allowed to decrypt. Thus, these schemes pose a huge computational burden on all receivers. On the other hand, with the property of gradual exposing of attributes, receivers can detect ciphertexts that are not intended for them earlier to save computational power.

Cryptographic Background

The construction of the presented solutions is based on several cryptographic schemes including *bilinear* pairing, Secret Sharing, and CP-ABE schemes, which are presented in chapter 1 Sections 1.2.2, 1.3.2, and 1.3.6, respectively. For simplicity, here we omit them.

9.3 SYSTEM AND MODELS

In this section, we present the system and models that are required to construct the proposed solution. We first present the definitions and new concepts of GIE. Then, the communication model and attack model are presented in sequence at the end of this section.

9.3.1 CONCEPTS

The GIE construction is based on set theory. We first describe several notations shown in below:

- y is an attribute.
- H is a one-way hash function $\mathcal{H} : \{0,1\}^* \to \{0,1\}^l$.

- S_y represents the set of users with attribute y. The cardinality of this set is represented as $|S_y|$. The intersection of two attributes y_i and y_j is represented as $S_{y_i} \cap S_{y_j}$ and the union of these two attributes is presented as $S_{y_i} \cup S_{y_j}$. ϕ represents an empty set and $|\bigcup_{i=1}^{n} S_{y_i}| = N$.

Now, we are ready to present several definitions. Despite the English definition of identity, i.e., a term uniquely pinpoints to a person, we extend the definition of identity as follows:

Definition 9.1 (Identity). An identity $I = \{y_1, \ldots, y_n\}$ is a set of attributes (i.e., terms) that can be used to identify a user or a group (or a set) of users. An attribute may not be unique to distinguish a user or a group of users.　■

Definition 9.2 (Anonymity). We define $\nabla_{\{y_1,\ldots,y_m\}}$ as the anonymity of a set of attributes $\{y_1, \ldots, y_m\}$, which is measured by the cardinality $|\bigcap_{\forall y_i \in \{y_1,\ldots,y_m\}} S_{y_i}|$.

Usually, one attribute y_i can create a certain level of anonymity, which is determined by the cardinality of its set S_{y_i}.

Definition 9.3 (Gradual Identity Exposure). The attributes describing an identity are gradually exposed in a one-by-one fashion, and the anonymity is monotonically decreasing. In other words, at any particular step, exposing attribute y_j will not increase the anonymity. We can have the following formula:

$$\nabla_{\{\ldots y_{j-1}\}} \geq \nabla_{\{\ldots,y_{j-1},y_j\}}, \qquad where \quad j \geq 1. \tag{9.1}$$

■

For example, if we have the following four attributes $y_1 =$ –ASU employee˝, $y_2 =$ –faculty at ASU˝, $y_3 =$ –female faculty at computer science at ASU˝, and $y_4 =$ –faculty teaching database at ASU˝. In this example, we have $|\nabla_{\{y_1\}}| \geq |\nabla_{\{y_1,y_2\}}| \geq |\nabla_{\{y_1,y_2,y_3\}}| \geq |\nabla_{\{y_1,y_2,y_3,y_4\}}|$.

Definition 9.4 (Optimal Identity Exposure). For a given set of l attributes $\{y_1, \ldots, y_l\}$, there are $l!$ possible exposure procedures. For a given procedure number j, we denote the anonymity at step i as ∇_i^j, i.e. the anonymity after exposing i'th attributes in the procedure. If the optimal identity exposure procedure exists, it must satisfy the following properties:

1　The sequence of exposed attributes set must satisfy (9.1).
2　For any step i, The anonymity ∇_i^* is always maximized:

$$\nabla_i^* = \max\{\nabla_i^1, \ldots, \nabla_i^{l!}\} \qquad \forall \quad 0 \leq i \leq n. \tag{9.2}$$

3　The optimality may not exist. However, if the optimality exists, there is a polynomial algorithm to find the optimal identity exposure sequence.

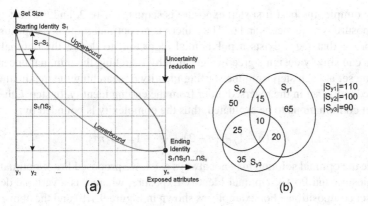

Figure 9.1: Uncertainty reduction with gradual identity exposure.

∎

Definition 9.4 describes a scenario that the overall system uncertainty (or remaining system uncertainty) is always the maximal after each uncertainty reduction step.

For example, in Figure 9.1(a), after y_1 and y_2 are exposed, the overall system uncertainty is $|S_1 \cap S_2|$. The higher the $|S_1 \cap S_2|$ value, the larger the value of the overall system anonymity. Thus, to achieve the best anonymity, we need to keep the system uncertainty at its maximum when selecting an attribute to expose.

Satisfying Definition 9.4 (property 2) guarantees that we can always achieve the maximal overall system anonymity after each step (except the last step, which can have maximal anonymity reduction). Based on the property 2, we can draw an upper-bound of the optimal identity exposure shown in Figure 9.1(a). Additionally, we can also draw lower-bound in the figure by restricting each step expose maximal information, which makes overall system anonymity at its minimal level. We must note that the upper bound and lower bound may not exist for all cases.

Property 3 states that the optimal solution of identity exposure may not always exist. However, if the optimal solution exists, we can always find a polynomial algorithm to derive the optimality. To prove the optimal solution may not exist, we can simply present an example. In Figure 9.1(b), three sets $S_{y_1}, S_{y_2}, S_{y_2}$, and their intersections are presented. The number in each area represents the size of corresponding set. There are 6 possible exposure sequences, which are listed in below:

No.	Exposure Sequence	Anonymity	Optimal at step 1	Optimal at step 2	Optimal at step 3
1	$y_1 \to y_2 \to y_3$	$110 \to 25 \to 10$	✓		✓
2	$y_1 \to y_3 \to y_2$	$110 \to 30 \to 10$	✓		✓
3	$y_2 \to y_1 \to y_3$	$100 \to 25 \to 10$			✓
4	$y_2 \to y_3 \to y_1$	$100 \to 35 \to 10$		✓	✓
5	$y_3 \to y_1 \to y_2$	$90 \to 30 \to 10$			✓
6	$y_3 \to y_2 \to y_1$	$90 \to 35 \to 10$		✓	✓

In this example, the best first step exposure is sequence 1 or 2, and the best second step exposure is sequence 4 and 6. Thus, there is no optimal sequence exposure.

To prove that there exists a polynomial algorithm to derive the optimal solution, we can simply design a greedy algorithm by searching the minimal uncertainty reduction set at each step from the starting identity or searching the maximal uncertainty reduction set in the reversed order from the ending identity. In step j, there are $n - j$ set combinations to be evaluated, thus the complexity is bounded by:

$$n + (n-1) + (n-2) + \cdots + 1 = O(n^2).$$

Since the optimal solution may not always exist, we proposed Pre-k-optimal Identity Exposure and Post-k-Optimal Identity Exposure, where k is a variable depends on the set compositions. For example, as shown in Figure 9.1(b) and the above table, there are pre-1-optimal arrangements, namely $y_1 \to y_2 \to y_3$ and $y_1 \to y_3 \to y_2$. There are post-2-optimal arrangements, namely $y_2 \to y_3 \to y_1$ and $y_3 \to y_2 \to y_1$.

9.3.2 COMMUNICATION MODEL

In our presented system, we need to confine an administrative domain, in which all users must trust the domain manager (i.e., trusted third party—TTP). Each user derives a set of attributes and corresponding private keys from the TTP based on the proposed solutions. TTP can be either online or offline depending on the types of applications using GIE. When the TTP is online, since it knows the exact number of users registered for each attribute, TTP can provide an online query service by accepting the requests from end users. The TTP can generate a correct sequence to satisfy the optimal identity exposure requirements. When users cannot reach the TTP, they need to construct the attribute tree by themselves and arrange the exposure sequence based on his/her own knowledge of the size of each attribute. Users' decisions may not accurately reflect the size of involved attribute size, but it is very flexible, and it can be applied in the scenario that the TTP is offline.

9.3.3 ATTACK MODEL

The attackers' goal is to compromise the anonymity features provided by GIE. Attackers can be either internal or external users of a given administrative domain. In order to compromise the proposed GIE scheme, attackers can collude to derive the extra information that each of them alone cannot derive.

9.4 CONSTRUCTIONS OF GRADUAL IDENTITY EXPOSURE

In this section, we present the detailed cryptographic construction to enable gradual exposure of an identity. In Section 9.4.1, we first consider how to convert an AND-gate access policy to an AND-gate chain that allows the system to expose attributes in a one-by-one fashion. Next, in Section 9.4.4, we present how to encrypt a message using the AND gate chain to allow a decryptor to discover the identity gradually in the decryption process.

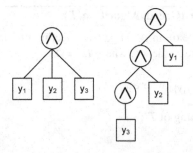

Figure 9.2: Converting an AND access policy tree to an AND gate chain.

9.4.1 CONSTRUCTION OF AND GATE CHAIN

An identity is created by an attribute tree through an AND gate, which requires the decryptor possesses private keys for each attribute in the tree to reveal the whole identity. Different attributes may have different anonymity levels, which is determined by the set cardinality. In Figure 9.2, we present a conversion from a one-level AND-gate chain to a multiple-level AND-gate chain. Attributes with low anonymity should be hidden, unless a decryptor possesses all attributes having a lower anonymity level. In other words, attributes in an AND-logic-gate tree are exposed to a decryptor gradually in the process of decryption. Only if the decryptor has the attribute and the corresponding private keys for current step, he/she can learn the attributes required for next step. On the other hand, if the decryptor does not have the attribute for the current step, the attribute for the next step cannot be revealed.

Since an optimal solution may not always exist, we need to produce an approximate solution to create an good-enough exposure sequence. To this end, we propose Post-k-optimal and Pre-k-optimal searching algorithms. To construct the pre-k-optimal and post-k-optimal, where k is maximized, we propose two greedy algorithms. The forward greedy algorithm can find a pre-k-optimal sequence with maximized k steps where the anonymity is maximized, and a backward greedy algorithm can find a post-k-optimal sequence with maximized k steps, where the anonymity is maximized. For example, as shown in Figure 9.1(b) there are pre-1-optimal arrangements, namely $y_1 \to y_2 \to y_3$ and $y_1 \to y_3 \to y_2$. There are post-2-optimal arrangements, namely $y_2 \to y_3 \to y_1$ and $y_3 \to y_2 \to y_1$.

Algorithm 9.1 ForwardGreedyAlgorithm(T).

Require: T is a structure of an AND gate connecting multiple attributes, and T' is the computed exposure sequence, where initially $T' = \emptyset$;
    while $|T'| \neq |T|$ **do**
        Find the largest $|S_y \cap S_{T'}|$ where $y \in T$ and $y \notin T'$;
        Append y to the end of T';
    end while
    return T'

Algorithm 9.2 BackwardGreedyAlgorithm(T).

Require: T is a structure of an AND gate connecting multiple attributes, and T' is
   the computed exposure sequence, where $T' = \varnothing$
   while $|T| \neq 0$ **do**
      Find the largest $|S_{T \setminus \{y\}}|$ where $y \in T$;
      Remove y from T;
      Append y to beginning of T';
   end while
   return T'

It is easy to prove that both forward greedy algorithms and backward greedy algorithms can find the best sequence to satisfy the overall system anonymity for maximized k steps.

9.4.2 SYSTEM SETUP

In this section, we describe how the trusted third party (TTP) set up the system. TTP
first defines the following functions:

- **parent**(x): return the parent node of x;
- The access tree $\mathscr{T}$ also defines an ordering between the children of every
  node, that is, the children of a node are numbered from 1 to *num*. The function **index**(x) returns such a number associated with the node x.

Then, TTP generates the following parameters. Note that the public key is a
system-wise public parameter used for encryption and decryption. The master key
is the system secret well-guarded by TTP.

- *Bilinear* map: $\mathbb{G}_0 \times \mathbb{G}_0 \to \mathbb{G}_1$ of prime order p with generator g.
- Choose two random $\alpha, \beta \in \mathbb{Z}_p$
- Public key $PK = < \mathbb{G}_0, g, g^\beta, g^{1/\beta}, e(g,g)^\alpha >$
- Master key is $< \beta, g^\alpha >$

9.4.3 KEY GENERATION

After the setup of system, each user need to be generate a set of private key components corresponding to his identity, i.e. the set of attributes. The key generation
algorithm is same to the BSW's scheme in [25]. The key generation algorithm will
take as input a set of attributes S and output a key that identifies with that set.

- The algorithm first chooses a random $r \in \mathbb{Z}_p$,
- Then choose random $r_j \in \mathbb{Z}_p$ for each attribute $y_j \in S$.
- Then compute the key as:

$$SK = (D = g^{(\alpha+r)/\beta}; \forall j \in S : D_j = g^r \times H(y_j)^{r_j}; D'_j = g^{r_j}).$$

9.4.4   ENCRYPTION AND DECRYPTION OF THE AND CHAIN

We now describe how to encrypt an AND-gate chain such that the attributes are exposed gradually. We modify the CP-ABE encrypt and decrypt primitive functions to achieve this goal (the original CP-ABE scheme is presented in Section 1.3.6). We assume that there is at least one attribute at each level of the AND-gate chain except the root.

Intuitively, the encryption and decryption algorithms work by using the BSW scheme as a sub-function. Each AND-gate acts as a subtree access policy to protect the attributes in the same level. The exception is at the lowest level, while each other levels' attribute is encrypted by the AND-gate subtree in the same level. For example, in Figure 9.3, the attribute y_2 is protected by the AND-gate subtree, so that only if the decrypor possess attribute y_1, can he reveal y_2 and proceed decryption. Moreover, instead of using attribute strings, we use hash values of attributes. A user can check whether the hash values match one of his attributes. On the other hand, this mechanism adds another level of protection since it is hard to guess attributes from hash values.

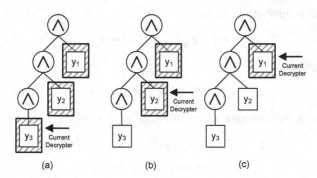

(a)   (b)   (c)

Figure 9.3: The process of decryption and attribute exposing. (a) Initially, all the attributes are hidden; (b) y_3 is exposed in the decryption process; (c) y_2 is exposed in the decryption process.

Encryption

The encryption algorithm encrypts a message M under the exposure schedule $\mathcal{T} = y_1 \to \dots \to y_n$. The encryption algorithm is performed in the top-down manner. Starting with level-0 AND gate x_0, the encryption algorithm chooses a random $s \in \mathbb{Z}_p$ and randomly chooses a 1-degree polynomial q_{x_0} with $q_{x_0}(0) = s$. Then it calculates:

$$C_{x_0} = Me(g,g)^{\alpha q_{x_0}(0)}; \qquad C'_{x_0} = g^{\beta q_{x_0}(0)}.$$

For the level-1 AND gate x_1, it randomly selects a degree-1 polynomial q_{x_1} with $q_{x_1}(0) = q_{\mathbf{parent}(x_1)}(\mathbf{index}(x_1)) = q_{x_0}(1)$. Then it calculates the hash value of level-1 attribute y_1 as $h_1 = H(y_1)$ and randomly chooses a k_1 to encrypt h_1 as $\{h_1\}_{k_1}$. Then, it computes:

$$C_{x_1} = k_1 e(g,g)^{\alpha q_{x_1}(0)}; \qquad C'_{x_1} = g^{\beta q_{x_1}(0)}.$$

For the level-1 attribute y_1, it sets:

$$q_{y_1}(0) = q_{\text{parent}(y_1)}(\text{index}(y_1)) = q_{x_0}(2).$$

Then, it computes:

$$C_{y_1} = g^{q_{y_1}(0)}; \qquad C'_{y_1} = H(y_1)^{q_{y_1}(0)}.$$

By repeating the process until the level n, for the level-m AND gate x_m, the encryption algorithm randomly selects a degree-1 polynomial q_{x_m} with $q_{x_m}(0) = q_{\text{parent}(x_m)}(\text{index}(x_m))$. Then it calculates the hash value of level-m attribute y_m as $h_m = H(y_m)$ and randomly chooses a random key k_m to encrypt h_m, i.e., $\{h_m\}_{k_m}$. Then, it computes:

$$C_{x_m} = k_m e(g,g)^{\alpha q_{x_m}(0)}; \qquad C'_{x_m} = g^{\beta q_{x_m}(0)}.$$

For each level-m attribute y_m, it sets:

$$q_{y_m}(0) = q_{\text{parent}(y_m)}(\text{index}(y_m)).$$

Then, it computes:

$$C_{y_m} = g^{q_{y_m}(0)}; \qquad C'_{y_m} = H(y_m)^{q_{y_m}(0)}.$$

Finally, we have the ciphertext CT as follows:

$$\begin{aligned}
CT = \{ & C_{x_0} = M e(g,g)^{\alpha q_{x_0}(0)}; C'_{x_0} = h^{q_{x_0}(0)}; \\
& h_n; C_{y_n} = g^{q_{y_n}(0)}; C'_{y_n} = H(y_n)^{q_{y_n}(0)} \\
& \forall i \in [1, n-1] : C_{x_i} = k_i e(g,g)^{\alpha q_{x_i}(0)}; \\
& C'_{x_i} = h^{q_{x_i}(0)}; \{h_i\}_{k_i}; \\
& C_{y_i} = g^{q_{y_i}(0)}; C'_{y_i} = H(y_i)^{q_{y_i}(0)} \}.
\end{aligned} \tag{9.3}$$

Decryption

Decryption algorithm is operated in a bottom-up fashion, as shown in Figure 9.3, starting from the level-n attribute, where the level-n attribute is initially exposed. The decryption algorithm calculates:

$$\begin{aligned}
\frac{e(D_i, C_{y_n})}{e(D'_i, C'_{y_n})} &= \frac{e(g^r \cdot H(i)^{r_i}, g^{q_{y_n}(0)})}{e(g^{r_i}, H(i)^{q_{y_n}(0)})} \\
&= \frac{e(g^r, g^{q_{y_n}(0)}) \cdot e(H(i)^{r_i}, g^{q_{y_n}(0)})}{e(g^{r_i}, H(i)^{q_{y_n}(0)})} \\
&= e(g,g)^{r q_{y_n}(0)}.
\end{aligned}$$

Since the level-$(n-1)$ AND gate has only one attribute, it can derive:

$$e(g,g)^{rq_{x_{n-1}}(0)} = e(g,g)^{rq_{y_n}(0)}.$$

Once the decryption algorithm derives the value $e(g,g)^{rq_{x_{n-1}}(0)}$, it calculates:

$$C_{x_{n-1}}/(e(C'_{x_{n-1}},D)/e(g,g)^{rq_{x_{n-1}}(0)}) = k_{n-1}.$$

Then the decryptor decrypts the h_{n-1} checks whether one of his/her attributes has the hash value h_{n-1}. If found, he/she can find the attribute y_{n-1} and then continues the decryption process through the following computations:

$$
\begin{aligned}
\frac{e(D_i, C_{y_{n-1}})}{e(D'_i, C'_{y_{n-1}})} &= \frac{e(g^r \cdot H(i)^{r_i}, g^{q_{y_{n-1}}(0)})}{e(g^{r_i}, H(i)^{q_{y_{n-1}}(0)})} \\
&= \frac{e(g^r, g^{q_{y_{n-1}}(0)}) \cdot e(H(i)^{r_i}, g^{q_{y_{n-1}}(0)})}{e(g^{r_i}, H(i)^{q_{y_{n-1}}(0)})} \\
&= e(g,g)^{rq_{y_{n-1}}(0)}.
\end{aligned}
$$

The level-$(n-2)$ AND gate is decrypted using the same method in level-$(n-1)$ AND gate operations except the level $n-2$ AND gate requires a degree-1 polynomial, whereas level-$(n-1)$ AND gate requires a degree-0 polynomial.

$$
\begin{aligned}
e(g,g)^{rq_{x_{n-2}}(0)} &= \prod_{z \in S_{x_{n-1}}} \left(e(g,g)^{rq_z(0)}\right)^{\Delta_{i,S'_{x_{n-1}}}(0)} \\
&= \prod_{z \in S_x} \left(e(g,g)^{r \cdot q_{\text{parent}(z)}(\text{index}(z))}\right)^{\Delta_{i,S'_{x_{n-1}}}(0)} \\
&= \prod_{z \in S_x} \left(e(g,g)\right)^{r \cdot qx(i) \cdot \Delta_{i,S'_{x_{n-1}}}(0)}.
\end{aligned}
$$

Where $i = \mathbf{index}(z)$, $S_{x_{n-1}} = \{x_{n-2}, y_{n-2}\}$ and $S'_{x_{n-1}} = \{\mathbf{index}(z) : z \in S_{x_{n-1}}\}$.

If the decryptor cannot find corresponding attribute, he/she cannot decrypt one more level to discover the next attribute. The same process for level-$(n-2)$ is repeated for each upper level to eventually recover message M. Brute force guessing the next level attribute won't help since the attacker can not check whether their guess is correct or not.

9.5 ANONYMITY EVALUATION MODELS AND SECURITY ANALYSIS OF GIE

In this section, we present several anonymity evaluation models for the new construction of GIE solutions. Security analysis of GIE is also provided.

9.5.1   SECURITY ANALYSIS

In this subsection, we analyze the security performance of our scheme under attacks to compromise the anonymity provided by GIE. Additionally, we will also present the security strength of GIE encryption scheme.

To evaluate the anonymity strength of GIE encryption scheme, we present the following lemma:

Lemma 9.1

At any given step from level n to level 1: (i) if attackers do not have the attributes for the current level, attackers cannot reduce uncertainty to the next level, and (ii) attackers cannot gain additional information by sharing their secret information. ∎

Proof of Lemma 9.1. *To prove the first property, given the strength of a hash function and meaningful terms used by attributes, attackers need to deploy dictionary attacks to map a given attribute to exposed hash value at the current level. Since attackers do not have the private key, it can only discover the attributes used by the next level. Thus, the compromised privacy is restricted by one level of the AND-gate chain.*

To address this vulnerability, the TTP can apply a keyed hash function $\mathscr{H}$ on a given attribute and give the hash value to the user. The secret key is only maintained by the TTP. In this way, the inputs of the hash function can be considered as a random number to prevent dictionary attacks. We must note that using the keyed hash function will reduce the flexibility of GIE. This is because every user needs to predict attributes that will be used and gets its corresponding secret key and hash values from the TTP in advance.

To prove the second property, we need to note that malicious attackers can combine their attributes and hash values to identify the hidden attributes. However, then GIE encryption algorithm restricts them at the currently level if they share their own secrets. For example, if attacker A has secrets for level-n attributes, and attacker B has secret for level-$(n-1)$ attributes. By share their secrets, they may reveal the attributes used at the level-$(n-1)$, however they cannot combine their secrets to correctly decrypt the level-$(n-1)$ gate. This is because the construction of the ABE decryption scheme requires using the same r value. However, the r values are not the same for different user when TTP distributing secrets to users. Thus, attackers cannot gain additional information through colluding attacks.

In the following context, we prove the presented GIE encryption scheme provides the same security strength of the original CP-ABE scheme. To prove the security of GIE, we reduce our scheme to CP-ABE using the following lemma.

Lemma 9.2

The security strength of GIE encryption scheme is equivalent to CP-ABE scheme. ∎

Proof of Lemma 9.2. *To prove that the proposed scheme is as secure as CP-ABE, we need to prove that the added components in the ciphertext, see Equation (9.3), namely C_{x_i}, C'_{x_i}, and h_{ik_i} where $i \in [1, n-1]$ do not reduce the security of the scheme. Other components of the ciphertext are identical to CP-ABE scheme.*

To prove that the additional components do not reduce the security of the proposed scheme, we need to prove that, given the C_{x_i} and C'_{x_i}, the possibility that an attacker can derive k_i or $e(g,g)^{\alpha q_{x_i}(0)}$ is negligible. Since k_i is randomly chosen and $e(g,g)^{\alpha q_{x_i}(0)}$ is randomized in $\mathbb{G}_2$, deriving each of them based on known $k_i e(g,g)^{\alpha q_{x_i}(0)}$ is hard.

Thus, assume that an attacker has ε advantage in deriving k_i or $e(g,g)^{\alpha q_{x_i}(0)}$, the attacker will have ε advantage in CP-ABE. This is because CP-ABE uses the same technique to protect the message. Thus, if CP-ABE scheme is secure, then our scheme is secure. Moreover, since the security of our scheme can be reduced to CP-ABE, it is collusion-resistant as CP-ABE.

9.6 PERFORMANCE EVALUATION OF GIE

In this section, we present the performance evaluation of GIE in the following aspects: (i) communication overhead, and (ii) computation overhead. In our performance comparison, we compare our scheme with BSW CP-ABE scheme [25], CN CP-ABE scheme [56], NYO scheme [165] and YRL scheme [230], which are described in the Related Work Section.

9.6.1 COMMUNICATION OVERHEAD

The communication overhead is incurred by the transmission of ciphertext. In our proposed solution, for each AND-gate chain with n attributes, $3n$ members in $\mathbb{G}_0$ and n members in $\mathbb{G}_1$ are required[2]. We compare our scheme with several CP-ABE schemes in the following table:

---

[2]Usually, the pairing takes the form $e : E(\mathbb{F}_{p^m}) \times E(\mathbb{F}_{p^{km}}) \to \mathbb{F}^*_{p^{km}}$, where p is a prime, m a positive integer, and k is the embedding degree (or security multiplier). Here, we use the classical algorithm of Weil pairing with $k = m = 1$.

Scheme Name	Support Anonymity	Attributes Supported	Ciphertext Size
BSW	No	∞	$(2n+1)\, G_0 + 1\, G_1$
CN	No	N	$(N+1)\, G_0 + 1\, G_1$
NYO	YES	N	$(2N+1)\, G_0 + 1\, G_1$
YRL	YES	N	$(2N+2)\, G_0 + 1\, G_1$
GIE	YES	∞	$3n\, G_0 + n\, G_1$

N: the number of predefined attributes and each attribute has 3 values, namely True, False, and Don't Care; n: the number of user chosen attributes.

9.6.2 COMPUTATION OVERHEAD

We analyze the computation overhead of our proposed solution in terms of the number of heavy cryptographic operations in the encryption and decryption process. For encrypting an AND-gate chain with n attributes, we need $3(n-1)$ exponentiations on G_0 and $n-1$ exponentiations on G_1.

For the decryption scheme, each attribute requires two *bilinear* pairings and each AND gate requires one *bilinear* pairing. Thus, as shown in Table 9.1, in an AND-gate chain with n attributes, the total number of pairing operations is $3n$. Comparing to BSW scheme, the total pairing operations of GIE are increased by n times with added anonymity features. The computation overhead of CN and YRL schemes depends on the predetermined value N, which is the total number of attributes in the system. Since the value of N can be very big, thus CN and YRL schemes are not efficient. In Table 9.1 we summarize the computation overhead evaluated based on the number of pairing operations.

Table 9.1

Comparative study for number of pairing operations

Scheme Name	Number of Pairing Operations
BSW	$2n$
CN	N
YRL	$2N$
GIE	$3n$

N: the number of predefined attributes; n: the number of user chosen attributes.

9.7 SUMMARY

In this chapter, we presented a new ABE-based approach, call Gradual Identity Exposure (GIE), to anonymize users' identities. We present a theoretical framework to model the GIE with several new proposed concepts. Compared to existing anonymizing techniques through hidden policy or pseudonyms, GIE is effective in that it provides a layered protection framework to protect users' identities through a step-by-step fashion. Moreover, GIE is more efficient and flexible since receivers do not need to try decrypting all ciphertexts as in hidden policy schemes.

Based on the proposed GIE construction, there are a few future research issues worth of investigating, including: the anonymity measurement and optimality condition in the gradual exposure process; the information theoretical optimality condition and efficient planning algorithms need to be further studied; in addition, GIE only exposes one attribute per step, how to extend the solution to expose multiple attributes in one exposure step to make the solution more flexible and how to increase the computation/communication efficiency; furthermore, the structure of multiple AND-gate chains combined by an OR gate needs to be investigated.

10 ABE for IoT and Mobile Cloud Computing

In a mobile edge cloud computing system, lightweight wireless communication devices extend cloud services into the sensing domain. A common mobile cloud secure data service is to inquiry the data from sensing devices. The data can be collected from multiple requesters, which may drain out the power of sensing devices quickly. Thus, an efficient data access control model is desired. To this end, a comprehensive security data inquiry framework for mobile cloud computing is presented based on ABE approaches. The presented solution has two focuses: first, we present how to use Privacy Preserving Cipher Policy Attribute-Based Encryption (PP-CP-ABE) that has been presented in Chapter 3 to protect sensing data. Using PP-CP-ABE, light-weight devices can securely offload heavy encryption and decryption operations to cloud service providers, without revealing the data content. Second, we present a new Attribute Based Data Storage (ABDS) system as a cryptographic group-based access control mechanism. The performance assessments demonstrate the security strength and efficiency of the presented solution in terms of computation, communication, and storage.

10.1 INTRODUCTION

With the fast development of wireless and IoT technologies, Mobile Edge Computing (MEC) has become an emerging cloud service model [108, 217], in which mobile devices and sensors are used as the information collecting and processing nodes for the edge cloud computing resources that are close to where the data collected. This application scenario describes a distributed or decentralized edge cloud computing setup, which requires highly correlated computing and networking interactions between mobiles (or IoT devices) to the edge (or cloud) computing/storage nodes. This new trend demands researchers and practitioners to construct a trustworthy architecture for mobile edge cloud computing, which includes a large number of lightweight, resource-constrained mobile devices.

Existing cloud provides two main services: storage and computation. Users' concerns about data security are the main obstacles that prevent the public cloud from widely adopted. These concerns origin from the fact that sensitive data are stored and processed in public clouds, which are operated by commercial service providers and shared by various other customers. Along with the other customers who can be potential competitors or malicious attackers, these service providers, esp., the storage and computing service providers, are usually not trusted by the data owner. Moreover, the multi-tenant data architecture directly results in the risk that a user's data being exposed to business competitors or malicious attackers, who may compromise the data server shared among tenants.

With the CP-ABE enabled cloud storage service, a new challenge is *how to incorporate wireless mobile devices, especially lightweight devices such as cell phones and sensors, into the cloud system.* This new challenge is originated from the fact that CP-ABE schemes always require intensive computing resources to run the encryption and decryption algorithms. To address this issue, an effective solution is to offload the heavy encryption and decryption computation without exposing the sensitive data contents or keys to the cloud service providers.

Another research challenge is *how to share encrypted data with a large number of users, in which the data sharing group can be changed frequently.* For example, when a user is revoked from accessing a file, he/she is not authorized to access any future updates of the file, i.e., the local copy (if exists) will get outdated. To this end, the updated data need to be encrypted by a new encryption key.

Furthermore, the third research challenge is *how to upload/download and update encrypted data stored in the cloud system.* For example, when changing certain data fields of an encrypted database, the encrypted data needs to be downloaded from cloud and then be decrypted. Upon finishing the updates, the files need to be re-encrypted and uploaded to the cloud system. Frequent upload/download operations will cause tremendous overhead for resource constrained wireless devices. Thus, it is desirable to design a secure and efficient cloud data management scheme to balance the communication and storage operational overhead incurred by managing the encrypted data.

To address the above described research challenges, in this chapter, we present a holistic secure mobile cloud data management framework that includes two major components:

1. A computing offloading-enabled ABE scheme that is based on the PP-CP-ABE presented in Chapter 3, in which users can securely offload computation intensive CP-ABE encryption and decryption operations to the cloud without revealing data content and secret keys. In this way, lightweight and resource constrained devices can access and manage data stored in the cloud data store.
2. An Attribute-Based Data Storage (ABDS) scheme that achieves information theoretical optimality in terms of minimizing computation, storage and communication overheads. Especially, ABDS minimizes cloud service charges by reducing communication overhead for data managements. The ABDS system achieves scalable and fine-grained data access control, using public cloud services. Based on ABDS, users' attributes are organized in a carefully constructed hierarchy so that the cost of membership revocation can be minimized. Moreover, ABDS is suitable for mobile computing to balance communication and storage overhead, and thus reduces the cost of data management operations (such as upload, updates, etc.) for both the mobile nodes and storage service providers.

In the performance evaluation, we demonstrate that the presented solution is computation efficient (i.e., saving 90% for encryption and 99% for decryption) for lightweight mobile devices and it is storage efficient of ABDS scheme, where both data inquirers and sensors only need to store $\log_2(N)$ private keys while N keys are required when using CP-ABE scheme.

The rest of this chapter is organized as follows. The related work is presented in Section 10.2. Section 10.3 presents system models. The new PP-CP-ABE scheme with offloading construction and ABDS design are presented in Section 10.4 and 10.5, respectively. In Section 10.6, we analyzed the security and discuss the performance of presented schemes with comparison to several related works. Finally, we summarize the solution in Section 10.7.

10.2 RELATED WORKS

Existing works related to the presented schemes includes (i) attribute-based encryption and (ii) cryptographic access control over untrusted storage.

Attribute Based Encryption (ABE) was first proposed as a fuzzy version of IBE in [186], where an identity is viewed as a set of descriptive attributes. There are two main variants of ABE proposed so far, namely Key Policy Attribute Based Encryption (KP-ABE [95]) and Ciphertext Policy Attribute Based Encryption (CP-ABE [25]). In KP-ABE, each ciphertext is associated with a set of attributes and each user's private key is embedded with an access policy. Decryption is enabled only if the attributes on the ciphertext satisfy the access policy of the user's private key. In CP-ABE [25, 56, 127, 220], each user has a set of attributes that associate with user's private key and each ciphertext is encrypted by an access policy. To decrypt the message, the attributes in the user private key need to satisfy the access policy. CP-ABE is more appealing since it is conceptually closer to the Role Based Access Control (RBAC) [188] model.

Cryptographic access control over untrusted storage is investigated in both cryptography community and networking community. In the cryptography community, Broadcast Encryption (BE) was introduced by Fiat and Naor in [81]. Compared with traditional one-to-one encryption schemes, BE is very efficient. Based on tradeoffs between key storage and ciphertext storage overhead, existing BE schemes can be generally categorized into the following classes: (i) constant ciphertext, linear public and/or private key on number of total receivers [33]; (ii) linear ciphertext on number of revoked receivers, constant (or logarithm) public and/or private key, [68, 163, 30]; (iii) sub-linear ciphertext, sub-linear public and/or private key [33]. In this work, a new construction of attribute-based data storage (ABDS) scheme is presented to address the deficiency of all 3 class existing works. Particularly, ABDS supports any arbitrary number of receivers with much lower complexity of storage and communication.

In a networking community, various encrypted file systems [124, 20, 70] were proposed to secure data over untrusted storage. Particularly, in [20], the authors proposed a distributed storage scheme where users offload encryption to a semi-trusted re-encryption server. However, if the server colludes with some malicious user, the data secrecy will be compromised completely. Compared with this scheme, the proposed PP-CP-ABE is secure even if service providers and malicious users collude. Recently, Yu et al. [231] proposed a security framework for cloud computing based on CP-ABE. This solution requires the users to disclose part of original private key to the cloud.

Data security in public cloud is an emerging research area [215, 236, 58, 210, 80, 143, 232, 89, 88, 106, 183, 128]. With the fast development of wireless technology, mobile cloud has become an emerging cloud service model [108], in which mobile devices and sensors are used as the information collecting and processing nodes for the cloud infrastructure. This new trend demands researchers and practitioners to construct a trustworthy architecture for mobile cloud computing, which includes a large number of lightweight, resource-constrained mobile devices.

While data integrity and retrievability in the cloud are also important security requirements, they are not the focuses of this dissertation. Readers can refer to research works in the provable data possession (PDP) [19, 80].

10.3 SYSTEM AND MODELS

10.3.1 NOTATIONS

The notations used in this chapter are listed in Table 10.1:

Table 10.1
Notations used in this chapter

Acronym	Descriptions
DO	Data Owner
DR	Data Requester/Receiver
ESP	Encryption Service Provider
DSP	Decryption Service Provider
SSP	Storage Service Provider
TA	Trust Authority
T	Access Policy Tree

10.3.2 OVERVIEW OF PP-CP-ABE SCHEME

In the presented solution, the notation DO is for Data Owner. A DO can be a mobile wireless device such as a smart phone or an environmental sensor that can request and/or store encrypted information from/in the cloud storage. The data are encrypted using the presented PP-CP-ABE scheme. Other than DO, there are many DRs (Data Requesters or Receivers) who can inquiry the information from the storage services of the mobile cloud. For example, a user may want to inquiry current pollution map of a particular city area. Since the data provided by DOs can be proprietary, it should be encrypted and only pollution map service subscribers can retrieve the data. In this case, the mobile cloud system only provides a service platform, and it should not be able to access the data content from the DOs. Here, the focus is on the encryption and decryption model to support the described application scenario; thus, due to the space limit, we do not describe how exactly the application is established in details. The presented system model should provide the following properties:

1. The data must be encrypted before sending to storage service provider (SSP);
2. The encryption service provider (ESP) provides encryption service to the data owner without knowing the actual data encryption key (DEK);
3. The decryption service provider (DSP) provides decryption service to data inquirers without knowing the data content;
4. Even ESP, DSP and SSP collude, the data content cannot be revealed;

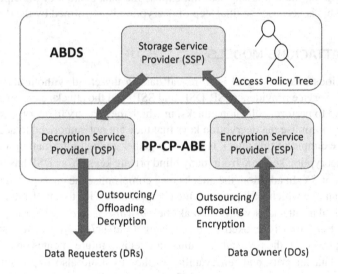

Figure 10.1: System architecture.

As shown in Figure 10.1, the SSP, ESP, and DSP form the core components of the system. A DR inquiries the data provided by a DO. ESP and DSP provide PP-CP-ABE services and SSP, e.g., Amazon S3, provides storage services. The cloud is semi-trusted, in which the cloud only provides computing and storage services with the assistance on data security; however, the data is blinded to the cloud. In particular, more powerful PCs and mobile phones can works as communication proxy for sensors that collect information.

Essentially, the basic idea of PP-CP-ABE to offload intensive but non-critical part of the encryption and decryption algorithm to the service providers while retain critical secrets. As we can prove later in this chapter, the offloading of computation does not reduce the security level compared with original CP-ABE schemes, where all computations are performed locally.

The encryption complexity of CP-ABE grows linearly on the size of access policy. During the encryption, a master secret is embedded into ciphertext according to the access policy tree in a recursive procedure, where, at each level of the access policy, the secret is split to all the subtree of the current root. However, the security level is independent on the access policy tree. In other words, even if the ESP possesses secrets of most but not all parts of the access policy tree, the master secret is still information theoretically secure given there at least one secret that is unknown to ESP. Thus, we can safely offload most part of encryption complexity to ESP by just retaining a small amount of secret information, which is processed locally.

As for the decryption, the CP-ABE decryption algorithm is computationally expensive since *bilinear* pairing operations over ciphertext and private key is a computationally intensive operation. PP-CP-ABE addresses this computation issue by securely blinding the private key and offloading the expensive pairing operations to the DSP. Again, the offloading will not expose the data content of the ciphertext to the DSP. This is because the final step of decryption is performed by the decryptors.

10.3.3 ATTACKING MODELS

The malicious attackers' goal is to reveal data in the cloud without authorization from DOs. Service providers (ESP, DSP, and SSP) and the attacks can combine their information to perform collusion attacks, in which they can try to decrypt the ciphertext and compromise the decryption keys that they are not authorized to access. One particular example of this attack is that they gather enough information to compromise the decryption keys SK from many blind private keys $\widetilde{SK}$ as DSP has the ability to get a lot of $\widetilde{SK}$. In addition, the attacks may compromise the encrypted data by use of the advantage which ESP provides the encryption service to gain from the DO.

In particular, attackers want to break the *Forward Secrecy*, which is defined as follows: After a user is revoked from accessing a file, he/she may have a local copy of the file; however, the revoked user must not get any future updates on this file.

While data integrity and retrievability in the cloud are also important security requirements, they are not the focuses of this solution. Readers can refer to research works in the provable data possession (PDP) [19, 80].

10.3.4 ACCESS POLICY TREE

In this section, we briefly describe the model of an access policy tree used in PP-CP-ABE as illustrated in Figure 10.2. The data access policy tree of PP-CP-ABE is composed by leaf nodes and internal nodes. Each leaf node represents an attribute, and each internal node is a logical gate, such as AND, OR, n-of-m. Several functions and terms are defined as follows to facilitate the presentation of the solution:

- **parent**(x): return the parent node of node x;
- **att**(x) denotes the attribute associated with the leaf node x in the data access tree;

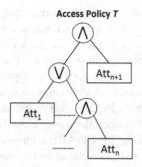

Figure 10.2: Illustration of a sample access policy tree.

- The access tree $\mathcal{T}$ composed by a set of leaf nodes (i.e., attributes) and internal nodes (i.e., logical gates) defines the data access policies, i.e., if a user *owns* a set of attributes that satisfy the logic operations of the tree to reach the root, it can access the secret secured by $\mathcal{T}$. Here *owns* means that the user has the private keys corresponding to the set of attributes. AND and OR are the most frequently used logical gates.
- num_x is the number of children of a node x. A child y of node x is uniquely identified by an index integer **index**(y) from 1 to num_x.
- The threshold value $k_x = num_x - 1$ when x is an *AND*, and $k_x = 0$ when x is an *OR* gate or a leaf node. k_x is used as the polynomial degree for node x using the threshold secret sharing scheme [194].

10.3.5 DEFINITIONS USED FOR SECURITY PROOFS

Definition 10.1 (Co-DBDH Assumption). *Given G_1, $G_1^a, G_1^b, G_2 \in G$ and $T \in G_T$, It is intractable to undistinguish $T = e(G_1, G_2)^{ab}$.*

Definition 10.2 (DLDH Assumption). *Given G, G_1, G_2, $G^a, G_1^b, T \in G$, It is intractable to undistinguish $T = G_2^{a+b}$.*

10.4 PRIVACY PRESERVING CP-ABE

10.4.1 OVERVIEW OF THE CONSTRUCTION

Essentially, the basic idea of PP-CP-ABE is presented in Chapter 3. How to offload intensive but non-critical part of the encryption and decryption algorithm to the service providers while retain critical secrets. As we can prove later in this chapter, the offloading of computation does not reduce the security level compared with original CP-ABE schemes, where all computations are performed locally.

The encryption complexity of CP-ABE grows linearly on the size of access policy. During the encryption, a master secret is embedded into ciphertext according to the access policy tree in a recursive procedure, where, at each level of the access policy, the secret is split to all the subtrees of the current root. However, the security level is independent of the access policy tree. In other words, even if the ESP possesses secrets of most but not all parts of the access policy tree, the master secret is still information theoretically secure given there at least one secret that is unknown to ESP. Thus, we can safely offload most parts of encryption complexity to ESP by just retaining a small amount of secret information, which is processed locally.

As for the decryption, the CP-ABE decryption algorithm is computationally expensive since *bilinear* pairing operations over ciphertext and private key is a computationally intensive operation. PP-CP-ABE addresses this computation issue by securely blinding the private key and offloading the expensive Pairing operations to the DSP. Again, the offloading will not expose the data content of the ciphertext to the DSP. This is because the final step of decryption is performed by the decryptors.

10.4.2 SYSTEM SETUP AND KEY GENERATION

The TA first runs **Setup** to initiate the PP-CP-ABE system by choosing a *bilinear* map: $e : G_0 \times G_0 \to G_1$ of prime order p with the generator g. Then, TA chooses two random $\alpha, \beta \in \mathbb{Z}_p$. The public parameters are published as:

$$PK = \langle G_0, g, h = g^\beta, e(g,g)^\alpha \rangle. \tag{10.1}$$

The master key is $MK = (\beta, g^\alpha)$, which is only known by the TA.

Each user needs to register with the TA, who authenticates the user's attributes and generates proper private keys for the user. An attribute can be any descriptive string that defines, classifies, or annotates the user, to which it is assigned. The key generation algorithm takes as input a set of attributes S assigned to the user, and outputs a set of private key components corresponds to each of attributes in S. The **GenKey** algorithm performs the following operations:

1. Chooses a random $r \in \mathbb{Z}_p$,
2. Chooses a random $r_j \in \mathbb{Z}_p$ for each attribute $j \in S$.
3. Computes the private key as:

$$SK = \langle D = g^{(\alpha+r)/\beta};$$
$$\forall j \in S : D_j = g^r \times H(j)^{r_j}; D'_j = g^{r_j} \rangle.$$

4. Sends SK to the DO through a secure channel.

10.4.3 PP-CP-ABE ENCRYPTION

To offload the computation of encryption and preserve the data privacy, a DO needs to specify a policy tree $\mathcal{T} = \mathcal{T}_{ESP} \wedge \mathcal{T}_{DO}$, where $\wedge$ is an *AND* logic operator connecting two subtrees $\mathcal{T}_{ESP}$ and $\mathcal{T}_{DO}$. $\mathcal{T}_{ESP}$ is the data access policy that will be performed by the ESP and $\mathcal{T}_{DO}$ is a DO-controlled data access policy. $\mathcal{T}_{DO}$ usually has a small number of attributes to reduce the computation overhead at the DO, in which it can be a sub-tree with just one attribute (see the example shown in Figure 10.3).

In practice, if $\mathscr{T}_{DO}$ has one attribute, DO can randomly specify an 1-degree polynomial $q_R(x)$ and sets $s = q_R(0)$, $s_1 = q_R(1)$, and $s_2 = q_R(2)$. Then DO sends $\{s_1, \mathscr{T}_{ESP}\}$ to ESP, which is noted as:

$$DO \xrightarrow{\{s_1, \mathscr{T}_{ESP}\}} ESP.$$

Here, we must note that sending s_1 and $\mathscr{T}_{ESP}$ will not expose any secrets of the solution. The proof will be given in Section 10.6.1.

ESP then runs the **Encrypt**$(s_1, \mathscr{T}_{ESP})$ algorithm, which is described below:

1. $\forall x \in \mathscr{T}_{ESP}$, randomly chooses a polynomial q_x with degree $d_x = k_x - 1$, where k_x is the secret sharing threshold value:
   a. For the root node of $\mathscr{T}_{ESP}$, i.e., R_{ESP}, Chooses a $d_{R_{ESP}}$-degree polynomial with $q_{R_{ESP}}(0) = s_1$.
   b. $\forall x \in \mathscr{T}_{ESP} \setminus R_{ESP}$ sets d_x-degree polynomial with $q_x(0) = q_{\mathbf{parent}(x)}$ (**index**(x)).
2. Generates a temporal ciphertext:

$$CT_{ESP} = \{\forall y \in Y_{ESP} : C_y = g^{q_y(0)}, C_y' = H(att(y))^{q_y(0)}\},$$

where Y_{ESP} is the set of leaf nodes in $\mathscr{T}_{ESP}$.

In the meantime, the DO performs the following operations:

1. Performs **Encrypt**$(s_2, \mathscr{T}_{DO})$ and derives:

$$CT_{DO} = \{\forall y \in Y_{DO} : C_y = g^{q_y(0)}, C_y' = H(att(y))^{q_y(0)}\}.$$

2. Computes $\tilde{C} = Me(g,g)^{\alpha s}$ and $C = h^s$, where M is the message.
3. Sends $CT_{DO}, \tilde{C}, C$ to the ESP:

$$DO \xrightarrow{\{CT_{DO}, \tilde{C}, C\}} ESP.$$

On receiving the message from the DO, ESP generates the following ciphertext:

$$CT = \langle \mathscr{T} = \mathscr{T}_{ESP} \bigwedge \mathscr{T}_{DO}; \tilde{C} = Me(g,g)^{\alpha s}; C = h^s;$$
$$\forall y \in Y_{ESP} \bigcup Y_{DO} : C_y = g^{q_y(0)}; C_y' = H(\mathbf{att}(y))^{q_y(0)} \rangle.$$

Finally, the ESP sends CT to the SSP.

10.4.4 OFFLOADING DECRYPTION

CP-ABE decryption algorithm is computationally expensive since *bilinear* pairing is an expensive operation. PP-CP-ABE addresses this computation issue by offloading the expensive Pairing operations to the DSP. Again, the offloading will not expose the data content of the ciphertext to the DSP.

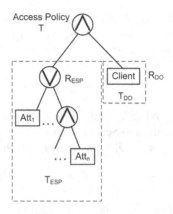

Figure 10.3: Illustration of access policy $\mathcal{T} = \mathcal{T}_{ESP} \bigwedge \mathcal{T}_{DO}$.

To protect the data content, the DO first blinds its private key by choosing a random $t \in \mathbb{Z}_p$ and then calculates $\tilde{D} = D^t = g^{t(\alpha+r)/\beta}$. We denote the blinded private key as $\widetilde{SK}$:

$$
\begin{aligned}
\widetilde{SK} \quad = \quad & \langle \tilde{D} = g^{t(\alpha+r)/\beta}, \\
& \forall j \in S : D_j = g^r \cdot H(j)^{r_j}, D'_j = g^{r_j} \rangle.
\end{aligned} \tag{10.2}
$$

Before invoking the DSP, the DO first checks whether its owned attributes will satisfy the access policy $\mathcal{T}$. If so, the DO sends $\{\widetilde{SK}\}$ to the DSP, and requests the SSP to send the ciphertext to the DSP. On receiving the request, the SSP sends $CT' = \{\mathcal{T}; C = h^s; \forall y \in Y_1 \bigcup Y_2 : C_y = g^{q_y(0)}; C'_y = H(\textbf{att}(y))^{q_y(0)}\}$ and $CT' \subset CT$ to the DSP:

$$
SSP \xrightarrow{\{CT'\}} DSP. \tag{10.3}
$$

Once the DSP receives both $\{\widetilde{SK}\}$ and CT', it then runs the **Decrypt**($\widetilde{SK}, CT'$) algorithm as follows:

1. $\forall y \in Y = Y_{ESP} \bigcup Y_{DO}$ the DSP runs a recursive function DecryptNode ($CT', \widetilde{SK}, R$), where R is the root of $\mathcal{T}$. The recursion function is the same as defined in [25] and DecryptNode($CT', \widetilde{SK}, y$) is proceeded as follows:

$$
\begin{aligned}
\text{DecryptNode}(CT', \widetilde{SK}, y) &= \frac{e(D_i, C_y)}{e(D'_i, C'_y)} \\
&= \frac{e(g^r \cdot H(i)^{r_i}, g^{q_y(0)})}{e(g^{r_i}, H(i)^{q_y(0)})} \\
&= e(g,g)^{r q_y(0)} \\
&= F_y.
\end{aligned}
$$

The recursion is processed as follows: $\forall y$ is the child of x, it calls $DecryptNode(CT'; \widetilde{SK}; y)$ and stores the output as F_y. Let S_x be an arbitrary k_x-sized set of children nodes y, the DSP computes:

$$
\begin{aligned}
F_x &= \prod_{y \in S_x} F_y^{\Delta_{i,S_x'}(0)} \\
&= \prod_{y \in S_x} \left(e(g; g)^{r \cdot q_y(0)} \right)^{\Delta_{i,S_x'}(0)} \\
&= \prod_{y \in S_x} \left(e(g; g)^{r \cdot q_{\text{parent}(y)}(\text{index}(y))} \right)^{\Delta_{i,S_x'}(0)} \\
&= \prod_{y \in S_x} \left(e(g; g)^{r \cdot q_x(i) \cdot \Delta_{i,S_x'}(0)} \right) \\
&= e(g, g)^{r q_x(0)}, \quad\quad\quad (10.4)
\end{aligned}
$$

where $i = \textbf{index}(z)$ and $S_x' = \{\textbf{index}(z) : z \in S_x\}$, $\Delta_{i,S_x'}(0)$ is the Lagrange coefficient. Finally, the recursive algorithm returns $A = e(g, g)^{rs}$.

2. Then, computes

$$
e(C, \widetilde{D}) = e(h^s, g^{t(\alpha+r)/\beta}) = e(g, g)^{trs} \cdot e(g, g)^{t\alpha s}.
$$

3. Sends $\{A = e(g, g)^{rs}, B = e(C, \widetilde{D}) = e(g, g)^{trs} \cdot e(g, g)^{t\alpha s}\}$ to the DO:

$$
DSP \xrightarrow{\{A,B\}} DO.
$$

On receiving $\{A, B\}$, DO calculates $B' = B^{1/t} = e(g, g)^{rs} \cdot e(g, g)^{\alpha s}$ and then it recovers the message:

$$
M = \frac{\widetilde{C}}{(B'/A)} = \frac{Me(g, g)^{\alpha s}}{(e(g, g)^{rs} \cdot e(g, g)^{\alpha s})/e(g, g)^{rs}}.
$$

10.5 ATTRIBUTE-BASED DATA STORAGE

In this section, we present an Attribute-Based Data Storage (ABDS) scheme that is based on PP-CP-ABE to enable efficient, scalable data management and sharing.

10.5.1 DATA MANAGEMENT OVERVIEW

The frequent data updates will cause additional expense for file managements. For example, to update existing files, e.g., changing certain data fields of an encrypted database, in which the encrypted data need to be downloaded from SSP to DSP for decryption. Upon finishing the updates, the ESP needs to be re-encrypted and to upload the data to the SSP. Thus, the re-encrypted process requires downloading and uploading the data, which may incur high communication and computation overhead, and as a result, will cost more for DOs.

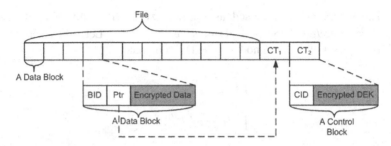

Figure 10.4: Illustration of a file organized into blocks with multiple control blocks.

To address the described cost issue, it is reasonable to divide a file into independent blocks that are encrypted independently. To update files, the DO can simply download the particular blocks to be updated. In this way, we can avoid re-encrypting the entire data. Moreover, data access control can be enforced on individual blocks using "lazy" re-encryption strategy. For example, when the data access memberships to a particular file are changed (i.e., the access tree is changed), this event can be recorded but no file changes are invoked. Until the data content needs to be updated, the re-encryption is then performed using the presented PP-CP-ABE scheme.

Partitioning the data into multiple small blocks also introduces addition overhead. This is because the extra control information needs to be attached for each data block for data management. For example, the control message should include a block ID and a pointer to its corresponding data access tree $\mathscr{T}$. In Figure 10.4, we depicted a sample file stored in SSP. As shown in Figure 10.4, each file is divided into blocks. A block is a tuple BID, Ptr, Encrypted Data, where BID is the unique identification of the block; Ptr is the pointer to the control block CT; and data is encrypted with a DEK. A control block CID, Encrypted DEK has a control block ID, i.e., CID and DEK encrypted by using PP-CP-ABE scheme.

The ABDS system should determine what is the appropriate data block size to be partitioned with a known file size. The design goal is to minimize the storage and communication overhead with the considerations of the following simple assumptions:

1. Every data update should only affect a small amount of data, e.g., updating certain data fields in the database;
2. In each unit time period, the number of blocks to be updated is known;
3. Each data block has the same probability to be updated.

Based on the above discussions, we can model the total cost C in a unit time period as follows:

$$C = 2nS_bC_c + \frac{F}{S_b}S_cC_s, \tag{10.5}$$

where n is the number of updated blocks in a unit time period, and $2n$ stands for an update includes one encryption and one decryption that require two transmissions; S_b is the size of block; C_c is the cost rate of data transmission that is charged by both cloud storage providers and wireless communication service providers; F is the size of file; S_c is the size of control data for each data block, and C_s is the charging rate of storage. To minimize cost C, DO can minimize (10.5) and derive the optimal block size:

$$S_b \geq 2\sqrt{2nC_cFS_cC_s}.$$

10.5.2 SETUP

PP-CP-ABE enables expressive policy with descriptive attributes to enforce data access control on the stored data. For example, if Alice wants to share a file to all CS students, she can specify the policy "CS *AND* Student". All the users whose attributes satisfy this policy can decrypt the data.

Besides the set of descriptive attributes enabled in the system, each user is assigned a unique binary ID: $b_0b_1\ldots b_{n-2}b_{n-1}$. We can define the term "*bit-assignment attribute*" that is represented as "B_i" or "$\overline{B_i}$" to indicate the binary value at position i in the ID. B_i indicates that the i'th bit of an ID is 1; $\overline{B_i}$ indicates that the i'th bit of an ID is 0. If the length of an ID is n, then the total number of bit-assignment attributes is $2n$. This means that two binary values are mapped to one-bit position (one for value 0 and one for value 1). Thus, a DO with ID u is uniquely identified by the set of bit-assignments S_u. Also, multiple DOs may have a common subset of bit-assignments. For example, a DO u_1's ID is 000 and a DO u_2's ID is 001, $S_{u_1} = \{\overline{B}_0, \overline{B}_1, \overline{B}_2\}$ and $S_{u_2} = \{\overline{B}_0, \overline{B}_1, B_2\}$ and $S_{u_1} \cap S_{u_2} = \{\overline{B}_0, \overline{B}_1\}$. Bit-assignment attributes can be used when the DO wants to share data to any arbitrary set of DOs. In this case, it may be hard to describe the set of DOs efficiently using descriptive attributes.

10.5.3 UPLOAD NEW FILES

Before uploading new files to the SSP, both ESP and DO are required to determine the encryption parameters, such as the block size. DO then invokes ESP with an access policy $\mathcal{T}_{ESP}$, which is the access policy to be enforced on the uploaded files. Here, we define some terms used in the following presentations:

- *Literal*: A variable or its complement, e.g., b_1, $\overline{b_1}$, etc.
- *Product Term*: Literals connected by AND, e.g., $\overline{b}_2b_1\overline{b}_0$.
- *Sum-of-Product Expression (SOPE)*: Product terms connected by OR, e.g., $\overline{b}_2b_1b_0 + b_2$.

Given the set of shared data receivers S, the membership functions $f_S()$, which is in the form of SOPE, specifies the list of receivers:

$$f_S(b_1^u, b_2^u, \ldots, b_n^u) = \begin{cases} 0 & \text{iff } u \in S, \\ 1 & \text{iff } u \notin S. \end{cases}$$

For example, if the subgroup $S = \{000, 001, 011, 111\}$, then $f_S = \overline{b}_0\overline{b}_1\overline{b}_2 + \overline{b}_0\overline{b}_1b_2 + \overline{b}_0b_1b_2 + b_0b_1b_2$.

Then, the DO runs the Quine-McCluskey algorithm [157] to reduce f_S to minimal SOPE f_S^{min}. The reduction can consider *do not care* values $*$ on those IDs that are not currently assigned to any DO to further reduce number of product terms in the membership function. For example, if $S = \{000, 001, 011, 111\}$, $f_S^{min} = \overline{b_0}\overline{b_1} + b_1 b_2$.

Finally, DO uploads the data blocks and the control block to SSP, where each data block is encrypted by the DEK and DEK is protected by the access policy in control block.

10.5.4 DATA UPDATES

Now, we investigate into how to efficiently handle the data updates, i.e., how to modify encrypted data with or without changing data access control policy.

Data Updates with Access Policy Change

In Section 10.5.1, we described the "lazy" re-encryption strategy adopted by DOs. Using the "lazy" re-encryption scheme, the DO continuously records the revoked data receivers. When there is a need to modify the data, the DO will choose a new data access tree that can revoke all previously recorded data receivers.

When DO updates a data block with access policy change, we need to consider the following cases:

- If there is no control block associated with the latest access policy, i.e., no data updates occurred after the latest access policy change event, the DO encrypt a new random DEK associated with the latest access policy with PP-CP-ABE and attach a new control block to the end of the file, see Figure 10.4.
- If there exists a control block associated with the latest access policy, i.e., at least one data block was encrypted with the newest access policy, the DO can simply re-direct the control block pointer, see Figure 10.4, to the control block associated with the latest access policy.
- If a control block is not pointed by any data block, this control block should be deleted.

Updates Without Access Policy Change

If no change is required to the access policy, DO can simply perform the PP-CP-ABE scheme and upload the updated data block in the SSP. The Block ID and the pointer to control the block are not changed.

10.6 PERFORMANCE EVALUATION

In this section, we first present the security assessments of the presented solution. Then, we present the computation, communication, and storage performance evaluation.

10.6.1 SECURITY ASSESSMENTS

We now briefly analyze the security of PP-CP-ABE scheme. We first describe the hardness assumptions used in this scheme: Given a *bilinear* map group system $S = (p, G, G_T, e(\cdot, \cdot))$, where two groups G, G_T have the prime order p. The security of this scheme is constructed on two basic assumptions: Co-Decision Bilinear Diffie-Hellman (Co-DBDH) assumption and Decision Linear Diffie-Hellman (DLDH) assumption.

The data structure of ciphertext and private key in PP-CP-ABE is the same as the original BSW CP-ABE [25]. Thus PP-CP-ABE can be viewed as a variation of CP-ABE. Particularly, in PP-CP-ABE, the access policy tree is constructed by two sub-trees $\mathscr{T} = \mathscr{T}_{ESP} \wedge \mathscr{T}_{DO}$. In general, $\mathscr{T}_{DO}$ contains a single attribute to reduce the computation and communication overhead. Thus, DO randomly specifies a 1-degree polynomial $q(x)$ and sets $s = q(0)$, $s_1 = q(1)$ and $s_2 = q(2)$. The tuple $\{s_1, \mathscr{T}_{ESP}\}$ is sent to ESP. It is easy to prove that, based on the threshold secret sharing scheme [194], for a given 1-degree polynomial $q(x)$, knowing s_1, secrets s and s_2 are informationally theoretically secure. In order to avoid the leakage of encrypted information for cloud service providers (including ESP and DSP), the following theorem proves that this scheme is secure against the adaptive chosen plaintext attacks (IND-CPA) based on Co-DBDH assumption.

Theorem 10.1. *Let $\mathscr{E}$ is a PP-CP-ABE scheme. If Co-DBDH is (t, q, ε)-hard on G, then the PP-CP-ABE scheme is (t', q', ε')-secure against the adaptive chosen plaintext attacks (IND-CPA), where $\varepsilon' > \varepsilon/4$, $q = q'$ and $t' > t$. Here q is the number of hash function queries made by the adversary, and t is the run time of attacks.*

Proof of Theorem 10.1. *Seeking a contradiction, we assume that there exists a PPT algorithm $\mathscr{A}$ that can break this scheme with a non-neglected probability ε, that is $Adv_{\mathscr{A}}^{IND-CPA}(\mathscr{E}) > \varepsilon$. Using $\mathscr{A}$, we can also construct a PPT algorithm $\mathscr{B}$ to break Co-DBDH problem: given $G_1, G_1^a, G_1^b, G_2 \in G$ and $T \in G_T$, it is intractable to undistinguish $T = e(G_1, G_2)^{ab}$, as follows:*

- *Setup: Let $G_2 = G_1^{\xi}$. $\mathscr{B}$ sets $\alpha = a\xi$ and chooses a random $\beta \in \mathbb{Z}_p$, such that the public key can be generated by using $PK = \langle G_0, g = G_1, h = g^{\beta} = G_1^{\beta}, e(g, g)^{\alpha} = e(G_1^a, G_2) \rangle$. $\mathscr{B}$ sends PK to the adversary $\mathscr{A}$.*
- *Learning: Given a policy $\mathscr{T}$, $\mathscr{A}$ makes use of PK to compute the corresponding ciphertext. The leaning will be stop after a polynomial time. Anytime, $\mathscr{A}$ asks a Random-Oracle query to obtain $H(att(x)) = G^r$ from $\mathscr{B}$, where r is a random integer.*
- *Challenges: $\mathscr{A}$ chooses two equal-length message M_0, M_1 and a valid policy $\mathscr{T}$ as the challenges and sends $\mathscr{B}$. $\mathscr{B}$ sets $s = b = q_R(0)$, such that the policy tree can be encrypted as follows: given a polynomial $q_x(x) = q_x(0) + \sum_{i=1}^{d_x} a_i x^i$, $C_x = g^{q_x(x)} = G_1^{q_x(0)} \cdot \prod_{i=1}^{d_x} G_1^{a_i x^i}$ and $C'_x = H(att(x))^{q_x(x)} = G_1^{q_x(0)r} \cdot \prod_{i=1}^{d_x} G_1^{r a_i x^i}$, where $H(att(x)) = G_1^r$ and r is a known random integer.*

And then $\mathcal{B}$ chooses a random bit σ and computes $\widetilde{C} = M_\sigma \cdot e(g,g)^{\alpha s} = M_\sigma \cdot$
$T, C = h^s = (G_1^b)^\beta$, finally outputs $CT = \langle \mathcal{T}, \widetilde{C}, C, \forall x \in Y_{ESP} \cup Y_{DO} : C_x, C_x' \rangle$.
* *Response: $\mathcal{A}$ returns a bit guess σ' to $\mathcal{B}$. If $\sigma = \sigma'$, $\mathcal{B}$ outputs Y, otherwise*
N.

When $T = e(G_1, G_2)^{ab}$, the ciphertext CT is a valid ciphertext due to $\widetilde{C} = M_\sigma \cdot$
$e(G_1, G_2)^{ab} = M_b \cdot e(g,g)^{\alpha s}$, where $s = b$. Otherwise, the ciphertext CT is a invalid
ciphertext.

$$\Pr[\mathcal{B}(G_1, G_1^a, G_1^b, G_2, T) = Y]$$
$$= \Pr[\mathcal{A}(PK, CT, \mathcal{T}) = \sigma]$$
$$= \Pr[\mathcal{A}(PK, CT, \mathcal{T}) = \sigma]\Pr[T = e(G_1, G_2)^{ab}] +$$
$$\Pr[\mathcal{A}(PK, CT_R, \mathcal{T}) = \sigma]\Pr[T \in_R \mathbb{G}_T]$$
$$= \frac{1}{2}(\Pr[\mathcal{A}(PK, CT, \mathcal{T}) = \sigma] + \frac{1}{2})$$
$$= \frac{1}{2} + \frac{1}{4}Adv_{\mathcal{A}}^{IND-CPA}(\mathcal{E}) > \frac{1}{2} + \frac{1}{4}\varepsilon.$$

Where $\Pr[\mathcal{A}(PK, CT, \mathcal{T}) = \sigma] = \frac{1}{2} + \frac{1}{2}Adv_{\mathcal{A}}(1^n)$ and $\Pr[T = e(G_1, G_2)^{ab}] =$
$\Pr[T \in_R \mathbb{G}_T] = \frac{1}{2}$. Hence, this contradicts the Co-DBDH hypothesis.

Based on the security assumptions presented in Section 10.3.3, ESP, DSP and
SSP are untrusted but honest service providers that will perform proper computation
according to PP-CP-ABE protocol and returns correct results. In order to compromise users' secret information, the ESP, DSP and SSP can perform collusion attacks.
In this scenario, an authorized user u' who satisfies the access tree $\mathcal{T}$ provides his
blinded private key $\widetilde{SK}$ to the DSP for decryption. Then, ESP and DSP can try to utilize the blinded private key of u' to derive M from $Me(g,g)^{\alpha s}$. ESP has s_1, and thus
it can easily derive $e(g,g)^{\alpha s_1}$. This is because $e(g,g)^\alpha$ is available from the public
parameters presented in (10.1). As the user u' satisfies the access policy $\mathcal{T}_{DO}$, DSP
can derive the following values $e(g,g)^{r's_1}$, $e(g,g)^{r's_2}$, $e(g,g)^{r's}$, and $e(g,g)^{t\alpha s + tr's}$
through the F_x function (see (10.4)) without knowing $alpha$ and r'. In the following
Table 10.2, we listed all rational terms that are available to ESP and DSP.

Table 10.2

Available rational terms to ESP and DSP

ESP	s_1	$e(g,g)^{\alpha s_1}$	$g^{\beta s_1}$	$g^{s_1/\beta}$
DSP	$e(g,g)^{r's_1}$	$e(g,g)^{r's_2}$	$e(g,g)^{r's}$	$e(g,g)^{t\alpha s + tr's}$

As we can see, ESP has the values s_1 and $e(g,g)^{\alpha s_1}$, but it is unaware of values s_2 or s. DSP possesses more terms as well as the blinded private key $\widetilde{SK}$ of u'

(see (10.2)). We must note that $\widetilde{SK}$ is not a valid CP-ABE private key, since the $\widetilde{D} = g^{t(\alpha + r')/\beta}$ is embedded with tr' and $t\alpha$, and the rest of all private key components $\{\forall j \in S : D_j = g^{r'} \cdot H(j)^{r_j}, D'_j = g^{r_j}\}$ are embedded with r'. Essentially, this blinded private key can be a valid CP-ABE private key when (i) the master key is $MK = \{\beta, g^{t\alpha}\}$; (ii) a colluding user contributes $D = g^{t(\alpha + r')/\beta}$, which is a valid component embedded with tr'; and (iii) a colluding user contributes $\{\forall j \in S : D_j = g^{r'} \cdot H(j)^{r_j}, D'_j = g^{r_j}\}$, which are binded by a random r', which is different from tr' in D. Since the t is the exponent of the generator g, deriving it is equivalent to solve the DLP problem, which is considered difficult. Thus, given the security of secret sharing and hardness of DLP on $\mathbb{G}_0$ and $\mathbb{G}_1$, ESP and DSP cannot derive $e(g,g)^{\alpha s_2}$ or $e(g,g)^{\alpha s}$ even if they collude.

Strictly speaking, the following theorem proves that this scheme holds the collusion key security against the blinded key attacks (KS-BKA) based on the DLDH assumption:

Theorem 10.2. *Let $\mathcal{E}$ is a PP-CP-ABE scheme. If DLDH is (t,q,ε)-hard on $\mathbb{G}$, then the PP-CP-ABE scheme is (t,q,ε)-secure against the blinded key attacks. Here q is the number of hash function queries made by the adversary, and t is the runtime of attacks.*

Proof of Theorem 10.2. *We assume that there exists a PPT algorithm $\mathcal{A}$ that can retrieve the original key SK from many blinded private keys $\{\widetilde{SK_i}\}$ with a non-neglected probability ε, that is, $Adv_{\mathcal{A}}^{KS-BKA}(\mathcal{E}) > \varepsilon$. Using $\mathcal{A}$, we can construct a PPT algorithm $\mathcal{B}$ to break the DLDH problem: Let $G_1 = G^{\xi_1}$ and $G_2 = G^{\xi_2}$, given some $G, G^{\xi_1}, G^{\xi_2}, G^a, G^{\xi_1 b}, T \in \mathbb{G}$, it is intractable to undistinguish $T = G^{\xi_2(a+b)}$, as follows:*

- *Setup: $\mathcal{B}$ sets $a = \frac{\alpha}{\beta}$, $b = \frac{r}{\beta}$ and $\beta = \xi_1$, such that $\alpha = a\beta = a\xi_1$ and $r = b\beta = b\xi_1$. Hence, the public key can be generated by using $PK = \langle \mathbb{G}_0, g = G, h = g^\beta = G^\xi = G_1, e(g,g)^\alpha = e(G^a, G^{\xi_1}) = e(G^a, G_1)\rangle$. $\mathcal{B}$ sends PK to the adversary $\mathcal{A}$.*
- *Learning: Given a policy $\mathcal{T}$, $\mathcal{A}$ makes use of PK to compute the corresponding ciphertext CT and queries the decryption key of $\mathcal{B}$. $\mathcal{B}$ returns a blinded private key $\widetilde{SK_i} = \langle \widetilde{D_i} = g^{t_i(\alpha + r)/\beta} = T^{t'_i}, \forall j \in S : D_j = g^r \cdot H(j)^{r_j}, D'_j = g^{r_j}\rangle$, where t'_i is a random integer $\mathbb{Z}_p$. The leaning will be stop after a polynomial time. Let $(\widetilde{SK}_1, \widetilde{SK}_2, \cdots, \widetilde{SK}_n)$ be some blinded keys learned by $\mathcal{A}$ Anytime, $\mathcal{A}$ asks a random-oracle query to obtain $H(att(x)) = G^k$ from $\mathcal{B}$, where k is a random integer.*
- *Response: $\mathcal{A}$ outputs a value Z. $\mathcal{B}$ checks whether or not $e(Z, G^{\xi_2}) = e(Z, G_2) = e(T, G)$. If this equation holds, it returns Y, otherwise N.*

When $T = G^{\xi_2(a+b)}$, the blinded private key $\widetilde{SK_i}$ is a valid key due to $\widetilde{D_i} = g^{t_i(\alpha + r)/\beta} = G^{t'_i \xi_2(a+b)} = T^{t'_i}$. Note that, $t_i = t'_i \cdot \xi_2$ is unknown for $\mathcal{B}$. Hence, $\mathcal{A}$ can output the original key $Z = g^{(\alpha + r)/\beta}$ and $e(Z, G_2) = e(T, G)$ with a non-neglected probability ε. However, when T is a random value in $\mathbb{G}_T$, $\mathcal{A}$ can also output a *value* $Z' = g^{(\alpha+r)/\beta} = T^{1/t_i''}$ *for a certain unknown* t_i'' *and* $t_i'' \neq \xi_2$. *Hence, we have* $e(Z', G_2) \neq e(T, G)$. *In summary, we have the equation*

$$\Pr[\mathscr{B}(G, G^{\xi_1}, G^{\xi_2}, G^a, G^{\xi_1 b}, T) = Y]$$
$$= Adv_{\mathscr{A}}^{KS-BKA}(\mathscr{E}) > \varepsilon.$$

Hence, this is a efficient algorithm to break DLDH problem and this contradicts the hypothesis.

10.6.2 PERFORMANCE EVALUATION

To evaluate the performance of the presented PP-CP-ABE scheme, we evaluate the computation overhead of service providers and users based on both theoretical analysis and experimental results. In the experimental analysis, we compared the computing overhead of various cryptographic operations in PC, Pocket PC and mobile sensors. The result showed that, without offloading, it is rather infeasible for the resource constrained devices to perform the operations.

Firstly, we analyzed the number of expensive cryptographic operations over G_0 and G_1, i.e., pairing, exponentiation, multiplication, performed by service providers and users' devices. In the analysis, we assume that the access policy T_{ESP} has a_1 attributes connected by an AND logical gate and T_{DO} only has 1 attribute. In addition, the root node is an AND gate.

Computation Overhead Analysis

In the Table 10.3, we compare the number of exponentiation, multiplications, and hash to G_0 operations incurred on ESP side and user side in the encryption offloading, where a_1 is the number of attributes in T_{ESP}.

Table 10.3

Number of cryptographic operations computed by ESP and user

	Exp G_0/G_1	Mul G_1	Hash to G_0
ESP	$2a_1/0$	0	a_1
User	$3/1$	1	1

We also provide a comparison of the number of exponentiation, multiplications, inversion, and pairing operations incurred by decryption offloading on DSP side and user side as shown in the Table 10.4, where a_1 is the number of attributes in T_{ESP}.

Evaluation of Computation Overhead

From the above analysis, we can see that the computation overhead is linear for service providers (ESP and DSP) and constant for the user. Among all operations,

Table 10.4

Number of cryptographic operations computed by DSP and user

	Exp G_1	Mul G_1	Inv G_1	Pairing
DSP	a_1	$2a_1$	a_1	$2a_1 + 1$
User	1	2	1	0

pairing and ECC operations are most computationally intensive. We conducted the experimental evaluation of cryptographic pairing and ECC operations on a wireless Mote sensor (8 bit-7.37 MHZ ATMega128L, 4KB RAM), a pocket PC (600 MHZ CPU), and a PC (1 GHZ CPU). The testing environments and results are listed in the Table 10.5.

Table 10.5

Computing time of cryptographic operations on embedded devices

	Pairing	Exp G_0	Mul G_0
PC (1GHZ CPU)	20 ms	5 ms	0.7 ms
Pocket PC (600 MHZ CPU)	550 ms	177 ms	26 ms
Sensor (8× 8MHZ)	31250 ms	10720 ms	196 ms

The result in Table 10.5 showed that, without offloading, it is rather infeasible for the resource-constrained devices to perform the operations. To show that PP-CP-ABE can offload most of the computation overhead from user to service providers., we implemented and evaluated the PP-CP-ABE on a PC with 1.6GHz Intel Atom processor running Linux 2.6.32. The computation time was measured using clock ticks returned by `clock_t clock(void)` function in standard C library. To illustrate that most of the computation overhead is offloaded to service providers, we run the user and server on the same platform and recorded the number of clock ticks. In the Figure 10.5, we compared computation overhead incurred on service providers and users in encryption and decryption offloading. The computation overhead was calculated in terms of 10 based logarithms, i.e., $\log_{10}$, of thousands (K) clocks ticks. As we can see from the figure, more than 90% of encryption and more than 99% of decryption computation are performed by the service providers.

Storage Performance of ABDS

We analyze the storage performance of ABDS and compare it with several related cryptographic access control solutions: broadcast encryption schemes (Subset-Diff) [81], BGW broadcasting encryption [34], access control polynomial (ACP) scheme [250].

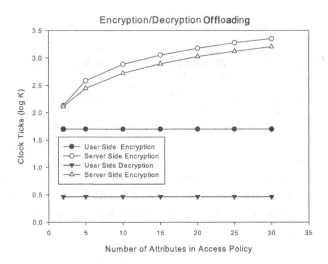

Figure 10.5: Performance evaluation of the encryption and decryption offloading.

The performance is assessed in terms of cipher-text storage overhead, key storage overhead (system parameters and public/private keys stored on the users and TA). We denote the total number of users in the system with N and a user wants to share a file to any given set of receivers in the system. The comparative results are presented in Table 10.6.

Ciphertext Storage Overhead In Subset-Diff scheme, the size of ciphertext is $O(t^2 \cdot log^2t \cdot logN)$, with t as maximum number of colluding users to compromise the ciphertext. For BGW scheme, the ciphertext size is $O(1)$ or $O(N^{\frac{1}{2}})$ as reported in [34]. In ACP scheme, the size of message depends on the degree of access control polynomial, which equals to the number of current receivers. Thus, the message size is $O(N)$. To control a set of receivers S using ABDS, the size of ciphertext depends on the number of product terms in the f_S^{min} (see 10.5.3). In [189], the authors derived an upper bound and lower bound on the average number of product terms in a minimized SOPE. Experimentally, the average number of message required is $\approx log(N)$ [55].

To investigate the average case, we simulated ABDS in a system with 512 users and 1024 users. In the simulation, we consider the cases of 0%, 5%, 25%, and 50% IDs are not assigned (i.e., *do not care* value). For each case, different percentages of receivers are randomly selected from the group. We repeat 100 times to average the results. Experimentally, the message size in CP-ABE starts at about 630 bytes, and each additional attribute adds about 300 bytes. Since the number of attributes in the access policy is bounded by $logN$, we can conclude that the ciphertext storage overhead of ABDS is in the order of $O(log^2 N)$.

Key Storage Overhead Compared with Broadcast Encryption schemes, ABDS greatly reduced the Key Storage Overhead of the TA and users' devices. In ABDS,

Table 10.6

Comparison of ciphertext storage overhead and key storage overhead in different cryptographic access control schemes

Scheme	Ciphertext Storage		Key Storage	
	single data receiver	multiple data receivers	TA	User
ABDS	$O(\log N)$	$\approx O(\log^2 N)$	$O(1)$	$O(\log N)$
Subset-Diff	$O(t^2 \cdot log^2 t \cdot \log N)$	$O(t^2 \cdot log^2 t \cdot \log N)$	$O(N)$	$O(t \log t \log N)$
BGW$_1$	$O(1)$	$O(1)$	N/A	$O(N)$
BGW$_2$	$O(N^{\frac{1}{2}})$	$O(N^{\frac{1}{2}})$	N/A	$O(N^{\frac{1}{2}})$
ACP	$O(N)$	$O(N)$	$O(N)$	$O(1)$

N: the number of group members; t: maximum number of colluding users to compromise the ciphertext.

the PK and MK is of constant size. Also, a user needs to store $\log(N)$ bit-assignment attributes. Thus, the storage overhead is $O(\log N)$, assuming a user does not store any IDs of the data receivers. Although the DO may need the list of data receivers' IDs along with the list of *do not care* IDs to perform Boolean function minimization, we can argue that this does not incur extra storage overhead.

- The data publishers do not need to store the receiver's IDs after the broadcast; thus, the storage space can be released.
- The TA can periodically publish the minimized SOPE of all *do not care* IDs, which can be used by data publishers to further reduce number of messages.
- If IDs are assigned to users sequentially, i.e., from low to high, TA can simply publish the lowest unassigned IDs to all users, who can use the all higher IDs as *do not care* values.
- Even if a user needs to store N IDs, the space is merely $N \log N$ bits. If $N = 2^{20}$.
- If a data publisher cannot utilize *do not care* values to further reduce the membership function in SOPE form, the ciphertext storage overhead might be a little higher.

10.7 SUMMARY

In conclusion, we present a secure data inquiry service architecture for mobile cloud computing. Especially, the solution enables lightweight wireless devices to securely store and retrieve their data in public clouds with minimal cost. To this end, we proposed an offloading enabled Privacy Preserving Cipher Policy Attribute-Based Encryption (PP-CP-ABE) to protect users' encrypted data. Using PP-CP-ABE, lightweight devices can securely offload intensive encryption and decryption operations to cloud service providers, without revealing the data content and used security keys. Also, we presented an Attribute-Based Data Storage (ABDS) system as a cryptographic access control mechanism. ABDSs achieve information theoretically optimal in terms of minimizing computation, storage, and communication overheads. Especially, ABDS minimize cloud costs charged by cloud service providers, as well as communication overhead for data managements. The performance assessments demonstrate the security strength and efficiency of the solution in terms of computation, communication, and storage.

11 ABE-Based Content Access Control for ICN

Information Centric Networking (ICN) is a new network architecture that aims to overcome the weakness of IP-based networking architecture. Instead of establishing a connection between the communicating hosts, ICN focuses on the content, i.e., data, transmitted in network. Thus, how to locate and access the desired content is crucial in ICN. Some existing solutions aim at resolving the content name through a name resolution service, which is similar to the DNS services. Other solutions are based on route-by-name scheme, which treats content names similar to existing routing protocols using IP addresses. Content copies in ICN can be cached at different locations. The content is out of its owner's control once it is published. Thus, enforcing access control policies on distributed content copies is crucial in ICN.

Attribute-Based Encryption (ABE) is a feasible approach to enforce such control mechanisms in this environment, in which access policies are embedded into the content regardless of the security mechanisms provided by the caches. However, applying ABE in ICN faces two challenges: from management perspective, it is complicated to manage attributes in distributed manners; from privacy protection perspective, unlike in traditional networks, the enforced content access policies are public to all the ICN users. Thus, it is desirable that unauthorized content viewers are not able to retrieve the access policy. To this end, a privacy-preserving access control scheme for ICN and its corresponding attribute management solution are presented in this chapter. The presented approach is compatible with existing flat name based ICN architectures.

11.1 INTRODUCTION

In the previous chapter, the relation anonymity issue is discussed in MANET. In such traditional networking schemes, if a network entity wants to access some information content, it has to locate and connect to the server that provides such service following network-routing protocols. As a result, the information is tightly associated with the location of the server. The entire network is centered around the connections between content consumers and content providers, making connection status an important factor to the network.

Witnessed by the fact that most of the network traffic is video sharing [59], various ICN architectures [48, 131, 66, 83, 2] are proposed. In ICN architecture, the core of networking is shifted from consumer-server connections to consumer-content connections. In this way, instead of identifying content owners' addresses, the network changes to identify authentic content copies scattered in network. Consumers do not need to know where copies of a content are located, i.e., the IP addresses of content owners. Content names are used to direct consumers to content copies. Content owners publish contents, which can be copied and distributed all over the network using network caches [177, 205]. Network caches are normally servers that are specifically designed for storage purpose or normal network entities with limited storage capabilities. This design enables contents being efficiently delivered to consumers with a higher efficiency. For example, it is able to retrieve the nearest (according to some metrics) copy of a content to a consumer. In contrast, in the traditional Internet networking framework, a consumer gets a content only from its original owner.

Though the design of ICN is efficient in retrieving contents, it brings great challenges to security issues during content caching and retrieving. One of them is that traditional access control mechanisms cannot be easily enforced in such environment. This is because, in ICN, content owners and consumers are not directly connected. Content owners have no control over the distributed network caches. To enforce access control to the content, several frameworks have been proposed [82, 201]. Most of them require additional authorities or secure communication channels in network to authenticate each content consumer. These schemes are sound but have too much reliance on traditional control schemes, making them inefficient in practice, especially in mobile ICN environment. Therefore, instead of enforcing the data access control mechanism on each caching server, a natural approach is to secure the content by enforcing the data access control through cryptographic approaches. If designed properly, only legitimate users who have proper cryptographic keys are able to access the data content. As each content is identified by the content name, it is easy for any network entity to access the content as long as such name is consistent among all the copies of the same content.

In this chapter, an attribute-based access control for ICN naming scheme is presented. In this scheme, attributes defined by different authorities can be synchronized more efficiently than traditional approaches. Content consumers do not need to negotiate their attribute keys when they request contents from other authorities. To facilitate the application in mobile environment, the presented approach aims to reduce the burden of a Trusted Third Party (TTP) and distribute part of its duties to several distributed attribute authorities.

The core of the presented solution is an ABE-based naming scheme. This approach is inspired by Attribute-Based Encryption (ABE) schemes [25, 233, 137]. Instead of incorporating a set of additional components, it only requires one additional trusted third party (TTP) in the network. In addition, it can be seamlessly incorporated into existing flat-name ICN architectures. In the presented approach, each network entity is assigned with a set of attributes with the help of a TTP according to their real identities. The access control policy for the content is based on combinations of the attributes in terms of *AND* and *OR* operations. This policy is enforced according to the content names instead of the contents. Moreover, privacy-preservation is provided for the content access policies, i.e., only legitimated content consumers are able to get part of the encryption policies and decrypt the data content. This feature can greatly improve the privacy protection on ICN data when they are distributed in the public domain. Especially, in wireless network, an access policy without privacy-preservation can be easily captured and monitored by any passive adversaries eavesdropping on the wireless channels. The presented approach also provides a user with the capability to identify its eligibility of the accessed contents through the encrypted names before actually accessing and processing the data content. To further support the use of ontology in attribute management, the presented scheme enables comparison between attributes, which gives the capability to rank attributes and associate different privileges accordingly. In summary, the expected contributions of this work can be listed as follows:

- It enables attribute rankings and access privilege management, making it flexible to construct a data access policy in real-world scenario. The content access policy is confidentially preserved. Ineligible consumers cannot derive the data access policies even if they collude together;
- It proposes a naming scheme for ICN network which combines the flexible attribute management solution with the privacy preserving access policy;
- It significantly reduces the computation and communication overhead for a potential consumer to determine his eligibility to access the content.

The remainder of this chapter is organized as follows. Section 11.2 reviews related work on ICN and ABE-related work. Section 11.3 presents the system models and preliminaries. Section 11.4 presents the detailed ABE-based ICN naming scheme. Performance evaluation of the presented solution and security analysis are provided in Section 11.5 and Section 11.6, respectively, and the summary of the presented solution is presented in Section 11.7.

11.2   RELATED WORKS

Before introducing details of the solution, related research results on ICN and ABE are presented in this section.

11.2.1   ICN SOLUTIONS

Several ICN architectures have been proposed in the past years. Although these approaches are different from each other in several aspects, the main idea is centered on information processing and management. Combined Broadcast and Content Based (CBCB) Routing [48] is a solution that runs on the application layer. It uses publish/subscribe scheme to publish contents. Each consumer broadcasts its interest in the form of attribute combinations. These interests are propagated through the network. At each router, the interests associated with an interface are updated in the form of predicates. When content is transferred through the network, the content is compared with the predicates on every interface to determine through which interfaces to forward the content.

Data-Oriented Network Architecture (DONA) [131] is deployed above the IP layer. The name of a content is in the form of $P: L$, where P represents the hash of the owner's public key, and L is a unique label the owner assigns to the content.

The owner registers the content into the name resolution system when it is ready to publish. Consumers use the name resolution system to find the nearest copy of the content. The system returns with the content copy or the IP address of the content location. Network of Information (NetInf) [66] follows a similar naming scheme as DONA. But instead of using the owner's public key to generate the digest, it uses a separate pair of public/private keys for the content. Multi-Level Distributed Hash Table (DHT) is used for name resolution purpose. A content owner needs to register its content in all the three levels of the DHT and content lookups are carried out from the lowest level upwards. If it is not successful, then a dedicated resolution system will be used for further assistance. Publish Subscribe Internet Technologies (PURSUIT) [83] is another solution that uses a similar naming scheme as DONA. However, it has a much different structure for retrieving content locations, which involves topology information and load balance. Besides, it uses Bloom filter for source-oriented routing to forward content copies to the consumers.

Named Data Networking (NDN) [2] doesn't specifically define the name structure. A name in NDN consists of multiple components, each of which can be a human-readable string or a digest of the content. Content providers are required to guarantee the uniqueness of name components. This solution uses names to execute a routing process that is similar to the current IP-based routing. Name tables, which are similar to route tables in IP network, maintain the prefix of names and the corresponding interfaces or data. In this way, a response to a content request can be the content itself. Also, this solution aims to provide a replacement to IP instead of being a layer above IP, which is different from approaches mentioned before.

Several research works have been conducted on applying NDN in a mobile network environment. In [17], the authors proposed a gossip algorithm to disseminate messages with a minimum number of transmissions. It is based on a modification from traditional NDN solution. In [234], a dedicated network architecture for mobile ad hoc ICN network is proposed. It supports both pull and push transport in multi-hop communications. Existing mobile ICN research works are mainly focused on lower-level networking mechanisms. For upper-level mechanisms, access control as an example, there is not much difference between a traditional ICN and a mobile ICN, except for the underlying networking related factors, such as mobility and mobility-related delay.

All these ICN solutions focus on the efficiency and security aspects of the network while access control to content and content privacy are not well addressed. In [82], an independent access control system is introduced to support the need in ICN. This system connects to the ICN structure through a component called the Relaying Party (RP). An additional component called Access Control Provider (ACP) is in charge of creating access policies and enforcing the policies to consumers' credentials. This system incorporates access control into ICN systems, but requires much more network interactions for a consumer to get the content. For content privacy purposes, [18] proposed a design in which each file is divided into blocks. A block from the file is mixed with blocks from "cover" content using randomizing transformations and the generated mixture is published to the network. In this way, adversaries could not retrieve the original file easily. To recover the file, an authentic consumer needs to get more information related to the file from a secure channel. With such information, the consumer requests related chunks from the network to generate the original file. This approach meets the security and privacy requirement to some extent, but through a complicated process. The requirement for a secure channel is very difficult to satisfy in many ICN application scenarios.

11.2.2   ABE SCHEMES

ABE schemes originate from Identity-Based Encryption (IBE), which aims to use the user's id as the public key for asymmetric encryptions. After that, an ABE scheme named Ciphertext-Policy ABE (CP-ABE) [25] was introduced by J. Bethencourt et al. This scheme assigns each user with a set of attributes according to their real-life identities and roles. There is one private key component corresponding to each attribute for each user. A policy specifying under what conditions the ciphertext can be successfully decrypted is constructed by the encryptor. This policy is transmitted together with the ciphertext, but in plaintext form. In other words, it is exposed to the network channel. Users who do not possess a satisfactory combination of attributes are not able to decrypt the ciphertext. This scheme enables providing access control to individual messages. A content owner is able to specify the required attribute combinations without knowing the receivers' key credentials. In addition, this scheme is secure against collusion attackers.

The original CP-ABE scheme is a possible candidate for enforcing access control in ICN, but it is not a good solution in such scenarios. The reason why CP-ABE is not suitable for ICN usage is that the policy is transmitted in clear text. In a traditional network, a user is authenticated before access is granted. However, once a content is published in ICN, the owner has no control on it. In this way, any network user who has access to the ciphertext is able to access the policy. Attackers can deduce the sensitivity of the message as well as inferring the identities of those who are involved in the message transmission. For example, a message encrypted with the policy $\{Chairman\}AND\{CEO\}$ from a hospital is highly likely to be more important and valuable than a message with policy $\{Nurse\}AND\{Intern\}$. Thus, attackers can easily identify the high-value users and concentrate attacks of different forms on them.

What is needed for CP-ABE is the capability to hide the policy into the ciphertext. For such purpose, several works [233, 166, 165] are proposed. An attacker cannot get any information about the policy even if it actually executes the decryption process. However, these solutions sacrifice efficiency for security in that any party that tries to decrypt the ciphertext will have to go through the entire decryption process which involves a heavy computation overhead. For instance, in [233], the decryption process includes a bit-by-bit decryption for the decrypting party's ID.

To save computation resources for the unsatisfactory users, D. Huang et al. proposed a scheme [117] to expose the policy attributes step by step. Only one attribute is exposed to the decryptor at one step. In this way, the decryptor is able to stop the decryption process as soon as it fails at a specific step. However, the price for such a feature is that one additional attribute, which is the one that fails the decryptor, is exposed. Besides, this approach supports AND-gates only, which limits the flexibility of the policy.

For attribute management purpose, it is desirable to enable the comparison between attributes so that nominal attributes can be mapped into ordinal values, e.g. $\{Nurse\} < \{Physician\}$. In [246, 219], Y. Zhu et al. proposed an encryption scheme using interval comparisons based on *bilinear* mappings. In this chapter, the idea for interval comparisons is adopted and applied to hidden-policy attribute-based encryption algorithms.

Comparatively, the proposed scheme achieves better flexibility by allowing the use of OR gates, and fully preserves the attribute policy in that ineligible nodes cannot get any information on any attribute in the policy that they do not have.

11.3   SYSTEM AND MODELS

In this section, we present the basic system and models to construct the ABE-based ICN naming scheme.

11.3.1   APPLICATION SCENARIOS

In a typical ICN system, there are three roles: content owner, content consumer, and content cache. A content owner creates the content and publishes it into the network. A consumer is a network entity that requests for the content. It gets the content with the help of the ICN infrastructure. A cache is an entity that keeps a copy of the content for a period of time in its own local storage so that whenever a request for the same content arrives, it directly responds to the request with a copy of the content to the consumer. All these three network roles are exchangeable for individual network entities. That is to say, a network entity can simultaneously be a publisher, a consumer, and a cache for different contents. In the following, an example in medical care is used through out the rest of this chapter to show how the presented scheme works. As shown in Figure 11.1, the content owner can be a patient, a content consumer can be a nurse or a physician, and the content caches are servers storing encrypted contents.

In an ICN network, users get content names from a Name Searching Service (NSS) and use the names to get the content through a Name-based Routing (NR) system. A user gets content names from the NSS and the NR is able to retrieve the content based on the names. Details on how these two systems are implemented is out of the focus of this work. Interested readers can refer to [131], [2], and [48] for more information. Additionally, the presented model includes a TTP that sets up Attribute-Based Access Control (ABAC) and Attribute-Based Encryption (ABE) related public parameters for the network. It also helps assigning and managing attributes to entities.

In the presented scheme, every network entity is associated with a unique identifier (UID) and a set of attributes. UID itself can be treated as a special attribute. A

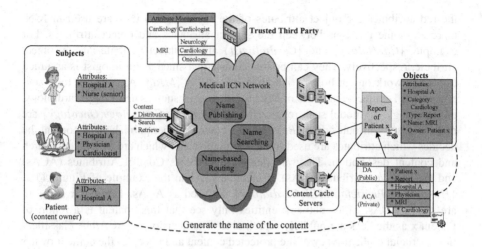

Figure 11.1: Basic ICN system model.

TTP is in charge of setting up global parameters for the entire network. An attribute (other than $UIDs$) can be defined and managed by any entity in network. But the definition and management process on an attribute should be carried out by the same entity. This entity is denoted as the authority of that attribute. As in this example, the attributes include: $\{HospitalA, Nurse, Physician, Cardiologist, MRI\}$. In the presented network model, multiple attribute authorities can be present at the same time. Thus, not only are all the network users organized in a distributed manner, the attribute authorities are also distributed. This property is supported by the specially-designed naming scheme presented in this chapter. Each of the authorities is in charge of an independent and non-overlapping set of attributes.

A content owner is able to set up an access policy for its content under this scheme. The policy is represented as a combination of related attributes with AND and OR gates. For example, if a content owner wants to create a file that should be accessible only to people working as a *Physician* or as a *Nurse* at Hospital A, then the policy could be $\{A\} AND \{\{Physician\} OR \{Nurse\}\}$. In this way, the owner does not need to know explicitly who should access the content. He can identify the attributes and the combination so that as long as a consumer satisfies the policy, he is able to access the content. Any entity that does not satisfy the policy will not be able to access the data in this content.

11.3.2 ATTRIBUTE-BASED NAMING AND ACCESS CONTROL

Attributes in an ICN network can be categorized into subject attributes and object attributes. As shown in Figure 11.1, attributes in green are subject attributes while the red attributes are object attributes of the report. When they are used in ICN, there are some relations between the subject attributes and object attributes. For example, $\{Cardiology\}$ and $\{Cardiologist\}$ are a subject attribute and an object attribute, respectively. They can be treated as equal since a cardiologist is assumed to always work on cardiology. Another example is $\{MRI\}$. As a useful tool, several medical subjects make use of MRI for diagnosis, such as neurology, cardiology, and oncology. To model such relationship, $\{Neurology, Cardiology, Oncology\}$ are defined as sub-attributes of $\{MRI\}$. When a content owner publishes the content, he decides which attributes are used for access control and which are for content search and content description. They are denoted as Access Control Attributes (ACAs) and Descriptive Attributes (DAs), respectively. As in the example of Figure 11.1, $\{Hospital\ A, Physician, MRI, Cardiology\}$ are used as ACAs. $\{Patient\ x, Report\}$ are used as DAs. Thus, network entities only see that this content is a report of Patient x as the DAs are publicly search-able. The decision on ACA/DA classification is crucial to the privacy of the protected content and it's up to the content owner to make such decisions.

11.3.3 COMPARABLE ATTRIBUTES AND ATTRIBUTE RANKINGS

In addition to the above-mentioned attribute setups, comparison between attributes is also supported. To illustrate this property, the example policy in 11.3.1, $\{A\}\ AND\ \{\{Physician\}\ OR\ \{Nurse\}\}$, is used. If for some reason, a modification to the policy is needed as: all the staff working at hospital A that rank higher than nurse are allowed to access the file, then in the traditional approach, it is necessary to enumerate all the attributes that are allowed and construct a very complex policy as $\{A\}\ AND\ \{\{Physician\}\ OR\ \{Nurse\}\ OR ...\}$. However, if a comparison relationship is set up between *Physician* and *Nurse* as *Physician* $>$ *Nurse*, meaning that a *Physician* attribute includes all the privileges of a *Nurse* attribute but with more that are not possessed by *Nurse*, then the original policy can be simplified. Suppose such a comparison relationship has been established with all the related occupation roles, then $\{A\}\ AND\ \{\{Physician\}\ OR\ \{Nurse\}\ OR ...\}$ can be reduced to $\{A\}\ AND\ \{Nurse\}$, which is easier for management purpose.

11.3.4 AN ILLUSTRATIVE EXAMPLE

In the example of Figure 11.1, there are three subjects: a nurse, a physician, and a patient. Their attributes are as shown in the figure. The patient publishes his MRI report in the network as the content. He, as the content owner, specifies an access policy as shown in Figure 11.2 for the MRI report. Its object attributes are listed in Figure 11.1. The content name is created following the procedure in Figure 11.3, which will be further illustrated in Section 11.4.1. When the nurse tries to access this content, she can successfully use her $\{Hospital\ A\}$ attribute to decrypt the first node but will get stuck at $\{Physician\}$, meaning this content is not prepared for her. When the physician accesses the content, she can successfully decrypt the entire decryption process from the leaf to the root level-by-level to reveal the random data encrypting key. Here, {*MRI*} is substituted with {*Cardiology*} since {*Cardiology*} is a sub-attribute. This is shown with the arrow in Figure 11.2. Also, {*Cardiology*} equals to {*Cardiologist*} in this case. Then, the physician uses the NR system to get the nearest copy of the content and uses the random data encrypting key derived from the name to decrypt the MRI report.

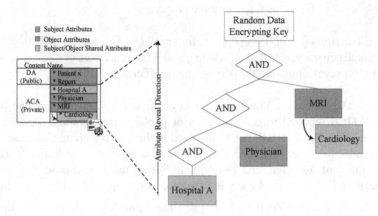

Figure 11.2: Creating a content name.

11.3.5 ATTACK MODEL

In order to guarantee the integrity of content, a digital digest signed by its owner is included in the content meta-data. Since data integrity is not the focus of this chapter, detailed information on this subject will not be provided. An attacker targets at the presented ABE scheme, in which an attacker is assumed to have two primary goals in compromising the ICN access control scheme:

- acquiring unauthorized privilege to the data protected under the presented ABE scheme;
- retrieving constitutional information of access policies to gain more information about the content, the owner, and the consumers. In other words, breaking the protection on the policies.

Sensitive information in this context includes but is not limited to the identity of the owner or consumers, the sensitivity of the content and the potential value of data in the content. For the first goal, attackers have to break the confidentiality mechanism of the protected data. Feasible methods include collusion attacks and vulnerability exploitation. The second attack goal is less important to an attacking party as a successful attack does not reveal information directly related to the protected secret. For the second goal, attackers need to analyze the proposed ABE-based scheme to identify possible ways to reveal the policy.

11.3.6   PRELIMINARIES OF ABE

The foundation of ABE-type algorithms is *bilinear* pairing computation. In this chapter, the design from [246] is adopted in terms of algebraic structure. Suppose there are two groups: an additive group G_0 and a multiplicative group G_1 with a same order $n = sp'q'$, where p' and q' are two large prime numbers. A *bilinear* map is defined as $e : G_0 \times G_0 \rightarrow G_1$. This map has three properties:

- Bilinearity: $e(aP, bQ) = e(P, Q)^{ab}$, for any $P, Q \in G_0$ and $a, b \in \mathbb{Z}_p$;
- Nondegeneracy: $e(g, h) \neq 1$, where g and h are generators of G_0;
- Efficiency: Computing the pairing can be efficiently achieved.

In CP-ABE, there are three types of keys: master key, public key and private key. A TTP is required to generate a set of public parameters and securely store the master key. The TTP will not be involved in the network communication. It can be offline all the time. The scheme of CP-ABE consists of four basic algorithms: **Setup**, **Encrypt**, **KeyGen**, and **Decrypt**. In **Setup**, the TTP chooses two random exponents $\alpha, \beta \in Z_p$. A public key is formatted as $< G_0, g, h, f, e(g,g)^{\alpha} >$ while the master key is (β, g^{α}). Here $h = g^{\beta}$, $f = g^{\frac{1}{\beta}}$. The public key is published by the TTP before deployment. When a party wants to encrypt a message M, it runs the **Encrypt** algorithm. The inputs of this algorithm are the public key, the message M and a policy tree T. The output is a ciphertext. The **KeyGen** algorithm is used to generate private keys based on its inputs: the master key and a set of attributes. For each network node, the TTP runs the **KeyGen** algorithm once to generate a private key according to attributes assigned to that node. When a node receives the ciphertext, it runs the **Decrypt** algorithm to get the encrypted data. This algorithm takes the ciphertext and the node's private keys as inputs.

In traditional CP-ABE schemes, the TTP is involved in both when a node joins in the system (node management) and when an attribute is created and assigned to a node (attribute management). When the scale of the system is large, the TTP will turn to a bottleneck for performance concerns. For this purpose, the presented scheme aims to isolate the duty of node management and attribute management. It off-loads the attribute management functions to other entities.

11.4   ABE-BASED ICN-NAMING SCHEME

In this section, the detailed design for the presented ABE-based naming scheme in ICN network is illustrated. This scheme is based on a previous work [117, 141].

11.4.1   CREATING A CONTENT

Initially, the TTP sets up global parameters for the entire network. Then, any entity in network can create attributes and assign them to other entities. Detailed process on how attributes are distributed is out of the scope of this work. Interested readers can refer to attribute allocation problem solutions for large-scale networks such as [28]. Once the attributes are assigned, network entities are able to create contents, i.e. start network communications. As shown in Figure 11.3, when an entity publishes a file, as the content owner, it creates an access policy for the content. The policy is represented as a combination of related attributes with AND and OR gates. For example, if a content owner wants to create a record that is accessible only to physicians and nurses working at hospital A, the policy can be constructed as: $\{A\}\,AND\,\{\{Physician\}\,OR\,\{Nurse\}\}$. In this way, content owners do not need to know explicitly who should access the content before constructing the policy. Instead, all they need to do is to identify the attributes and the combinations of attributes for a qualified content user so that as long as a consumer satisfies the policy, it is able to access the content. Any entity who does not satisfy the policy will automatically be deprived of the privilege to access the information in this content. No additional network participants are needed during this entire process to monitor the access control enforcement.

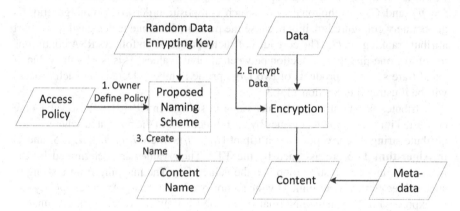

Figure 11.3: Creating content.

After creating the policy, the owner generates a random data-encrypting key and uses it to encrypt the file. This encryption process can be any type of cryptosystem, the choice of which is not directly related to the presented scheme. The encryption result is set as the data part of the content item. The meta-data part includes public parameters used for data integrity assurance and data decryption, like the type of cryptosystem used for the random data encrypting key.

The content owner creates a name for the content. He uses the presented scheme to encrypt the random key under the policy he has specified. The result is used as the content name. Here, it is necessary to emphasize that the generated name hides the content access policies so that no one can get the entire policy from the name. In fact, the content name is a ciphertext after a series of encryption operations. It exhibits as a random sequence of bits to any viewer.

A consumer who needs this file is able to get a copy of the content by its name through the ICN network. Before he retrieves the content, he can use his own attributes to decrypt the name. If his attributes satisfy the hidden policy embedded in the name, he can get the random data-encrypting key protected in the name. The data of the content then can be decrypted using the random key to recover the original file. If a consumer cannot successfully decrypt the content name, it implies the consumer is not allowed to access the original file. Thus, even if he downloads the content, he still does not have the random data encrypting key to decrypt it. A benefit of the propose scheme is that a normal user can delay the downloading process of the content till he successfully decrypts the content name, which helps reduce the workload of underlying network.

11.4.2 ABE-BASED NAMING SCHEME

In this section, a composite order group G_0 with an order $n = p^2q^2$ is used, where p and q are two large prime numbers. In other words, the composite value s in Section 11.3.6 is set to pq. Two subgroups G_s and G_t of G_0 are chosen such that $s = pq$, $t = pq$, and G_s is orthogonal to G_t. Such composite-order group configuration is deliberately configured mainly because the presented scheme is designed to support attribute rankings in G_s. The core idea of such configuration follows RSA conditions to enforce one-direction deduction between attribute values. This is why the value of s and t are set to be products of two large prime numbers. Details on such process will be illustrated in Section 11.4.3.

Attributes of an entity can be any value in strings. In CP-ABE, these values are converted into mathematical values by hash functions. In the presented scheme, each attribute string A_i corresponds to a triplet (I_i, k_i, h_i), where $I_i, k_i, h_i \in \mathbb{Z}_{n'}^*$. S_i and T_i in **Algorithm 11.3** are assigned by the TTP. Their values are determined by the generators under each sub-group and the value of h_i. The mapping from a string to such a three-tuple is determined by the authority of attribute A_i. An access policy can be expressed in Disjunctive Normal Form (DNF) of attributes. In each conjunctive clause of the DNF, the sequence of attributes is determined by the encrypting party, i.e. the content owner. The sequence of encrypting a conjunctive clause (encryption sequence) is opposite to the decryption sequence. To help identify the decryption thread, a public attribute A_{Pub} is defined in the scheme. Unlike other attributes, A_{Pub} is associated with a triplet $(S_{Pub}, T_{Pub}, I_{Pub})$, which are publicly known. For each conjunctive clause, the encryptor adds A_{Pub} at the end of the encryption sequence. In other words, the special attribute A_{Pub} is always the last attribute in encryption and the first attribute in decryption process. Additionally, the encryptor is required to simplify the DNF so as to reduce the size of attribute policy.

In the presented scheme, a **GlobalSetup** algorithm is run by the TTP to generate global parameters for the system. For each node joining in the network, the TTP runs **NodeJoin** algorithm once to generate a unique secret for the node. The input of **NodeJoin** is the node's UID and the outputs are $-D_{UID}, X_{Pub,UID}, Y_{Pub}, Z_{Pub,UID}''$. For each attribute, the authority in charge runs the **AuthoritySetup** algorithm to generate secrets associated with that attribute. Besides, this naming scheme includes three more basic algorithms: **KeyGen**, **Encrypt**, and **Decrypt**. Once set up, the authority of an attribute runs **KeyGen** for each node carrying this attribute to allocate the inherent attribute secrets. **Encrypt** and **Decrypt** are used by encryptors and decryptors respectively for message processing.

The **GlobalSetup** algorithm generates global parameters $-G_s$, G_t, φ, ψ, φ^β, $e(\varphi, \psi)^\alpha$, $Enc_k(\cdot)$, $Dec_k(\cdot)$, $(P_{Pub}, S_{Pub}, T_{Pub})$, $ROOT''$, and global secrets $-\beta$, $g^{\alpha''}$, where α and β are random values and $Enc_k(\cdot)$, $Dec_k(\cdot)$ are a pair of symmetric encryption/decryption functions.

Algorithm 11.1 GlobalSetup

1: Choose two *Bilinear* groups G_0 and G_1 with a composite order $n = p^2 q^2$, where p and q are two large prime numbers. g is the generator of G_0;

2: Choose two subgroups G_s and G_t of G_0 such that: the order of G_s and G_t are both $n' = pq$; G_s and G_t are orthogonal to each other;

3: Choose two generators $\varphi \in G_s$ and $\psi \in G_t$;

4: Choose two random values $\alpha, \beta \in \mathbb{Z}_{n'}^*$;

5: Define a constant $ROOT \in G_1$ as identification of the secret message;

6: Choose a pair of symmetric encryption functions $Enc_k(\cdot)$ and $Dec_k(\cdot)$ in G_1;

7: Define a public attribute, $(S_{Pub}, T_{Pub}, I_{Pub})$, $S_{Pub} \in G_s$, $T_{Pub} \in G_t$, $I_{Pub} \in \mathbb{Z}_{n'}^*$;

8: The global parameters are $-G_s$, G_t, φ, ψ, φ^β, $e(\varphi, \psi)^\alpha$, $Enc_k(\cdot)$, $Dec_k(\cdot)$, $(S_{Pub}, T_{Pub}, I_{Pub})$, $ROOT''$, global secrets are $-\beta$, $\psi^{\alpha''}$.

The **NodeJoin** algorithm is defined as in **Algorithm 11.2**.

Algorithm 11.2 NodeJoin

1: For each node with UID in network, generate a random number $r_{UID} \in \mathbb{Z}_{n'}^*$;

2: Calculate $D_{UID} = \psi^{(\alpha + r_{UID})/\beta}$;

3: Calculate:

$$X_{Pub,UID} = \varphi^{r_{UID}} S_{Pub}^{r_{Pub}},$$

$$Y_{Pub} = \varphi^{r_{Pub}},$$

$$Z_{Pub,UID} = e(\varphi, \psi)^{r_{UID} I_{Pub}}.$$

where $r_{Pub} \in \mathbb{Z}_{n'}^*$ is a random number for each node;

4: Choose a random value $P_{UID} \in \mathbb{Z}_{n'}^*$;

5: Assign to the node $-D_{UID}$, $X_{Pub,UID}$, Y_{Pub}, $Z_{Pub,UID}$, P_{UID}''.

Each individual authority that manages an attribute A_i will have to run **AuthoritySetup** to set up attribute secrets.

Algorithm 11.3 AuthoritySetup

1: For each attribute A_i, choose random numbers $I_i, k_i, h_i \in \mathbb{Z}_{n'}^*$;

2: For each attribute A_i, generate $S_i \in G_s$ and $T_i \in G_t$, where $S_i = \varphi^{h_i}$ and $T_i = \psi^{h_i}$.

The **KeyGen** algorithm generates private keys corresponding to each attribute for each node holding this attribute. It is defined in **Algorithm 11.4**. When the node

---

Algorithm 11.4 KeyGen

1: The authority passes I_i, S_i and T_i to TTP;
2: TTP computes and sends back to the authority:

$$X_{i,UID} = \varphi^{r_{UID}}S_i^{r_i},$$

$$Y_i = \varphi^{r_i},$$

$$Z_{i,UID} = e(\varphi, \psi)^{r_{UID}l_i},$$

$$L_{UID} = T_{Pub}^{1/P_{UID}}.$$

where $r_i \in \mathbb{Z}_{n'}^*$ is a random number;
3: The authority assigns $X_{i,UID}$, Y_i, $Z_{i,UID}$, and L_{UID} to the node together with I_i, h_i and k_i.

---

receives keys from the authority, it checks if $L_{UID}^{P_{UID}} = T_{Pub}$ is true. If it's true, it updates P_{UID} with P_{UID}^2 and accepts the keys. This update is intended to defend against replay attack on L_{UID}. If not true, it will discard the keys.

The **Encrypt** algorithm works following the encryption sequence of each clause. In the following, each attribute is denoted from I_1 to I_m, m is the number of attributes in the clause. In the example of Figure 11.2, $I_1 = MRI$, $I_2 = Physician$, $I_3 = Hospital\ A$, $I_4 = A_{Pub}$, $m = 4$. Any encryptor needs to choose a random value $s \in Z_p$, set $I_0 = s$ and follow **Algorithm 11.5**.

Algorithm 11.5 Encrypt

1: Calculate $C = Ke(\varphi, \psi)^{\alpha s}$, $C' = \varphi^{\beta s}$ and $C'' = Enc_K(ROOT)$;
2: For each attribute A_n, **if** a triplet $(C_{1,n}, C_{2,n}, C_{3,n})$ has already been calculated, move to the next attribute A_{n+1} and restart step 3 with A_{n+1}; **else, goto** step 4;
3: Choose a random number $l_n \in \mathbb{Z}_{n'}^*$;
4: Calculate:

$$C_{1,n} = \psi^{(I_{n-1}-I_n)l_n},$$

$$C_{2,n} = T_n^{(I_{n-1}-I_n)l_n},$$

$$C_{3,n} = (k_n l_n)^{-1}.$$

$1 \leq n \leq m$;
5: Calculate $C_{1,m+1} = \psi^{(I_m-I_{Pub})}$, $C_{2,m+1} = T_{Pub}^{(I_m-I_{Pub})}$.

The **Decrypt** algorithm works following the decryption sequence. Note that the first attribute in decryption sequence is always A_{Pub}. A decryption process follows **Algorithm 11.6**.

When **Decrypt** algorithm succeeds, S_k is the random data encrypting key embedded in C.

Algorithm 11.6 Decrypt

1: Start from the public attribute A_{Pub};
2: For each attribute A_n that the decryptor possesses, compute:

$$\frac{Z_{n,UID_{dec}} \cdot e(X_{n,UID_{dec}}, (C_{1,n})^{k_n C_{3,n}})}{e(Y_n, (C_{2,n})^{k_n C_{3,n}})}$$

$$= e(\varphi, \psi)^{r_{UID_{dec}}(I_{n-1})};$$

3: **If** $e(\varphi, \psi)^{r_{UID_{dec}}(I_{n-1})}$ is the decryptor's private key, go to step 2 with attribute A_{n-1}; **else** go to step 4;
4: Calculate

$$S_k = C/(e(C', D_{UID})/e(\varphi, \psi)^{r_{UID_{dec}}(I_{n-1})}).$$

if $Dec_{S_k}(C'') == ROOT$, **Success**; **else Failure**.

11.4.3 ATTRIBUTE RANKINGS

The presented ABE scheme extends capabilities of traditional ABE schemes and is able to support comparison between values of the same attribute. In real world scenario, this means, for instance, two attribute values *Physician* and *Nurse* of attribute **Occupation** can be compared and have the relationship *Physician* > *Nurse*. In other words, it means that the *Physician* attribute subsumes all the privileges the *Nurse* has, but the *Nurse* does not have any of the additional privileges the *Physician* has. Such capability is applicable and desirable when the privilege of the lower-ranking role (*Nurse*) is a subset of that of the higher-ranking role (*Physician*). In traditional ABE solutions, each attribute value (*Physician* and *Nurse* in the above example) corresponds to a set of cryptographic components that are designated for that specific attribute (**Occupation** in the example) of a specific user. Components for different values of the same attribute are not related. In other words, the key components of *Physician* are independent to those of *Nurse*. To establish ranking relations between attribute values, certain connections need to be established between the corresponding key components. Specifically, a one-direction relation between values of the same attribute is supported in the presented scheme. It allows a higher-ranking user (*Physician*) to be able to legally derive the corresponding lower-ranking role (*Nurse*) key components for herself. However, the lower-ranking role cannot derive anything useful regarding the higher-ranking role.

Such capability can be achieved by deliberately assigning appropriate values in **KeyGen** algorithm. Specifically, as in the previous example, the scheme assigns h_P for *Physician* and h_N for *Nurse* such that $h_P = h^{\alpha_P}$, $h_N = h^{\alpha_N}$, $h \in \mathbb{Z}_{n'}^*$, and $\alpha_P < \alpha_N$. Thus, it is easy to derive $S_P = \varphi^{h_P}$ and $S_N = \varphi^{h_N}$. This is different from traditional ABE scheme, where both S_P and S_N are randomly chosen. Such difference is the connection that is established between comparable values (*Physician* and *Nurse*) of the same attribute (**Occupation**).

Recall when the order of $\mathbb{G}_s$ is defined, it is written as $n' = pq$, where p and q are two large prime numbers. In other words, n' is a composite number satisfying

RSA algorithm requirements. If a user U_P is assigned with $S_P = \varphi^{h^{\alpha_P}}$, i.e. the key for *Physician*, the user is able to calculate the corresponding key S_N for *Nurse* as long as $\alpha_P < \alpha_N$. This process can be done as:

$$S_N = \varphi^{h^{\alpha_N}} = (\varphi^{h^{\alpha_P}})^{h^{\alpha_N - \alpha_P}} = (S_P)^{h^{\alpha_N - \alpha_P}} \tag{11.1}$$

This means when attributes are assigned to U_P, it is optional to assign the value $h^{\alpha_N - \alpha_P}$ to the user together with S_P. Thus, when the user needs to decode some message dedicated for *Nurse*, he can easily calculate S_N following equation (11.1). However, if another user U_N has the attribute *Nurse*, he cannot deduce S_P following the same equation in a similar way. This is because in this case, $\alpha_P - \alpha_N < 0$. Under RSA assumption, h^{-1} cannot be efficiently computed due to the secrecy of n'.

A benefit of such extension to the original scheme is that it allows the ranking relations among attributes without incurring too much workload on TTP. Only eligible users, *Physician* owners in this example, can use such capability and the value $h^{\alpha_N - \alpha_P}$ is only useful to eligible users.

It is necessary to clarify that the attribute authority can decide whether to assign the value $h^{\alpha_N - \alpha_P}$ to a specific *Physician* owner or not. In other words, a *Physician* owner does not automatically have the capability to derive his *Nurse* components unless it acquires such value. Such derivation capability is carried out under the control of TTP.

With such knowledge, the TTP can assign two more values Δh and Δr to user U_P in **KeyGen** algorithms. When needed, the user can derive his key values corresponding to attribute *Nurse* afterwards. The modified step 3 of **KeyGen** is as:

$$X_{P,UID} = \varphi^{r_{UID}} S_P^{r_P},$$
$$Y_P = \varphi^{r_P},$$
$$Z_{P,UID} = e(\varphi, \psi)^{r_{UID} I_P},$$
$$L_{UID} = T_{Pub}^{1/P_{UID}},$$
$$\Delta h = h^{(\alpha_N - \alpha_P) r_P},$$
$$\Delta r = \Delta h I_N / I_P.$$

Thus, the r_{UID} for U_P's *Nurse* attribute is changed to $r'_{UID} = r_{UID}\Delta h$. Correspondingly, the following can be computed:

$$X_{N,UID} = (X_{P,UID})^{\Delta h} = \varphi^{r_{UID}\Delta h} S_N^{r_P} = \varphi^{r'_{UID}} S_N^{r_P},$$
$$Y_N = Y_P,$$
$$Z_{N,UID} = (Z_{P,UID})^{\Delta r} = (e(\varphi, \psi)^{r_{UID} I_P})^{\Delta h I_N / I_P}$$
$$= e(\varphi, \psi)^{r_{UID}\Delta h I_N} = e(\varphi, \psi)^{r'_{UID} I_N},$$
$$L'_{UID} = L_{UID}.$$

Here, it is necessary to point out that to make sure the values of h for two comparable attributes are the same, comparable attributes need to be managed by the same authority. This means one single authority defines the relative order between these attribute values. This requirement is reasonable in real-world scenario since in most cases a single authority (the hospital in this example) defines values of the same attribute (job position). It is rare to require two separate authorities to define separate values for the same attribute.

11.4.4 APPLY ABE-BASED NAMING SCHEME IN ICN

With the above naming scheme, the following capabilities can be achieved in ICN scenarios:

* A content owner is able to specify the access control policy without knowing the consumers' keys;
* The policy confidentiality can be fully protected from being leaked to adversaries;
* Step-by-step attribute exposure is enforced for consumers to determine their eligibility efficiently in computation;
* Flexible attribute management is supported.

Using this scheme, any entity who wants to publish data contents needs to create the content following the procedures shown in Figure 11.3.

The owner firstly creates a random symmetric key K. Then the data to be published is encrypted using K. The resulting ciphertext C is then used to generate a metadata of C. Both the metadata and C are parts of the final content. Then the owner needs to specify an access policy P of attributes, which identifies what attribute requirements an authentic consumer should satisfy. After that, the owner uses this policy to encrypt K following **Encrypt** algorithm. The result is used as the content name.

In this way, the owner does not need to know individual public keys of all the potential consumers in advance, which is required in traditional methods.

11.5 PERFORMANCE ANALYSIS AND EVALUATION

The performance improvement provided by the presented scheme is evaluated in this section. In the following, the computation and communication performance is presented in two parts: real-world implementation and complexity analysis.

11.5.1 EVALUATION OF THE NAMING SCHEME

In this section, the ABE-based naming scheme is evaluated from performance aspect. This includes analysis on its computation and communication (storage) overheads. The computation consumption analysis is carried out by comparing the presented scheme with existing ABE schemes. The communication comparison is carried out on both the content name and the content itself respectively since they both are transmitted in the network.

From computation perspective, the time consumptions for key generation, encryption and decryption processes are tested. In real world application, the time consumption for a consumer to decrypt the content's name is much more important than that for other functions. This is because each content is encrypted once, but decrypted by multiple users for multiple times. In addition to testing the real-world time consumption for each function, a comparison is conducted on the decryption overhead with existing ABE solutions: CP-ABE [25], CN scheme [57], NYO scheme (the 2nd construction in [165]), YRL scheme [233] and GIE scheme [117]. The idea is to compare the number of most time-consuming operations needed in each scheme. Such comparison is carried out in complexity analysis.

11.5.2 REAL-WORLD IMPLEMENTATION

For real-world implementation, a machine with a four-core 2.80 GHz processor and 4 GB memory running Ubuntu 10.04 is used for experiment. Pairing-Based Cryptography (PBC) Library [153] is used to handle the pairing computations. A type-A1 curve [145] is generated using the parameter generating tools included in this library for the following tests. It randomly generates the prime numbers used for the curve, with a length of 512 bits for each of them.

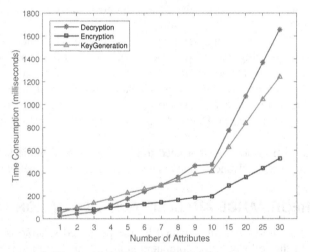

Figure 11.4: Computation performance

Each operation is run for ten times for key generation, encryption and decryption (Figure 11.4). Here the policies are set to be a conjunctive clause of different number (shown in x-axis) of attributes. This is because given a fixed number of attributes, this form requires the most time for computation. In other words, it directly represents the correlation between the number of attributes involved and the time needed for computations. The reason why the encryption function consumes more time when the number of attributes is small is that the cost for computing C in Algorithm 11.5 requires an additional pairing operation, which is independent to the number of attributes. When few attributes are involved, this additional pairing takes a high portion of the entire time consumption. This portion reduces as the attribute number grows, which explains why the time consumption for encryption eventually becomes the smallest among the three functions.

In theory, the time consumption should be linear to the number of attributes involved. The curve in Figure 11.4 is not perfectly linear, but it meets the expected growing trend. There are several reasons why it is not strictly linear. Before decrypting the message attribute by attribute, in the implemented program, there are some necessary steps to initialize global parameters, read files and allocate memory space. Similarly, at the end of the algorithm, there are some clean-up work involved, such as writing files and releasing memory space. Such time consumption is related to the number of attributes involved but not strictly proportional. Also, at step 4 of Decryption algorithm, there is one additional pairing operation. Thus, when the number of attributes is small, this additional operation takes more portion of the total time than when the number of attributes is large. If the possible variance introduced by system level factors, for instance the resource consumption from other processes, are also considered, the variance in the figures is reasonable in practice.

11.5.3   COMPLEXITY COMPARISON

For comparison purpose, every atomic operation is tested for fifty times and the average values are chosen as benchmarks for further comparison. Results of the experiment (Table 11.1) show that pairing operation takes longer than any other operations. Therefore, the comparison metric is set to be the number of pairing operations in decryption process.

Table 11.1

Time-consumption of different operations (in milliseconds)

	Pairing	Exponentiation	Multiplication	Inversion
Time	7.675	0.491	0.029	0.024

Following the above-mentioned idea, there are some terms that need to be defined: N_{attr} is used to denote the number of attributes a consumer has, N_{all} refers to the total number of attributes defined in the network ($N_{all} \gg N_{attr}$). The presented naming scheme is denoted as ICN-ABE in the rest of this manuscript. Since the policy is publicly known in CP-ABE and CN, decrypting parties are able to decide what attributes to use in decryption. Therefore, for those who satisfy the policy, the time taken for decryption in CP-ABE is proportional to the number of attributes involved, which is denoted as N_{invo}, $N_{invo} \leqslant N_{attr}$. The time taken for a successful decryption in CN is related to the number of attributes defined in the entire system. This is because each user is assigned with a value (Positive, Negative, and Wildcard) for every attribute defined in CN. It is obvious that unauthorized users would not bother to try decryption, which is why an alternative result in both schemes is that it takes 0 in time as the user would halt the decryption process.

An unauthorized user in GIE or ICN-ABE is not able to proceed with the decryption process if it cannot satisfy the next attribute. In this situation, N_{part} is used to denote the number of attributes that the consumer has already decrypted, where $N_{part} \leqslant N_{invo}$. Therefore, there are two possibilities for the computation cost in GIE and ICN-ABE, one for a successful decryption and the other for a failed one. Since OR-gate is not widely supported by all the ABE-schemes mentioned before, the performance is tested with policies consisting of attributes and AND-gates. Test result is shown in Table 11.2. It is necessary to point out that in real world, N_{all} is much larger than N_{attr}. Therefore, CN scheme has the largest cost. Among all the anonymity schemes, GIE and the presented scheme cost less than NYO and YRL. As a matter of fact, the cost of the presented scheme is around 2 thirds of that of GIE.

Table 11.2

Comparison of computation cost in decryption

Scheme	Hidden Policy	Number of Pairings
CP-ABE	No	$2N_{invo} + 1$ or 0
CN	No	$N_{all} + 1$ or 0
NYO	Yes	$2N_{attr} + 1$
YRL	Yes	$2N_{attr} + 2$
GIE	Yes	$3N_{invo}$ or $3N_{part}$
ICN-ABE	Yes	$2N_{invo} + 1$ or $2N_{part}$

To evaluate the communication costs, the size of content names is compared. The purpose to compare network names is to make sure that names generated by the presented scheme do not consume much more storage space than existing solutions. In PBC library [153], a data structure element't with size of 8 bytes is used to represent an element. For the presented scheme, a block of 24 bytes is needed to store the network name. Compared with this name size, a content in CBCB [48] is identified by a set of attributes with corresponding values. The size of this attribute set is determined by the content owners. Thus, it is reasonable to model the names as a human-readable string of an undetermined size. NDN [2] shares a similar problem with the name size since the names in NDN also consists of a number of human-readable strings. As mentioned before, DONA [131], NetInf [66] and PURSUIT [83] share the same naming scheme. Therefore, only the size of DONA's name is used for comparison. In [131], the size of a name is confined to 40 bytes in its protocol header. Thus, the size of network names in the presented scheme is small enough to fit in existing ICN solutions.

The number of attributes used in ciphertext is denoted as N_{ciph}. For each attribute in the policy, the corresponding ciphertext consists of two elements from $\mathbb{G}_0$ and one element from $\mathbb{Z}_p$ in ICN-ABE. The total size of a ciphertext is $1\mathbb{G}_1 + (2N_{ciph} + 4)\mathbb{G}_0 + N_{ciph}\mathbb{Z}_p$. This means the ciphertext consists of 1 element from $\mathbb{G}_1$, $2N_{ciph} + 4$ elements from $\mathbb{G}_0$ and N_{ciph} elements from $\mathbb{Z}_p$. Comparison results are shown in Table 11.3. Here the sizes of attribute policy in CP-ABE and CN are not considered. CP-ABE has the smallest ciphertext size. Among the four schemes supporting anonymity, the ciphertext sizes in NYO and YRL are much larger than those in GIE and ICN-ABE. This is because these two schemes encrypt the ciphertext for all the attributes in the network. GIE and ICN-ABE are of the same order of magnitude with ICN-ABE performing better.

Table 11.3
Comparison of ciphertext size

Scheme	Ciphertext Size
CP-ABE	$1\mathbb{G}_1 + (2N_{ciph} + 1)\mathbb{G}_0$
CN	$1\mathbb{G}_1 + (N_{all} + 1)\mathbb{G}_0$
NYO	$\geqslant 1\mathbb{G}_1 + (2N_{all} + 1)\mathbb{G}_0$
YRL	$1\mathbb{G}_1 + (3N_{all} + 3)\mathbb{G}_0$
GIE	$N_{ciph}\mathbb{G}_1 + 3N_{ciph}\mathbb{G}_0$
ICN-ABE	$1\mathbb{G}_1 + (2N_{ciph} + 4)\mathbb{G}_0 + N_{ciph}\mathbb{Z}_p$

11.6   SECURITY ANALYSIS

From security perspective, the strength of the presented scheme is analyzed based on the attack model presented in Section 11.3.5. For the first attack goal, a security theorem is provided with its corresponding security proof as in Section 11.6.2. For the second goal, the scheme is analyzed based on details of the algorithms.

Theorem 11.1. *Let G_0 and G_1 defined as in Section 11.4.4. For any adversary A, the advantage it can gain from the interaction with the security game defined in Section 11.6.1 is negligible.*

Proof of Theorem 11.1 (Proof Sketch). *The proof for this theorem is provided in Sections 11.6.1 and 11.6.2. In the proof, it is verified that the attacker cannot break the encryption algorithm to get any data exposed. Furthermore, it is also proved that attackers cannot conduct collusion attacks onto the system. This is because if collusion attacks are feasible, the adversary in the security game of Section 11.6.1 can overcome the constrain that no single user can satisfy the policy and still get the secret information decrypted. Thus, the attacker is able to gain a non-negligible advantage in this game.*

For the second attack goal, the attacker will stop at the first attribute, A_k, that he doesn't own in the decryption process. If he can get to know this additional attribute, he must get it from step 3 in Algorithm 11.6. This means that the attacker possesses the secret key $Z_{i,UID}$ of the attribute A_k, which contradicts to the assumption that he does not possess such an attribute.

The rest of this section focuses on the proof of Theorem 11.1. Before going into details of the proof, the security model in terms of a security challenge game is presented in Section 11.6.1.

11.6.1 ABE SECURITY MODEL

In this section, the focus is placed on the naming scheme, which can be modeled in the form of a game between a challenger and an adversary. The challenger simulates the operations of the TTP and the attribute authorities, while the adversary tries to impersonate as a number of normal network nodes. The game consists of the following five steps:

Game 1:

- **Setup.** The challenger runs the **GlobalSetup** algorithm and returns to the adversary the global parameters.
- **Phase 1.** The adversary can ask for a certain number of attribute keys in the name of a number of different users from the challenger. The number of allowed keys and users are arbitrary. The challenger runs the **NodeJoin** algorithm for each user involved in the requests and returns the corresponding secret information. The adversary then plays in the roles of these users to request for attributes from the challenger. The challenger runs the **AuthoritySetup** algorithm to create parameters for authorities and runs the **KeyGen** algorithm to generate the corresponding attribute keys that are requested by the adversary on behalf of the authorities and the TTP. In other words, **Key-Gen** in this game is conducted all by the challenger itself. The challenger creates new authorities only when it is necessary.
- **Challenge.** The adversary provides two messages M_0 and M_1 to the challenger together with an access policy A. A satisfies that none of the users created by the challenger has attributes satisfying A. It is possible that a combination of attributes belonging to different users who are impersonated by the adversary can satisfy policy A. The challenger flips a coin b and encrypts M_b using A as:

$$C = \begin{cases} e(\varphi, \psi)^{\alpha s} & \text{if } b = 1, \\ e(\varphi, \psi)^{\theta} & \text{if } b = 0. \end{cases}$$

It then sends the ciphertext back to the adversary.

- **Phase 2.** The adversary can ask for more attributes and users from the challenger. But if any single user can gain satisfactory attribute combinations for A, the challenger aborts the game. Up to now, all the attributes or attribute keys mentioned in the game description refer to private keys. The adversary can always request for any public keys, which is only for encryption purpose.

- **Guess.** The adversary makes a guess b' on the real value of b.

The adversary's advantage in this game can be defined as $ADV = P[b' = b] - \frac{1}{2}$. The presented scheme is secure if for all the polynomial time adversaries, the advantage is at most negligible in the game.

11.6.2 SECURITY PROOF

In this section, the sketch for security proof is provided following the structure in [25]. Before going into details of the proof, the security game described in Section 11.6.1 is modified. This modification follows the same idea as in [25] and it is intended to change from differentiating two random messages M_0, M_1 to differentiating $e(\varphi, \psi)^{\alpha s_j}, e(\varphi, \psi)^{\theta_j}$ so that the generated intermediate results can be represented using the four mappings that are to be introduced in this section. The goal of such modification is essentially to facilitate the subsequent security proof. To differentiate these two games, the one in Section 11.6.1 is referred to as **Game1** and the modified game as **Game2**.

Modified Game (Game 2):

Game2 consists of five steps similar to **Game1**. The steps **Setup, Phase1**, and **Phase 2** are the same as in **Game1**. The **Challenge** step is different in that the challenger does not choose one message to construct the ciphertext C. Instead, its outputs C_j are:

$$C_j = \begin{cases} e(\varphi, \psi)^{\alpha s_j} & \text{if } b = 1, \\ e(\varphi, \psi)^{\theta_j} & \text{if } b = 0. \end{cases}$$

Here, all the θ_j are randomly chosen from $Z_{n'}^*$ following independent uniform distribution.

Suppose an adversary **adv1** in **Game1** has the advantage of ε, his corresponding adversary **adv2** in **Game2** can be constructed according to the following strategy:

- Forward all the messages between **adv1** and the challenger during **Setup, Phase1**, and **Phase 2**;
- In the **Challenge** step, **adv2** gets two messages M_0 and M_1 from **adv1** and the challenge C from the challenger. **adv2** flips a coin δ and sends $C' = M_\delta C$ to **adv1** as the challenge for **adv1** in **Game1**. **adv2** generates its guess based on the output δ' from **adv1**. If $\delta' = \delta$, then the guess is 1; otherwise, it is 0.

The advantage that **adv2** has in this game can be calculated as $\frac{\delta}{2}$.

In the following, it will be shown that no polynomial adversary can distinguish between $e(\varphi, \psi)^{\alpha s}$ and $e(\varphi, \psi)^{\theta}$. Therefore, no adversary can have non-negligible advantage in the security model.

Security Guarantee in the Modified Game

In this section, the proof sketch follows the generic group model introduced in [199] and uses a simulator to model the modified security game between the challenger and the adversary. The simulator chooses random generators $\varphi \in G_s$ and $\psi \in G_t$. It then encodes any member in G_s and G_t to a random string following two mappings: $f_0, f_1 : \mathbb{Z}_{n'} \to \{0,1\}^{\lceil \log n' \rceil}$. It also encodes any member in G_1 to a random string in a similar way: $f_2 : \mathbb{Z}_n \to \{0,1\}^{\lceil \log n \rceil}$. One additional mapping f_3 is used to convert elements in $\mathbb{Z}_{n'}^*$ to string representations: $f_3 : \mathbb{Z}_{n'}^* \to \{0,1\}^{\lceil \log n' \rceil}$. These four mappings should be invertible so that the simulator and the adversary can map between the strings and the elements of corresponding algebraic structures in both directions. Four oracles are provided to the adversary by the simulator to simulate the group operations in G_s, G_t, G_1, and the pairing respectively. Only the string representations can be applied to the oracles. The results are returned from the simulator in such string representations as well. These oracles will strictly accept inputs from the same group, i.e. strict enforcement on the input from the same group for the respective group operations. The simulator plays the role as the challenger in the modified game.

- **Setup.** The simulator chooses G_s, G_t, G_1, e, φ, ψ, and random values α, β. It also defines the mappings f_0, f_1, f_2 and the four oracles mentioned above. The simulator chooses the public attribute parameters $I_{Pub} \in \mathbb{Z}_{n'}^*$, $S_{Pub} = f_0(\mu) \in G_s$, $T_{Pub} = f_1(\lambda) \in G_t$, and $ROOT \in G_1$, where λ and μ are random strings. The public parameters are G_s, G_t, $\varphi := f_0(1)$, $\psi := f_1(1)$, $\varphi^\beta := f_0(\beta)$, $e(\varphi, \psi)^\alpha := f_2(\alpha)$, $(S_{Pub}, T_{Pub}, I_{Pub})$, and $ROOT$.
- **Phase 1.** When the adversary runs **NodeJoin** for a new user with UID, the simulator generates a random number $r_{UID} \in \mathbb{Z}_{n'}^*$. It returns to the adversary with $D_{UID} = f_1((\alpha + r_{UID})/\beta)$, $X_{Pub,UID} = f_0(r_{UID})f_0(\mu r_{Pub,UID}) = f_0(r_{UID} + \mu r_{Pub,UID})$, $Y_{Pub} = f_0(r_{Pub})$, and $Z_{Pub,UID} = f_2(r_{UID}I_{Pub})$, here $r_{Pub,UID} \in \mathbb{Z}_{n'}^*$ is a random number chosen by the simulator. When the adversary requests for a new attribute A_i that has not been used before, the simulator randomly chooses $I_i, k_i, h_i \in \mathbb{Z}_{n'}^*$ and $S_i = f_0(h_i) \in G_s$, $T_i = f_1(h_i) \in G_t$ to simulate the process for setting up an attribute authority for this new attribute. For each attribute key request made from the adversary, the simulator computes $X_{i,UID} = \varphi^{r_{UID}} S_i^{r_i} = f_0(r_{UID} + h_i r_i)$, $Y_i = \varphi^{r_i} = f_0(r_i)$, and $Z_{i,UID} = e(\varphi, \psi)^{r_{UID} I_i} = f_2(r_{UID} I_i)$, where r_i is a random number chosen from $\mathbb{Z}_{n'}^*$. The simulator passes all these values to the adversary as the attribute keys associated with A_i.
- **Challenge.** When the adversary asks for a challenge, the simulator flips a coin b and chooses a random value $s \in \mathbb{Z}_{n'}^*$. If $b = 1$, the simulator calculates $C = f_2(\alpha s)$; if $b = 0$, it picks a random value $s' \in \mathbb{Z}_{n'}^*$ and calculates $C = f_2(s')$. In addition, it calculates $C' = \varphi^{\beta s}$ and $C'' = Enc_K(ROOT)$. It also computes other components of the ciphertext following **Encrypt**: $C_{1,n} = f_1((I_{n-1} - I_n)l_n)$, $C_{2,n} = f_1(h_n(I_{n-1} - I_n)l_n)$, and $C_{3,n} = f_3((k_n t_n)^{-1})$, where $h_n \in \mathbb{Z}_{n'}^*$ is a random number chosen by the simulator.

- **Phase 2.** The simulator interacts with the adversary in a similar way as in **Phase 1** with the exception that the adversary could not acquire attribute keys enabling a single user to satisfy the access policy $\mathbb{A}$. The output of this step is similar to that of Phase 1 except that the simulator obtains more user IDs and attributes in this step.

From this game, it can be seen that the adversary only acquires string representations of random values in $\mathbb{Z}_n^*$, $\mathbb{Z}_n$ and combinations of these values. All the queries can be modeled as rational functions. It can further be assumed that different terms always result in different string representations [25]. As shown in [25], the probability that two terms share the same string representation is $O(q^2/n)$, where q is the number of queries made by the adversary. It is assumed in the rest of the proof that no such collision happens.

Now an argument can be made that the adversary's views are identically distributed between the two cases when $C = f_1(\alpha s)(b = 1)$ and when $C = f_1(s')(b = 0)$. As a matter of fact, what the adversary can view from the modified game with the simulator are independent elements that are uniformly chosen and the only operation that the adversary can do on these elements is to test if two of them are equal or not. Thus, the situation that the views of the adversary differ can only happen when there are two different terms v_1 and v_2 that are equal when $b = 1$. Since αs and s' only occur in group G_1, the results from f_1 cannot be paired. Queries by the adversary can only be in the form of additive terms. Then it can be derived: $v_1 - v_2 = \gamma\alpha s - \gamma's'$, where γ is a constant. By transformation, it can be written as: $v_1 - v_2 + \gamma's' = \gamma\alpha s$. This implies that by deliberately constructing a query $v_1 - v_2 + \gamma's'$, the adversary may be able to get the value of $e(g, g)^{\gamma\alpha s}$. Now it needs to be proved that such a query cannot be constructed by the adversary based on the information it gets from the modified game.

In fact, the information that an adversary can acquire from this game is listed as in Table 11.4. This table excludes values related to L_{UID} as it is not related to αs. To construct the desired value, the adversary can map two elements from G_s and G_t into one element in G_1. He can also use elements in Z_n to change the exponentials. From this table, it can be easily seen that to obtain a value containing αs, the adversary can pair βs and $(\alpha + r_{UID})/\beta$ to get $\alpha s + r_{UID}s$ in G_1. In fact, this is the only way to get a term containing αs. But it is not feasible. Both βs and $(\alpha + r_{UID})/\beta$ belong to G_t, while the pairing requires one element from G_s and one from G_t, respectively.

A more detailed illustration for the above argument is that: by conducting the query on behalf of the users that the adversary has established in **Phase 1**, the adversary can get a polynomial $\gamma\alpha s + \sum_{UID \in U_{query}} \gamma r_{UID}s$, where U_{query} is the set of UIDs used by the adversary. To eliminate the second part in this polynomial, the adversary can use items in the table containing $I_{n-1} - I_n$ and r_{UID} to construct desirable polynomial. But this is impossible for the adversary under the game assumption because:

- Firstly, the adversary cannot reconstruct s from either $t_n(I_{n-1} - I_n)h_n$ or $(I_{n-1} - I_n)h_n$ since the h_ns are chosen as random values for each attribute that it is impossible to get $s = \sum_{n \in P_a}(I_{n-1} - I_n) + I_{Pub}$ from them without

Table 11.4

Query information accessible to the adversary

μ	β	$r_{UID} + \mu r_{Pub,UID}$
r_{Pub}	h_i	$r_{UID} + h_i r_i$
r_i	$(I_{n-1} - I_n)h_n$	$t_n(I_{n-1} - I_n)h_n$
λ	$(\alpha + r_{UID})/\beta$	βs
h_i		
α	$r_{UID}I_{Pub}$	$r_{UID}I_i$
I_{Pub}	I_i	k_i
$(k_n t_n)^{-1}$	h_i	

peeling off the h_ns. Here, P_a represents the set of attributes satisfying the policy;

- Secondly, the adversary cannot reconstruct s from I_{Pub} and I_i in Z_p. This is because no single user is assumed to satisfy the attribute policy that the adversary cannot reconstruct a valid attribute combination satisfying the policy. Thus, he cannot find the constitution of P_a for the equation $s = \sum(I_{n-1} - I_n) + I_{Pub}$.
- Thirdly, the item with r_{UID} cannot be canceled.

Therefore, based on the information an adversary can get from the presented scheme, the attacker can not differentiate a random ciphertext from an authentic one. The security of the presented scheme is proved. ■

11.7 SUMMARY

In this chapter, a comprehensive access control solution for ICN network is presented. This solution is based on a privacy-preserving ABE-based naming scheme. This scheme greatly reduces the communication and computation overhead compared to existing ABE solutions. Also, this scheme is designed in a public-key pattern, making it more flexible for attribute management. From security and privacy perspective, the ABE-based naming scheme achieves a high security level as CP-ABE, but with attribute anonymity protection for policy privacy and flexible attribute rankings. Experiments and analysis confirm the effectiveness of the proposed solution.

12 ABE for Vehicular Network Security Policy Enforcement

Vehicular Ad Hoc Networks (VANETs) are usually operated among vehicles moving at high speeds, and thus their communication relations can be changed frequently. In such a highly dynamic environment, establishing trust among vehicles is difficult. To solve this problem, we propose a flexible, secure and decentralized attribute based secure key management framework for VANETs. The presented solution is based on attribute-based encryption (ABE) to construct an Attribute-Based Vehicular Network Security Policy Enforcement (AVN-SPE) framework. AVN-SPE considers various road situations as attributes. These attributes are used as encryption keys to secure the transmitted data. AVN-SPE is flexible in that it can dynamically change encryption keys depending on the VANET situations. At the same time, AVN-SPE naturally incorporates data access control policies on the transmitted data. AVN-SPE provides an integrated solution to involve data access control, key management, security policy enforcement, and secure group formation in highly dynamic vehicular communication environments. The presented performance evaluations show that AVN-SPE is efficient, and it can handle large amount of data encryption/decryption flows in VANETs.

12.1 INTRODUCTION

In vehicular ad hoc networks (VANETs) , which include both vehicles and roadside units, security and privacy research is mainly based on the entity level or data level (a.k.a., entity trust [170, 169, 90, 91, 180] vs. data trust [181]). At the entity level, previous research mainly focused on how to ensure the genuineness of the data source, i.e., providing origin integrity. The entity trust requires validation of an entity (e.g., an identity, a license number, or a pseudonym), which is usually performed by using authentication techniques. The data trust requires to evaluate the trustworthiness of data contents. The evaluation technologies of data trust can be generally classified as data integrity checking, probability-based statistic modeling techniques, and majority rule-based evaluation. In VANETs, transmitted data should be accessed by their intended receivers. However, due to fast movements of vehicles, most existing key management solutions only consider setting up entity trust without considering who should access the data. Moreover, due to the broadcasting nature of VANETs, it is desirable to enforce a group-based key management solution to improve the communication efficiency with strong security and privacy policies on who are eligible for the corresponding group communications. This factor has been largely ignored in existing solutions for both the entity trust and data trust. Thus, it is highly desired to allow the data source to specify the security access control policies on the broadcasted data.

To enable the security data access control policy in VANETs, we use *attributes* as the basic properties of vehicles for access control and secure group communications in VANETs. Particularly, attributes are used to describe the roles of VANET communication participants. Attributes abstract entity and data trust at a certain level, and they can be used to identify a group of entities. For example, attributes can be described as follows: (i) ownership of vehicles: taxes are associated with a company, police cars in a city, (ii) type of events: accidents, congestions, and (ii) property of events: location-based services, road traffic updates, etc. Attributes can be further classified as dynamic and static attributes, depending on whether attributes change frequently or remain the same during a relatively long period in comparison to ephemeral connections of VANETs. Vehicles that fulfill a set of descriptive attributes form a group. Considering attributes as policies associated with a group, we introduce a new concept *policy group*. A policy group is a group of vehicles confined by their attributes, such as common interests, security or service requirements, or environmental constraints (e.g., street name, time, driving direction, etc.). A policy group is defined by the message source and is organized automatically without relying on an on-line trust party to manage the group. This means as long as a vehicle "satisfies" the specified attributes by the message source, it will be able to decrypt the message encrypted by using the given attributes. To enable such a capability, we propose a novel Attribute-based Security Policy Enforcement (AVN-SPE) framework by using the basic formation of Attribute-based Encryption (ABE) scheme [25], which utilizes Identity-Based Encryption (IBE) [39] and threshold secret sharing scheme [194].

The main focuses of the presented solutions are in three-fold: (i) AVN-SPE provides an architectural solution that enforces policy control in highly dynamic communication environments. The policies defined are based on vehicles' surrounding situations and can be modified to achieve different security and privacy goals for VANETs. We describe the AVN-SPE architecture by defining policies for data access control through vehicular group communications. (ii) We show how AVN-SPE policies can be extended to perform subgroup vehicular communications with minimum communication and computation overhead. (iii) We present an optimization of ABE for vehicular networks, which helps AVN-SPE run more efficiently. The performance evaluations demonstrate the soundness of AVN-SPE for large-scale vehicular networks.

The rest of the chapter is organized as follows: Section 12.2 details previous work in area of vehicular group communication; in section 12.3, we present the AVN-SPE model; Section 12.4 describes AVN-SPE in detail; performance evaluation of AVN-SPE is detailed in Section 12.5; finally, we summarize this chapter in Section 12.6.

12.2 RELATED WORKS

In VANETs, the group formation, key distribution, and group maintenance are difficult tasks considering the ephemeral vehicle-to-vehicle communications. In CARAVAN [187], the group formation assumes that vehicles forming a group are moving with the similar speed and maintaining relatively constant distance from each other. This assumption is very restricted in that it assumed a very ideal communication scenario in VANETs. Moreover, the methodology adopted by CARAVAN faces following issues: (1) All vehicles moving at a similar velocity within a certain distance forms a group. Thus, the group formation is very restricted. (2) The selection procedure of group leader invokes additional communication overhead. (3) The communication overhead within the group (V2V) and outside the group (V2I/I2V) is very high due to frequent exchanges of PKI certificates for every communication session.

In [97, 147], the authors propose to utilize group communication and group signature based schemes to achieve security and privacy. The solutions in [147] assume that the underlying group has already been formed. The group communications thereafter are performed by using group signature schemes [51] to protect the privacy of the transmitting vehicle. Similar to CARAVAN, GSIS [147] has the similar issues of rigid group formation in dynamic VANET communication environments. The group signature scheme proposed by Guo et. al. [97] provides a solution for group formation by grouping vehicles based upon their location and roles. However, the communications among vehicles belonging to different categories are not addressed.

The solutions proposed by Raya et. al. [179] presented a location-based group formation in VANETs. The group formation is performed based on the location of the vehicles, i.e., based on *where* the vehicle is rather than *whom* the vehicle is. This solution has several restrictions. For examples, the road is assumed to be dissected into small cells as groups. The vehicle closest to the center of a cell is declared as group leader. A group leader encrypts the group key with every group member's public key to establish secure group communications. As a result, the key issue needs to be address is how to adjust the cell size according to the density of vehicles on the road to maximally reduce the group management and communication overhead. This approach requires intensive collaborations among vehicles, which are very difficult in highly dynamic VANETs.

12.3 AVN-SPE SYSTEM AND MODELS

In this section, we present the network model, policy tree formation, and the attack model.

12.3.1 NETWORK MODEL

The network model (Figure 12.1) of VANETs comprises on-road units, off-road units, and the interfacing layer. On-road units consist of vehicles, Road Side Units (RSUs) and communication networks such as the cellular networks. The on-line trusted parties, e.g., RSUs, are usually managed by a local transportation department office. Vehicles can use wireless LAN technologies to establish short distance communications or use Internet-based security services through RSUs or cellular

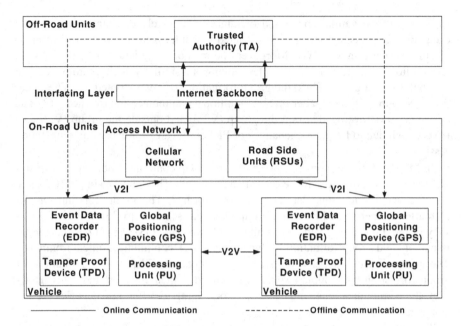

Figure 12.1: Network model for AVN-SPE.

networks to establish remote communications. We assume that vehicles are equipped with following hardware units:

- **Event Data Recorder (EDR)**: EDR records activities of the vehicle such as engine overheating, and road events such as accidents observed by the vehicle in motion.
- **Tamper-Proof Devices (TPD)**: TPD contains information about the vehicle that cannot be modified, such as Vehicle Identification Number (VIN), certificates, private keys.
- **Processing Units (PU)**: Processing units are responsible for performing V2I, I2V or V2V communications. For example, creating event messages, enforcing security and privacy policies, encrypting/decrypting data, etc.
- **Global Position System (GPS)**: Every vehicle is assumed to be equipped with GPS devices.

Off-road units consist of trusted authorities (TA) that provides the standard key management services for users to derive their private keys according to their dedicated attributes. Communications between off-road and on-road units are enabled through the interfacing layer, i.e., the Internet.

12.3.2 POLICY TREE FORMATION

Before describing policy tree formation, we define what is a *policy* in this chapter.

Definition: *A policy is defined as a rule R described over a set of attributes Y, where Y are linked together by a tree structure PT. The rule governs the operations over the data by providing access control, if and only if the access structure PT is satisfied with the requesters' attributes.*

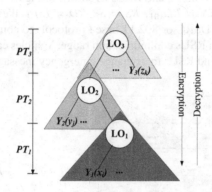

Figure 12.2: Policy tree example based on attributes.

In AVN-SPE, an access tree structure is refereed as a Policy Tree (PT). A PT regulates the policies by means of specifying attributes in the PT. These policies are defined over a set of attributes that describe the access control rules to the data. These policies can be broadly categorized into static and dynamic attributes. An example of such a classification is represented in Table 12.1, where Y_s and Y_d represent static and dynamic attribute set, respectively. The security policies are represented as a combination of attributes and the associated logical operators (LO), as shown in Figure. 12.2. For example, $PT_1 = (Y_1(x_1) \wedge Y_1(x_2)) \vee Y_1(x_3)$ represents logical operations among three attributes $\{Y_1(x_1), Y_1(x_2), Y_1(x_3)\}$, where PT_1 is satisfied if either the combination of attribute $Y_1(x_1) \wedge Y_1(x_2)$ is true or if $Y_1(x_3)$ is true. In this example, a PT is represented as a tree with attributes as leaf nodes and logic operators as internal nodes, where logic operators $LO = \{LO_i | \wedge, \vee, <, \leq, >, \geq, k$ out of $n\}$.

Table 12.1
Classification of attributes

Road Attributes	Environment Attributes	Vehicle Attributes
Y_d: Road Name (RN)	Y_d: Date (ED)	Y_s: Vehicle Category (VC)
Y_d: Road Segment Number (RS)	Y_d: Time Stamp (ET)	Y_d: Vehicle Application Or Service (VS)
Y_d: Road Direction (RD)	Y_s: City Name (EC)	
Y_d: Road Intersection (RI)	Y_s: State Name (ES)	

Example scenario:

The example scenario discussed next provides a base for understanding how policy trees assist secure group/subgroup communications. Here, we present an illustrative example, which will be utilized in the following contexts. Consider a VANET scenario consisting of vehicles and RSUs as shown in Figure 12.3(a). The dynamic attributes set $Y_d = \langle RN = RD_{101}, RS = S_i, RD = East/West, ED = 03/15/09, ET = T_s \rangle$. RSUs uses DSRC or 802.11 based protocols to communicate with vehicles when they come in RSUs communication range. Vehicles can communicate with neighboring vehicles and RSUs to exchange emergency messages or normal data.

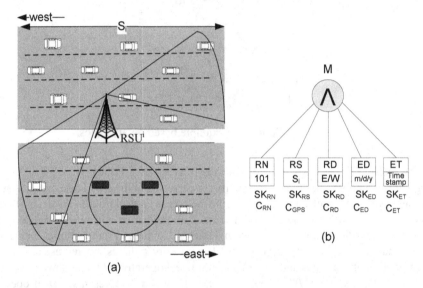

Figure 12.3: (a) Example group formation in VANETs. (b) Dynamic attribute-based policy tree.

Private Key Generation and Distribution

Once attributes are determined, TAs are responsible to generating corresponding private key component[1] for each attribute possessed by a VANET user. Thus, a set of private key components with respect to a set of attributes form a private key for a user. It must be noted that each private key component is derived for a public attribute. Although multiple users can share the same attribute, their corresponding private keys are different. Moreover, they cannot collude to gain additional attributes without generating corresponding private key components from the trusted authority. In essence, the attributes and logic operators construct the policies, and users share the same set of attributes in the policy tree to form a *policy group* for secure communications.

---

[1] A private key component is a private key corresponding to an attribute. It is equivalent to the private key of an identity in IBE scheme [39].

AVN-SPE uses a hybrid and decentralized trust framework to distribute private key components for VANET users. For dynamic attributes, decentralized servers can be deployed through road side units (RSUs) or through well deployed cellular networks. The private keys (SK) of static attributes can be derived in advance using off-line methods via a TA. The private key components of dynamic attributes can be derived from a local on-line trusted server, such as an RSU or using cellular networks.

As shown in Figure 12.2, the encryption process in PT is performed through a top-down approach and the decryption process is performed in a bottom-up fashion. In this example, the PT also shows a hierarchical structure containing three sub-policy trees $(PT_1, PT_2,$ and $PT_3)$. For example, if vehicles possess attributes at PT_1 level, they can derive the secret at level LO_1. However, if LO_2 is an AND gate, they must also possess attributes at the PT_2 level to retrieve the secret at LO_2. From this example, we can see that the policy tree approach is very flexible by constructing the policy tree structure with different logical gates, which provide us a powerful tool to construct group and subgroup communications. In the following context, we refer to the policy tree created by the off-line trusted authority as static $s - PT$ and the policy tree created by the on-line trusted parties as dynamic $d - PT$.

Policy Group Formation

The policies for group formation are defined as a set of rules that grant accesses to the data by restricting who can satisfy the policy tree. In other words, policy group formation in APSE is based upon the policy trees, which are regulated by the attributes (see Table 12.1) common to all vehicles. Although using common attributes can result in very large group size; in AVN-SPE, this group size can be confined by a specific location. The group size in Figure 12.3(a) is limited to only vehicles present at the location, as they share common dynamic attributes RS.

Policy Sub-Group Communications

Utilizing static attributes in conjunction with dynamic attributes provides the flexibility of performing subgroup communications. Here, the policy is defined as a set of rules that grant data access for the policy group. For example, if the attribute vehicle category (VC) in Y_s is used with dynamic attributes in PT shown in Figure 12.3(b), the access to the data is restricted to subgroup members that belong to VC. Here, VC can be a civilian vehicle, a government vehicle, or can belong to a specific organization or company. Consider the example scenario, we assume that black vehicles are police cars. As all group members sharing dynamic attributes in that location can communicate securely using PT, the messages exchanged among the police vehicles are also available to other group members. We can prevent the information leakage by extending the policy tree. For example, we can enforce another attribute: vehicle category $(VC = police)$. As civilian vehicles will not have the private key component for the attribute $VC = police$, they cannot decrypt the messages exchanged among police vehicles. It is important to note that the CP-ABE scheme cannot be directly applied by adding another leaf-node in PT shown in Figure 12.3(b). This is because CP-ABE requires a universal trusted authority to generate attributes for each vehicle. The message cannot be decrypted by combining private key components issued from different authorities. We solve this problem by combining multiple policy trees (Figure 12.2). The detailed procedure is explained in Section 12.4.

12.3.3 ABE PROTOCOLS

Here we present ABE protocols that will used in later presentations.

Protocol 12.1 (Setup). *This algorithm takes a security parameter* **k** *and returns a public key PK and master secret key MK. This setup operation is performed by selecting:*

- Bilinear *map:* $G_0 \times G_0 \to G_1$ *is a map from addition group* G_0 *to multiplicative group* G_1 *of prime order p with a group generator g.*
- *Two random numbers* $\alpha, \beta \in \mathbb{Z}_p$ *are selected.*
- *A hash function* $H : \{0,1\}^* \to G_0$ *that maps the bit string to a point in group* G_0.

The setup algorithm generates:

$$PK = \langle G_0, g, h = g^\beta, f = g^{1/\beta}, e(g,g)^\alpha \rangle, \tag{12.1}$$

$$MK = \langle \beta, g^\alpha \rangle. \tag{12.2}$$

Protocol 12.2 (Key Generation (MK,Y)). *The key generation protocol takes a set of attributes Y and the master secret key as input. It outputs a set of private key components, in which each corresponds to an attribute* $y \in Y$. *The algorithm is operated by the TA and it works in following three steps:*

- *For a user v, TA chooses a random* $r \in \mathbb{Z}_p$,
- *TA chooses a random* $r_y \in \mathbb{Z}_p$ *for each attribute* $y \in Y$.
- *TA finally computes the key as*

$$SK = \langle D = g^{(\alpha+r)/\beta}; \forall y \in Y, D_y = g^r \times H(y)^{r_y}; D'_y = g^{r_y} \rangle. \tag{12.3}$$

Protocol 12.3 (Encryption(PK,M,PT)). *Message M is encrypted under the tree PT with the public key PK. We only describe the encryption procedure for one-level tree for simplicity. The algorithm chooses a polynomial* q_k *for each node k (including the leaves) in the tree PT starting from the root node R. For each node k in the tree, the degree* d_k *of the polynomial* q_k *is set to be one less than the threshold value* T_k *of that node, that is,* $d_k = T_k - 1$. *Starting from R, the algorithm chooses a random key* $s \in \mathbb{Z}_p$ *and sets* $q_R(0) = s$. *Then another point* d_R *is chosen on polynomial* q_R *to define it completely. Finally, for all leaf nodes* $y \in Y$ *of tree PT, the ciphertext CT is constructed over PT by computing:*

$$CT = \langle PT; \tilde{C} = M \cdot e(g,g)^{\alpha s}; C = h^s = g^{\beta s}; \forall y \in Y, C_y = g^{q_y(0)}; C'_y = H(y)^{q_y(0)} \rangle.$$
$$\tag{12.4}$$

Protocol 12.4 (Decryption(CT, SK, k)).

- *We first define a recursive algorithm DecryptNode(CT, SK, k) that takes inputs such as a ciphertext CT, private key components SK associated with attribute set $Y = \{y\}$, and an attribute y from PT.*
- *If the node k represents a leaf node, then $y = \mathbf{att}(k)$ is the attribute value of node k. If $y \in Y$, then,*

$$DecryptNode(CT, SK, k)$$

$$= \frac{e(D_y, C_y)}{e(D'_y, C'_y)} = \frac{e(g^r \cdot H(\mathbf{att}(k))^{r_y}, g^{q_x(0)})}{e(g^{r_y}, H(\mathbf{att}(k))^{q_x(0)})}$$

$$= \frac{e(g^r, g^{q_x(0)}) \cdot e(H(\mathbf{att}(k))^{r_y}, g^{q_x(0)})}{e(g^{r_y}, H(\mathbf{att}(k))^{q_x(0)})} = e(g, g)^{r q_x(0)}.$$

If $y \notin Y$, define DecryptNode$(CT, SK, k) = \perp$.

- *The decryption algorithm begins by calling the function DecryptNode on the root node R of the tree PT. If the tree is satisfied by Y, set*

$$A = DecryptNode(CT, SK, R) = e(g, g)^{r q_R(0)} = e(g, g)^{rs}.$$

- *M is reconstructed by computing*

$$M = \frac{\widetilde{C} \cdot A}{e(C, D)}.$$

12.3.4 ATTACK MODEL

The attack model considers existence of both passive and active attackers. Attackers' goals are to hinder group formation and gain unauthorized access to the data exchanged within the group. We assume that attackers can be VANET participants. Although, RSUs and the TA cannot be compromised, the attacker can try to impersonate an RSU or a TA. Attackers can intercept all the traffic transmitted by RSUs and vehicles, and they can inject fake messages.

12.4 DESCRIPTIONS OF AVN-SPE

In this section, we discuss AVN-SPE in detail. When vehicles entering an area controlled by an RSU, the RSU and vehicles perform mutual authentication by exchanging certificates. The certificates being issued by the trusted authority (TA) can be verified by both vehicles and the RSU. As the RSU is connected to the Internet, RSU also checks vehicles' certificate against Certificate Revocation List (CRL). The RSU then generates the private key components for the dynamic attributes with respect to road conditions, location, and time. As the example shown in Figure 12.3(b), there are five dynamic attributes that are monitored by *RSU i* (denoted as *RSU(i)* if road direction (RD) attribute is ignored (see the dynamic attributes descriptions in Table 12.1). In the follows, we describe the two phases that are involved during the AVN-SPE procedure.

12.4.1   PHASE I: GROUP KEY DISTRIBUTION

In this phase, group keys are established between vehicles and they are used in
Phase II to perform group and subgroup communications. Phase I is performed by
$RSU(i)$ by executing Protocol 12.5 after successful mutual authentication has been
performed.

Protocol 12.5 (Group Key Distribution).
$RSU(i)$ and vehicles perform the following operations:
    *1. $RSU(i)$ chooses random numbers $\alpha, \beta \in \mathbb{Z}_p$ and $r^v \in \mathbb{Z}_p$ for vehicle
$v \in V$, where V is the vehicle set.*
    *2. $RSU(i)$ executes SETUP protocol (presented in protocol 12.1 with
vehicles, and generates public key $PK_{RSU(i)} = \langle G_0, g, h, f, \zeta^{\alpha^v} \rangle$ and private
master key $MK_{RSU(i)} = \langle \beta, g^{\alpha} \rangle$.*
    *3.   For all vehicles in range, compute their private key components $SK_v =
\langle D; D_y, \forall y \in Y_d; D'_y \rangle$, where $v = 1, 2, 3, \dots$ using (12.3).*
    *4.   For all vehicles, perform encryptions: $E_{PK_v}(SK_v)$, where PK_v is the
public key of the vehicle derived from its certificate.*
    5. $RSU(i)$ transmits encrypted data ($E_{PK_v}(SK_v)$) to vehicle v.

The private key components are generated for all dynamic attributes in Y_d. Apart
from the road and vehicular attributes, the environmental attributes such as ET
changes more frequently. Hence, if the message is encrypted with ET at t_1, message decryption at t_2 should not reconstruct the original message M. However, if ET
is excluded, the system will be vulnerable to replay attacks. Thus, to include ET,
$RSU(i)$ generates a time stamp T_s for a specified time interval. This time stamp T_s
is a constant value for the time interval $t_2 - t_1$. For example, if the time interval is
for *five* minutes ranging from 3:00PM - 3:05PM, the $RSU(i)$ generates a time stamp
T_s by choosing a random number $\gamma \in \mathbb{Z}_p$ and hashing it with the RSU's ID. As a
new random number γ is chosen for every interval, the uniqueness of time stamp
T_s is guaranteed. Successful completion of protocol 12.5 guarantees that every vehicle has obtained dynamic attributes along with the corresponding private key components that vehicles will use in Phase II to perform group communications using
policy trees.

12.4.2   PHASE II: GROUP AND SUB-GROUP COMMUNICATION

During phase II, each vehicle can create a PT by defining its own policies. One such
policy example is illustrated in the example scenario shown in Figure 12.3. Since the
PT in Figure 12.3(b) contains only dynamic attributes, all vehicles satisfying the PT
will be able to decrypt the message. Hence, a group is formed with no clear bound-
aries. This flexibility significantly reduces the overhead involved in adding/deleting
members, rearranging and updating group keys. However, this flexibility can be further extended to form subgroups within the group at the cost of minor group key management overhead. The advantage of subgroup communications has been detailed in
Section 12.3.2. Traditionally, the frequent group and subgroup changes will require

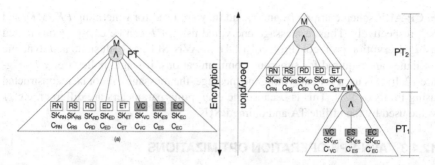

Figure 12.4: Sub-group communication using (a) Single tree PT containing both static and dynamic attributes; (b) Two separate trees, PT_1 containing static and PT_2 dynamic attributes.

setting up separate group keys between the specified group members leading to significant overhead.

In AVN-SPE, group/subgroup formation is through the policy tree specified by the message originator, thus there is no group member addition/deletion operations and key updates are not required. Following the PT discussed in the example scenario (Figure 12.3(b)), the communications performed between any two vehicles can be heard by all group members. To restrict the communication within a subgroup, e.g., only police vehicles, the sender can perform subgroup communication protocol as shown as follows:

Protocol 12.6 (Sub-Group Communication Protocol).
A vehicle:

    1. Creates two policy trees: static policy tree PT_1 and dynamic policy tree PT_2.

    2. Calls Protocol 12.3, ENCRYPTION$\langle PK, M, PT_2 \rangle$ and obtains

$$CT_2 = \langle PT_2; \tilde{C}; C; \forall y \in Y_d, C_y; C_y' \rangle$$

    3. Randomly selects C_y as a secret from CT_2.
    4Calls Protocol 12.3, ENCRYPTION$\langle PK, C_y, PT_2 \rangle$ and obtains

$$CT_1 = \langle PT_1; \tilde{C}'; C'; \forall x \in Y_s : C_x'; C_x'' \rangle$$

    5. The ciphertext is in the form of $\langle CT_1; CT_2 \setminus C_y; PT = \{PT_1; PT_2\} \rangle$.

Note that the selected C_y is considered as a secret and it will not be transmitted with CT_2. The encryption is enforced in a top-down fashion, i.e., first operates on PT_2, and then is followed by PT_1. The policy tree PT_1 is constructed over static attributes whereas PT_2 over dynamic attributes as shown in Figure 12.4(b). In CP-ABE, the private key components for all attributes form an access tree that is generated using key generating parameters for a group. Key generating parameters used in CP-ABE scheme are α, β, and r, and they are used for generating PK, MK, and SK, respectively. Thus, a message encrypted using PT can be easily reconstructed using decryption protocol (Protocol 12.4). In AVN-SPE, as both static and dynamic attributes are required for subgroup communications, if single policy tree PT (Figure 12.4(a)) is used for encrypting the message, the message cannot be reconstructed using Protocol 12.4. This is because the encrypting parameters are different, e.g., r value used by the offline TA and online RSUs.

12.4.3   AVN-SPE OPERATION OPTIMIZATIONS

In VANETs, due to vehicles' mobility, long communication delay is not acceptable for safety-related applications. It is desirable to minimize the delay of policy-tree-based encryption procedures discussed in previous sections. As AVN-SPE is constructed based on CP-ABE, the optimization techniques (described in CP-ABE[25]) such as combining similar attributes can be directly applied to AVN-SPE. Additionally, we propose two techniques to expedite the AVN-SPE operations. The first operation optimization is to use a Key Encryption Key (KEK), which can significantly reduce the time of performing encryption/decryption protocol. The data can be encrypted by running any symmetric key encryption algorithm using the key and then the key is encrypted using a policy tree. As the key size will usually be less then data size, encrypting KEK through PT is a cost-effective procedure. The second optimization is to standardize some common situations in VANETs. As CP-ABE requires transmitting PT in plaintext along with the CT, the ciphertext size can be very big. To reduce this overhead, a standard table with indexes having a generalized set of tree structures can be embedded into the vehicle's processing unit[2]. If a generic policy tree is used, the vehicle can send the index of the tree by looking up the table. It is important to note that, although vehicles can create their own policy trees, this optimization is especially useful when communication regarding standard messages like accident report has to be performed. In next section, we present a comparison between traditional and optimized CP-ABE schemes.

12.5   PERFORMANCE ASSESSMENTS

In this section, we provide performance assessments of AVN-SPE. Particularly, the assessment focuses on computation overhead, communication overhead, and security strength.

12.5.1   PERFORMANCE EVALUATION SETUP

The evaluation has been performed on a 64-bit Pentium IV with a 3.2GHz processor. The protocol implementation uses 160-bit elliptical curve cryptography. The vehicles are assumed to be able to transmit data over a distance up to 1,000 meters. The

---

[2]The table can be installed at the time of vehicle manufacturing. We assume that all manufacturers follow same standard indexing table

packet payload size as per SAE J1746 standard [185], GPS, NTCIP hazard codes [121], and standard protocol headers [61] are set to 100, 200, or 400 (Bytes). We assume the V2V communications use 802.11 based technologies and V2I uses DSRC technology [63]. Following 802.11 standards, the maximum allowable payload size is 2312 bytes (with WEP header) that we have used in the simulation to provide the worst-case performance results. In addition, as the data rate for DSRC varies from 6 mbps to 27 mbps, the round trip time taken for V2I/I2V communication using DSRC in exchanging 400 bytes of data is very small and can be neglected. As per US traffic standards [223], the maximum allowable speed for vehicles on 4-8 lane highway is 90mph and the average inter-vehicle safe distance is assumed to be 10m in jammed and 30m in smooth traffic conditions. The benchmark taken for performing cryptography related operations is listed in Table 12.2. Particularly, we use the CP-ABE implementation provided at [123].

Table 12.2

Notation and values

Time Notation	Operation Time (ms)	Description
T_{CV}	0.07	PKI certificate verification time.
T_{SV}	0.07	Signature verification time.
$T_{G(K)}$	0.00025	256-bit symmetric key generation time.
$T_{Sig(V)}$	1.42	Time for signing value V.
T_E	0.07	Time to encrypt messages using PKC.
T_D	1.52	Time to decrypt messages using PKC.

12.5.2 COMPUTATION OVERHEAD

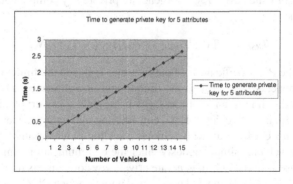

Figure 12.5: Key generation time.

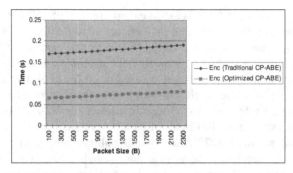

(a) Encryption Time comparison between Traditional and Optimized CP-ABE for Group Communication.

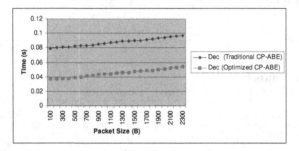

(b) Decryption Time comparison between Traditional and Optimized CP-ABE for Group Communication.

Figure 12.6: Computation overhead.

Figure 12.5 shows the time to generate private key components for vehicles in the region controlled by an RSU. For secure group communications, minimum of four attributes (RN, RS, ED and ET) are required when the road direction is ignored. The total time taken by the RSU (T_{RSU}) to generate private key components for vehicles is calculated as:

$$T_{RSU} = (T_{CV} + T_{G(SK)} + T_{Sig[G(SK)]} + T_E) \times N_v,$$

where T_{CV} is the PKI certificate verification time, $T_{G(SK)}$ is the time to generate private key components, $T_{Sig[G(SK)]}$ is the time to sign the generated private key components, T_E is the time to encrypt messages using public key cryptography (PKC), and N_v is the number of vehicles. Let's consider the maximum capacity on an eight lane (single-side) US freeway with vehicles traveling at maximum allowable speed of 90mph. If 8 vehicles simultaneously enter the communication range of an RSU, the RSU will use 1.42 seconds to generate private key components for all the vehicles within $D_t = 0.034$ miles. If an RSU is mounted with two such DSRC each monitoring a single direction, then the distance traveled in generating private key components is 0.0678 miles, which equals to handle 172mph road traffic.

Figures 12.6(a) and 12.6(b) show a comparison of the amount of time taken to perform encryption/decryption in a group. The time is taken to perform encryption and decryption using AVN-SPE with optimized operations are calculated as:

$$T_{Enc-Opt} = T_{G(K)} + T_{E(AES-K)} + T_{E(CPABE(PT)-KEK)} + T_{Sig(E(KEK))}$$

$$T_{Dec-Opt} = T_{SV} + T_{D(CPABE(PT)-KEK)} + T_{D(AES-K)}.$$

$T_{E(Protocol-Key)}$ and $T_{D(Protocol-Key)}$ represent the time to perform encryption and decryption operations, respectively. Figures 12.6(a) and 12.6(b) show the amount of time is taken to perform encryption/decryption by traditional and optimized CP-ABE schemes with only one policy tree comprising of 5 attributes. The messages are first encrypted by performing 256-bit Advanced Encryption Standard (AES) encryption. Then the key is encrypted using CP-ABE encryption algorithm. The decryption process is in the reverse order. It can be observed that the time taken by a vehicle to perform encryption and decryption using optimized CP-ABE scheme is significantly less than using CP-ABE to encrypt the entire message. Figures 12.6(a) and 12.6(b) show that the measurement for a maximum allowable packet size for 802.11 standard, i.e., 2312 bytes for inter-vehicle communications. Even for a packet size of 2313 bytes, the time taken to encrypt the message with optimized scheme is 0.081 sec, which is much less than 0.191 sec taken by the traditional CP-ABE scheme. Similarly, the decryption time for optimized is 0.052 sec whereas for the traditional scheme is 0.097 sec.

12.5.3 COMMUNICATION OVERHEAD

The measurements for subgroup communication in VANETs are shown in Figures 12.7(a) and 12.7(b). In the performance evaluation, we considered two policy trees PT_1, PT_2 as shown in Figure 12.4. The encryption was first performed with PT_2 (5 attributes) followed by PT_1 (i.e., 3 attributes). We evaluate the optimized procedures. In addition, a symmetric KEK is first encrypted with PT_2 and then PT_1. The time taken to encrypt (0.322 seconds for non-optimized solution and 0.12 seconds for optimized solutions for a data with 2312 bytes) a message using two policy trees in both non-optimized and optimized CP-ABE scheme is higher than the single tree structure; however, it is still less than the time required to generate new keys for all vehicles in a subgroup. We can observe the same properties for the evaluations of decryption process in Figure 12.7(b).

12.5.4 SECURITY ANALYSIS

In this section, we discuss the security vulnerabilities and the countermeasures of AVN-SPE to against attacks. As the security of group and subgroup communication is based on the private key components generated by the RSU, a compromised RSU can completely disrupt AVN-SPE operations. Although we assume that RSU cannot be compromised, however an attacker can deploy an adversarial RSU. The certificate based mutual authentication procedure performed when a vehicle enters the RSU's

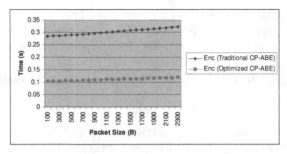

(a) Encryption Time comparison between Traditional and Optimized CP-ABE for Subgroup Communication.

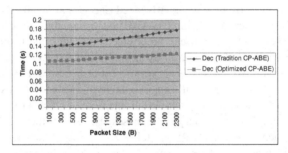

(b) Decryption Time comparison between Traditional and Optimized CP-ABE for Subgroup Communication.

Figure 12.7: Computation overhead.

coverage range hinders the attackers from deploying this attack. For an adversary impersonating other vehicles, the adversary cannot generate valid signatures for the ciphertext transmitted to other vehicles. This is because the attacker does not have the genuine vehicle's private key, and thus the attacker will fail to impersonate another valid user.

We must note that AVN-SPE does not prevent attackers from encrypting a message using a set of attributes. This is because both attributes and encrypting parameters are publicly known. Thus, signatures of transmitted messages are required when authentication is demanded. To reduce the number of times on computing signatures, using the optimized solution, we can just attach a signature for the first transmitted message. The following traffic can be authenticated using the KEK encrypted in the first message.

AVN-SPE may suffer from denial of service (DoS) attacks, i.e., a sender can send messages encrypted with a number of attributes to overburden the receiver. Although DoS attacks is difficult to prevent as it is difficult to prevent any vehicle from sending messages using AVN-SPE, the attackers can be revoked. As vehicles have to sign the ciphertext before transmitting, other vehicles can report this mischievous activity to RSUs. RSUs then can generate new dynamic attributes to revoke the misbehaved vehicles. The detailed approaches for revoking misbehaved vehicles is out of scope of this chapter.

12.6   SUMMARY

In this chapter, we presented a novel attribute-based solution for policy enforcement on VANET data access control. The solution provides a general framework for defining policies enforced by means of policy trees, which can be modulated to achieve secure V2V, V2I, or I2V communications in VANETs. We also provided performance analysis and measurements to show that AVN-SPE is practical for VANET communications. However, other security and policies related issues like anonymity in VANETs, efficient authentication based on AVN-SPE, collusion resistance under strong security requirements, misbehavior detection and revocation, etc., need to be further explored.

13 Using ABE to Secure Blockchain Transaction Data

Blockchain technology is increasingly being adopted as a trusted platform to support business functions including trusted and verifiable transactions, tracking, and validation. However, most business use-cases require privacy and confidentiality for data and transactions. As a result, businesses are forced to choose private blockchain solutions and unable to take full advantage of the capabilities, benefits and infrastructure of public blockchain systems. To address this issue, we present an Attribute-Based Encryption (ABE) security solution built on a private-over-public (PoP) blockchain approach. The policy based distributed operation of ABE conforms well to the blockchain concept. The cross-chain PoP approach provides the benefits from both public blockchains and private blockchains. Businesses will be able to restrict access, maintain privacy, and improve performance, while still being able to leverage the distributed trust of public blockchains. This solution presents the ABE-based security framework and protocol for securing data, transactions as well as smart contracts. Security analysis and performance evaluation show the presented solution to be effective, efficient and practical. It can greatly reduce the cost and complexity for businesses compared to running isolated private blockchain solutions.

13.1 INTRODUCTION

Public and private blockchains have many similarities [122]: (a) both are decentralized peer-to-peer networks, where each participant maintains a replica of a shared append-only ledger of digitally signed transactions; (b) both maintain the replicas in sync through a protocol referred to as consensus; and (c) both provide certain guarantees on the immutability of the ledger, even when some participants are faulty or malicious. The main distinction between public and private blockchain is related to who is allowed to participate in the network, execute the consensus protocol and maintain the shared ledger. A public blockchain network is completely open, and anyone can join and participate in the network. One of the drawbacks of a public blockchain is the substantial amount of computational power to maintain a distributed ledger at a large scale to achieve consensus, in which each node in a network must solve a complex, resource-intensive cryptographic problem called a Proof of Work (PoW) [162] to ensure all are in sync.

Many people believe private blockchains could provide solutions to many financial enterprise problems that public blockchain solutions do not, such as abiding by regulations such as the Health Insurance Portability and Accountability Act (HIPAA)

[100], anti-money laundering (AML) [191] and know-your-customer (KYC) laws [27], etc. Another disadvantage is the openness of public blockchain, which implies little to no privacy protection for transactions and only supports a weak notion of security. Both of these are important considerations for enterprise use cases of blockchain.

From business standpoint of views, private blockchains provide many salient features for business purposes [167]. For examples, they provide interesting opportunities for businesses to leverage their trustless and transparent foundation for internal and business-to-business use cases. With the advent of smart contracts, this technology could eventually replace many centralized businesses. Moreover, private blockchain is usually much faster, cheaper and respects the company's privacy. Private blockchains also provide more control power over the participants in the blockchain. For examples, banks and financial institutions have to worry heavily about regulations, and usually they cannot use the public blockchains due to their open and permission-free nature allowing anyone to participate, in which using public blockchains is contradictory to the regulations, to which they must abide.

The Hyperledger project [45] from the Linux Foundation, R3CEV's Corda [208], and the Gem Health network [176] are several of the different private blockchain projects under development. They solutions, while purposefully designed for enterprise applications, lose out on many of the valuable attributes of the public blockchains that are permissionless systems, simply because they are not widely applicable, but are instead built to accomplish specific tasks and functions.

Cross-chain functionality aims to combine the best features of different blockchain systems [146], both private and public, for the purposes of exchanging value across disconnected ecosystems. Ripple [44] has made notable strides to this effect, with Inter-ledger already testing transactions across multiple ledgers simultaneously in different currencies. ZCash [105] provides privacy protection for Bitcoin [162] users. Hawk [132] and Ekiden [54] have been proposed using off-chain approaches to provide data privacy protection. However, none of existing solutions clearly addressed the problem of applying access control policies to enforce data privacy protection on transaction secrets. For example, when using smart contract solutions, e.g. Ethereum [1], for procurement in supply-chain, transaction parameters such as product name, quantity, price, purchasing terms, shipping options, address, etc. could all be sensitive business secrets. They should be only viewable for relevant stakeholders. Hyperledger [45] addresses this problem by relying on a TA approach to build permission groups for data access control. However, data access has to be predefined. It is not suitable for complex and dynamic businesses logic that require dynamic access control. Moreover, traditional infrastructure-based data access control model, e.g., Role-Based Access Control (RBAC) [188], is incompatible with the distributed nature of blockchain operations where transaction data are mobile and shared by multiple blockchain participants.

To address the data access control and privacy protection issues in public blockchain, this chapter presents a distributed Attribute-Based Encryption (ABE) solution applied to private block-chains over public blockchains (PoP blockchain, PoP block, or PoP for short) approach. The PoP architecture is presented in

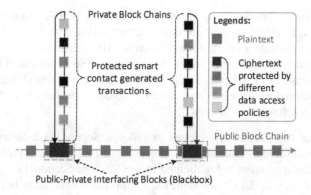

Figure 13.1: Illustration of PoP blockchain architecture.

Figure 13.1, where we use ABE-protected state channel and attach multiple private blockchains on a public blockchain. The PoP approach for deploying the ABE solution provides the benefits of both worlds.

Applying ABE on an off-chain basis means it can inter-operate with the public blockchain without interference. Private blockchains transactions can be much less computationally intensive and provide superior performance [129] since they do not have to be verified by all participants. Businesses are able to choose the private blockchain solution that best suits their needs independently from the public blockchain. Each private blockchain can be viewed as a protected state channel. The integrity of a private blockchain can be validated and checked in ciphertext and in aggregate by all public blockchain participants.

The public blockchain infrastructure is leveraged to provide validation and immutability for the entirety of the private blockchain state channel. This can take the form of the final private blockchain transaction result, or a hash of the entire private blockchain. Therefore, distributed trust on the public chain is not necessary for the private blockchain. At the same time, ABE provides data privacy for the private blockchain state channel. Only participants with the appropriate permissions and corresponding ABE attribute private keys can view and validate their relevant blocks in the private block chain. It provides the benefits of private block chains in terms of privacy without requiring the deployment of trusted nodes or multiple verification nodes. It essentially minimizes the entry cost businesses in adopting blockchain solutions.

In summary, the presented PoP solution has the following main features:

- It is a decentralized trust model for key management of ABE-based data access control. Using this approach, it can incorporate access control policies into ciphertext to protect content of smart contracts.
- It is a privacy-preserving messaging protocol to allow private blockchain participants to interact with the smart contract that can generate a private blockchain. This chapter illustrates how to use this protocol based on a supply-chain procurement application.

- The solution provides two smart contracts: PPP (Public Parameters and Policies) to establish attribute based security trust model and ppSCM to provide secure data access control based on ABE scheme.
- A comprehensive security and performance analysis is presented based on the presented PPP scheme. The presented solution is practical that can significantly reduce the effort and cost to establish dedicated and isolated private blockchains.

The rest of this chapter is arranged as follows: Section 13.2 describes system and models that serve as foundation for this presented solution, and an supply-chain based blockchain solution is highlighted in this section; Section 13.3 presents the details of the presented PoP solution; the performance evaluation is presented in Section 13.4; finally, we summarize the presented solution in Section 13.5.

13.2 SYSTEM AND MODELS

In this section, we present several system models that construct the PoP solution. First, we present an example application scenario using IoT-based blockchain for a supply-chain procurement procedure; then, the background of smart contract is presented; how to use attribute-based data privacy protection is described; and the system security model is presented at the end.

13.2.1 BLOCKCHAIN TECHNOLOGY FOR SUPPLY CHAIN

To illustrated the presented solution, in Figure 13.2, we use a supply chain example based on Block-Chain Technology (BCT), which involves multiple parties, i.e., suppliers, buyers, carriers, IoT companies, and banks. In the figure, the middle box maintains the constructed blockchains.

The potential of having all the information written in a blockchain allows the creation of an authoritative record that can be used to automatically establish smart contracts. Without such an authoritative record, smart contracts written on a Blockchain could hardly be executed, because parties need to agree on data and information that, like smart contracts themselves, are agreed to by a whole network through a consensus mechanism. The one-layer blockchain solution sees as such a fully integrated and automated trade network where documents and goods are transparently identified and tracked along the supply chain. Because the information is registered on a distributed database, it makes it tamper-resistant and fosters greater trust in the trade network. The left side of the figure present a BCT-supported purchase related transaction by using Ethereum's Decentralized App (DApp) [29] solution involves 4 main procedures based on supply-chain operation procedures:

- *Order Processing*: The order-processing workflow starts with a PO from the buyer. Within the blockchain, once created, the PO is time-stamped and can become a valid document whose clauses can be executed only if valid, due to the programming features of smart contracts. Assuming delivery documents can also be registered on it, the metadata of the invoice, PO and

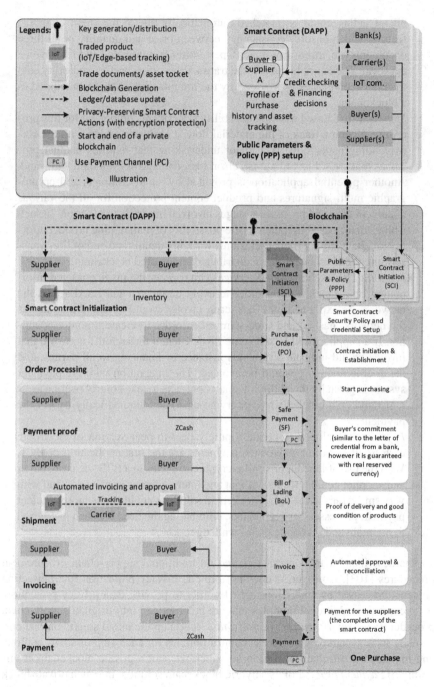

Figure 13.2: A supply chain scenario using IoT devices, blockchain, and data encryption protections.

bill of lading could be matched automatically due to the smart contracts feature, which ensures consistency between price and quantity in all three documents (i.e. three-way-match), permitting an automated and fast invoice approval. The entire history of the transactions offers perfect audibility, and trust between parties is provided by the immutability of the data entered in a blockchain.

- *Shipment*: IoT-based tracking capability is a critical component for this procedure. Keeping track of the material flow at each step, along with the corresponding paper flow, is a major undertaking that requires manual processes that are subject to human error, loss, damage or even theft and fraud. Another potential application is provided by smart contracts and cryptographic multi-signatures and product content protection for all the various documentation and processing stages involved in a trade transaction. In such a blockchain-based IoT, there is the possibility of maintaining product information, its history, product revisions, warranty details and end of life, transforming the blockchain into a distributed and trusted blockchain.

- *Invoicing*: Blockchain-based services can register the invoice-related information on a blockchain in order to avoid duplicates and fraud across the network. As explained by [103], each invoice would be distributed across the network, hashed and time-stamped in order to create a unique identifier. If a supplier tried to sell same invoice again through the network, that invoice would indicate a previous instance of financing to all parties, and the double financing would be avoided. The integration with the payment system is given by the ability of smart contracts to take control over an asset registered on a blockchain (e.g. crypto-cash) and automatically trigger the payment.

- *Payment*: Developed to create a purely peer-to-peer version of electronic cash to allow online payments, payments are the first application of BCT. With the use of Bitcoin or similar cryptocurrencies in a B2B scenario, buyer and supplier could transact without any intermediaries (e.g. banks) and with very small transaction fees. Blockchain solutions could create more efficient payment processes between banks, eliminating the need for each institution to maintain and reconcile their own ledger.

The described smart contract is based on traditional supply-chain procurement procedures [103]. However, it does not provide privacy protection for transaction contents processed by smart contracts. In order to provide data privacy protection, we present two additional modules that are incorporated into original supply-chain procedures: (a) *Smart contract initialization*: it sets up the initial smart contract credentials such as agreed data access control policies for each step of smart contract and initiates the off-chain operation, in which we start a private blockchain at this point. (b) *Payment proof*: PoP is a hybrid blockchain solution, in which private block chains are interfaced into public chains. Moreover, the private chain can also incorporate public blockchain evidence into the private blockchain. The addition of the payment proof procedure is to utilize the payment channel [174] feature of public blockchains to prove the buyer has sufficient money to pay for the purchased product. The buyer first pays for the product to a Escrow account, and once the product is landed, the cashed money will be delivered to the supplier to close the blockchain-based purchase.

In the presented model, both IoT companies and Banks play a crucial role. Relying on IoT companies' tracking capability, Banks can use the traceability nature of blockchains. By gaining access control to protected business transaction data, Banks can monitor the healthiness of business entities for credit evaluation loan decision making. Due to the page limits, this research focuses on the data privacy protection and skips the details of Banks and IoT companies involved smart contract and blockchain related activities.

13.2.2 SMART CONTRACT

In biticoin, the concept of "scripting" has already existed, which is actually a weak version of smart contract. With a script, one can set a condition on when a transaction can spend the "Unspent Transaction Outputs" (UTXO) However, the script in Biticoin suffers from the following issues: first, it lacks Turing-completeness, thus does not nearly support everything; second, it is value-blinded; third, it lacks state, UTXO can either be spent or not, there is no way to keep other states except for these two; fourth, it is blockchain blinded.

Ethereum smart contract is to build a decentralized application to create a blockchain with a build-in Turing complete programming language. Therefore, smart contract means is defined to be a cryptographic "boxes" that contain value and only unlock it if certain conditions are met. The same as a transaction, a smart contract will also be stored in the blockchain and can be retrieved by its address and integrity can be guaranteed as well. To trigger a smart contract is just like a remote processor call. The input would be included in the transaction. That is, smart contract creation, smart contract function call and smart contract destroy are all included in a transaction. With smart contract, one can express logics such as "only after April 17th, 2018, can the document be sent to A". The smart contract is running in an Ethereum Virtual Machine (EVM), and the smart contract involved user interactions and data processing modules are usually running by the DApp.

In the presented supply-chain example shown in Figure 13.2, there are two smart contracts are involved: (1) *public blockchain smart contract*: the smart contract on the right side box includes multiple stake holders providing supply-chain services to settle down a PPP (Public Parameters and Policies). A PPP describes what encryption public parameters will be used for data privacy protection, who may serve as a trusted party for data access control management for running private blockchains, and what security policies to be enforced in the private blockchain. We can treat a PPP as a "template" that can be reused to build a private blockchain. Thus, multiple PPPs can be generated for different use cases of private blockchains. (2) *Private blockchain smart contract*: the second smart contract on the left side box in Figure 13.2 represents a one purchase between a supplier and a buyer. In addition, an IoT company can be involved to provide product tracking and inventory.

13.2.3 ABE-ENABLED ABAC

In the literature, a large number of Attribute-Based Encryption (ABE) solutions have been proposed. ABE is a way to implement attribute-based access control, in which, data will be encrypted and a data owner could define an access policy describing what attributes the data users need to own in order to get access of the data. Only users who own satisfiable set of attributes and gets the corresponding private key from a trusted third party, namely TA could decrypt the data. Among existing ABE schemes, decentralized ABE provides some attractive features, especially that any entity could play the role of a trusted authority or attribute authority to distribute attributes and generate private keys for users. In this way, all the organizations can work in a federated way naturally even with no need of knowing the existence of each other.

In this chapter, we present an extended Lewko's scheme [137] by adding distributed trust management to allow multiple parties to collaboratively establish the trust and distribute secret keys. The following described federated Authority Setup and Federated KeyGen protocols are newly proposed. In the following sections, Lewko's scheme is briefly described. Interested reader can refer to Lewko's [137] for its security proofs. The presented approach has the following salient features compared to existing blockchain data privacy protection solutions:

- It is distributed and mobile, i.e., every participant in the system can serve as a trust authority to issue attributes and corresponding private keys for private blockchain participants; the access control policy is associated with ciphertext, which can be freely shared among blockchain stakeholders without needing an access control infrastructure for data management.
- It is federated, i.e., attributes can be shared among private blockchain participants. This also means that the scheme allows a coalition to be established for a private blockchain for attributes and corresponding private keys generation. The coalition can prevent single point failure issue as well resisting to $n - 1$ collusion problem, where n is the size of the coalition.
- It provides interoperability feature, i.e., attributes and corresponding private keys generated from different trust authorities can be used together to form a data access control policy. For example, Alice can use her own generate private key for attribute A_1 and another attribute A_2, which is generated by Bob to decrypt a data protected by data access policy enforced by the policy $\{A_1 \ AND \ A_2\}$.

A typical security policy should include multiple descriptive terms (i.e., attributes) such as:

$$\mathscr{P}1 = \text{The } \underline{pricing} \text{ and } \underline{quantity} \text{ can be accessed by}$$

$$\text{the } \underline{supplier} \text{ and the } \underline{buyer}.$$

In this policy, 'pricing' and 'quantity' are accessing objects, and 'supplier' and 'buyer' are attributes describing accessing subjects. These attributes can be used as public encryption keys. In each block created by the private blockchain, data are encrypted by using one or multiple data access policies.

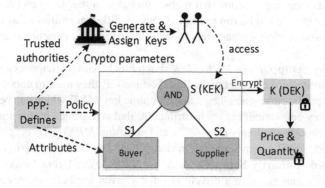

Figure 13.3: Attribute (or policy)-based access control setup.

The policy $P1$ is presented as a tree structure in Figure 13.3, which is called Policy Tree (PT). A PT is constructed by attributes at the leaves and intermediate nodes are logical gates. Using secret sharing scheme, a tree-root level secret s can be used as a Key-Encrypting-Key (KEK) to protect at symmetric key K as the Data Encrypting-Key (DEK) to protect data such as the values of price and quantity. The smart contract generated PPP defines attributes and policies for a particular application, and trusted parties, and crypto parameters used for key generation and encryption in the private blockchain.

Scheme Construction

The data privacy protection is based on attribute-based encryption scheme presented by Lewko at. al [137]. Using Lewko's scheme for PoP, each blockchain participant can generate private keys for attributes. As a result, each participant is a trusted authority for key generation. In rest of presentation, we use the term authority and blockchain participants interchangeably. However, Lewko's scheme requires users to derive their private key from one trusted authority to allow them to use the same attribute and corresponding private key to decrypt a ciphertext. The trust authority is fully trusted since he/she will know all the generated private keys. We extend Lewko's solution by adding multi-authority key generation scheme, called *federated authority setup* and *federated key generation*, for an attribute and corresponding private key is generated by multiple authorities. As a result, only if all involved trusted authorities got compromised, e.g., using collusion attack, can they derive the private key for a user. The extended Lewko scheme is presented as follows: Considering the heavy computation overhead, part of the encryption and decryption computation is outsourced to the edge nodes. The following is the outsourced version of the scheme.

Global Parameters Setup$(\lambda) \rightarrow GP$: The Global Parameters (GP) can be established in advance by a well-known organization, e.g., in the supply-chain industry. Since the GP is publicly known, it is not critical for which party to generate the

GP. The organization selects a composite *Bilinear* group G or order $N = p_1p_2p_3$. $GP = \{N, g_1, H : \{0,1\}^* \to G\}$, where g_1 is a generator of group G_{p_1} and the hash function H is mapping function that maps a global identifier to an element of group G. This algorithm might be run multiple times by different entities so as to generate multiple candidate global parameters in the candidate public parameters and policies (PPP).   □

Authority Setup$(GP) \to MPK, MSK$: Each blockchain participant can serve as a trusted authority for key generation. Based on the GP, they need to choose and publish a set of public parameters, i.e., a master public key MPK, which can be later used for private key generation. For each attribute A_i that is managed by the authority, the authority j chooses randomly $\alpha_i, y_i \in \mathcal{Z}_N$ and publishes $MPK_j = \{e(g_1, g_1)^{\alpha_i}, g_1^{y_i}, \forall i\}$ as the public key. The corresponding master private key is $MSK_j = \{\alpha_i, y_i \forall i\}$.   □

Federated Authority Setup$(GP, AAS) \to M\hat{P}K, M\hat{S}K$: Using Lewko's scheme, each authority can generate a private key for a given attribute. However, for each attribute, it requires every user to derive their private keys from the same authority. Thus, the authority must be fully trusted. To relax this requirement, we need to involve multiple authorities for private key generation to prevent single point failure issue. If there are n federated authorities, then this scheme is resistant to $n - 1$ authority collusion problem. The federated authority setup algorithm is run when multiple attribute authorities need to generate the public key and private key for their shared attribute(s). For simplicity, we assume there are n attribute authorities in the set AAS, i.e., $AAS = \{AA_1, AA_2, \cdots, AA_n\}$. AA_i will generate $\alpha_i, y_i \in \mathcal{Z}_N$. Each AA_i will generate an individual master private key $MSK_i = \{\alpha_i, y_i \forall i\}$. AA_{i-1} will send the individual master public key to AA_i as:

$$MPK_{i-1 \to i} = \{e(g_1, g_1)^{\sum_{j=2}^{i} \alpha_{j-1}}, g_1^{\sum_{j=2}^{i} y_{j-1}}\},$$

AA_i will calculate

$$MPK_{i \to i+1} = (e(g_1, g_1)^{\sum_{j=2}^{i} \alpha_{j-1}})^{\alpha_i}, (g_1^{\sum_{j=2}^{i} y_{j-1}})^{y_i}.$$

The final federated master public key and private key for an attribute is defined as follows:

$$M\hat{P}K = \{e(g_1, g_1)^{\sum_{j=1}^{n} \alpha_j}, g_1^{\sum_{j=1}^{n} y_j}\},$$

$$M\hat{S}K = \{\sum_{j=1}^{n} \alpha_j, \sum_{j=1}^{n} y_j\}.$$

          □

Encrypt$(M, (A, \rho), GP, \{MPK\}, M\hat{P}K) \to CT$: M is a message, A is an $n \times \ell$ access matrix, and ρ maps its rows to attributes. For each row in A, the algorithm chooses a random number $r_x \in \mathcal{Z}_N$. A random vector $w \in \mathcal{Z}_N^{\ell}$ with 0 being the first entry is chosen randomly. ω_x denotes $A_x \cdot \omega$. The data owner chooses $s \in \mathcal{Z}_N$ and a vector $v \in \mathscr{Z}_N^\ell$ randomly where s is its first entry. $\lambda_x = A_x \cdot v$ with A_x being the xth row of the matrix A. The ciphertext is as follows:

$$CT =< C_0, C_{1,x}, C_{2,x}, C_{3,x} >, \text{ where}$$

$$C_0 = Me(g_1,g_1)^s, C_{1,x} = e(g_1,g_1)^{\lambda_x} e(g_1,g_1)^{\alpha_{\rho(x)} r_x},$$

$$C_{2,x} = g_1^{r_x}, C_{3,x} = g_1^{y_{\rho(x)} r_x} g_1^{\omega_x}, \forall x.$$

CT is the ciphertext of DEK, which is used to encrypt data object o in the blockchain protocol. In this encryption protocol, the encryptor needs to identify which master public key parameters are used for each involved attribute. Later, a decryptor can use private keys generated from corresponding public keys.   □

KeyGen$(GID, i, \{MSK\}, GP) \to SK_{i,GID}$: In PoP, the GID can be an address that is used to identify the blockchain participant. For a global identifier GID with attribute i belonging to an authority, the authority generates the following private key

$$SK_{i,GID} = g_1^{\alpha_i} H(GID)^{y_i}.$$

Using the *KeyGen* scheme, an authority can generate private keys for other blockchain participants.   □

Federated KeyGen$(GID, i, \{MSK\}, \hat{MSK}, GP) \to SK_{i,GID}$: The federated key generation algorithm is run when multiple attribute authorities need to generate the private key for an attribute shared among multiple users. Assume that $AAS = \{AA_1, AA_2, \cdots, AA_n\}$. AA_i will generate $g_1^{\alpha_i} H(GID)^{y_i}$. AA_{i-1} will send AA_i the private key component:

$$SK_{i-1 \to i} = g_1^{\sum_{j=2}^{i} \alpha_{j-1}} H(GID)^{\sum_{j=2}^{i} y_{j-1}},$$

then, AA_i will calculate

$$SK_i = (g_1^{\sum_{j=2}^{i} \alpha_{j-1}})^{\alpha_i} (H(GID)^{\sum_{j=2}^{i} y_{j-1}})^{y_i}.$$

The final secret key of the shared attribute for the user GID is as follows.

$$SK_{i,GID} = g_1^{\sum_{j=1}^{n} \alpha_j} H(GID)^{\sum_{j=1}^{n} y_j}.$$

□

Decrypt$(CT, \{SK_{i,GID}\}, GP) \to M$: Assume that the ciphertext is encrypted under an access matrix (A, ρ). If the decrypt holds the private key $\{SK_{\rho(x),GID}\}$ for a subset of rows A_x of A satisfying that $(1, 0, \cdots, 0)$ is in the span of these rows, then the plaintext message M can be obtained in the following way:

$$C_{1,x} \cdot e(H(GID), C_{3,x})/e(SK_{\rho(x),GID}, C_{2,x})$$

$$= e(g_1,g_1)^{\lambda_x} e(H(GID), g_1)^{\omega_x}.$$

The decryptor chooses constants $c_x \in \mathcal{Z}_N$ so that $\sum_x c_x A_x = (1, 0, \cdots, 0)$ and computes

$$\prod_x (e(g_1, g_1)^{\lambda_x} e(H(GID), g_1)^{\omega_x})^{c_x} = e(g_1, g_1)^s.$$

To verify the transaction including encrypted data, the decryption algorithm will be called.   □

Offloaded Encryption and Decryption

Considering participant running DApp on mobile devices, ABE based computation can be potentially intensive for mobiles. Here, we presented an ABE offloading model that can significantly reduce the computation overhead on mobiles. In the study, we assume that an edge cloud node can assist mobiles to perform offloading functions. **Encrypt**$(M, (A, \rho), GP, \{MPK\}, M\hat{P}K) \rightarrow CT$: M is a message, A is an $n \times \ell$ access matrix and ρ maps its rows to attributes. For each row in A, the algorithm chooses a random number $r_x \in \mathcal{Z}_N$. A random vector $w \in \mathcal{Z}_N^\ell$ with 0 being the first entry is chosen randomly. ω_x denotes $A_x \cdot \omega$. The data owner chooses $s \in \mathcal{Z}_N$ and a vector $v \in \mathcal{Z}_N^\ell$ randomly where s is its first entry. $\lambda_x = A_x \cdot v$ with A_x being the xth row of the matrix A. The ciphertext is as follows where C_0, C_1 are calculated by the data owner, e.g., a mobile device, and C_2, C_3 are calculated by the edge node:

$$CT = <C_0, C_{1,x}, C_{2,x}, C_{3,x}>, where$$
$$C_0 = Me(g_1, g_1)^s, C_{1,x} = e(g_1, g_1)^{\lambda_x} e(g_1, g_1)^{\alpha_{\rho(x)} r_x},$$
$$C_{2,x} = g_1^{r_x}, C_{3,x} = g_1^{y_{\rho(x)} r_x} g_1^{\omega_x}, \forall x.$$

CT is the ciphertext of DEK, which is used to encrypt data object o in the blockchain protocol. In this encryption protocol, the encryptor needs to identify which master public key parameters are used for each involved attribute. Later, a decryptor can use private keys generated from corresponding public keys.   □

Decrypt$(CT, \{SK_{i,GID}\}, GP) \rightarrow M$: Assume that the ciphertext is encrypted under an access matrix (A, ρ). If the decrypt holds the private key $\{SK_{\rho(x), GID}\}$ for a subset of rows A_x of A satisfying that $(1, 0, \cdots, 0)$ is in the span of these rows, then the plaintext message M can be obtained in the following way. The edge node will calculate $C_{1,x} \cdot e(H(GID), C_{3,x})$ and the private key holder only needs to calculate $e(SK_{\rho(x), GID}, C_{2,x})$.

$$C_{1,x} \cdot e(H(GID), C_{3,x}) / e(SK_{\rho(x), GID}, C_{2,x})$$
$$= e(g_1, g_1)^{\lambda_x} e(H(GID), g_1)^{\omega_x}.$$

The decryptor chooses constants $c_x \in \mathcal{Z}_N$ so that $\sum_x c_x A_x = (1, 0, \cdots, 0)$ and computes

$$\prod_x (e(g_1, g_1)^{\lambda_x} e(H(GID), g_1)^{\omega_x})^{c_x} = e(g_1, g_1)^s.$$

To verify the transaction including encrypted data, the decryption algorithm will be called.   ☐

13.2.4 SECURITY MODEL

PoP assumes that blockchain participants are curious, selfish, and greedy, and they want to learn business secrets incorporated into blockchains. They may collude to share their secrets to gain additional data access capabilities that should not assigned to them. Moreover, they may drop off from blockchain creating procedure to take the goods without paying for it.

13.3 POP SYSTEM MODELS

13.3.1 ACCESS SYSTEM CONSTRUCTION

Definition 13.1 (Contract). A smart contract C is defined by a procedure $C = \{T\}$ that specifies a set of interdependent transactions among contract subjects (or participants) in group G.   ☐

Definition 13.2 (Transaction). A transaction defines a sequential atomic data actions $\{a\} \in A = \{read, write, change\}$, and each action a is restricted by a privilege $\alpha(a, T) : \mathscr{P}(a, T) \mapsto S_{a,T}$, where the privilege $\alpha(a, T)$ are defined by capabilities such as $\{can, cannot, restricted\ by/to\}$ of the action a in the transaction T. The privilege is described in the security policy $\mathscr{P}(a, T)$, and the privilege can be mapped/translated to $S_{a,T}$ denoting the subset of subjects in the overall participating group G.   ☐

Definition 13.3 (Subject). A subject $s \in S$ or G (here S is a subset of subjects and G is the overall group includes all the subjects) is an entity involved in smart contract that can perform actions, and each subject has a data access privilege described by a policy $\mathscr{P}(s) : \cup_{\forall(a,T) \rightarrowtail s} \mathscr{P}(a, T)$, in which $\cup_{\forall(a,T) \rightarrowtail s}$ denotes the union for all action and transaction pair (a, T) it involves ($\rightarrowtail$ is an "involve" operator) the subject s. It is the collection of policies for a smart contract involving the subject s. Here, we denote $\mathscr{P}(s) = \cup_{\forall(a,T) \rightarrowtail s} \mathscr{P}(s, a)$ and $\mathscr{P}(s, a)$ is the data privacy policy for action a. Correspondingly, $\mathscr{P}(S)$, $\mathscr{P}(S, a)$ are defined for S as a subset of subjects.   ☐

Definition 13.4 (Object). An object $o \in O$ represents a data (or a file, a piece of information) that subjects want to access to perform actions such as *read, write* and *change*. The access policy to an object o for a smart contract $P = \{T\}$ is represented as $\mathscr{P}(o) = \cup_{\forall(a,T) \rightarrowtail o} \mathscr{P}(o, a)$, which represents the collection of data access policies involved with all actions on an object o.   ☐

Definition 13.5 (Data Access Capability)**.** The following access policies are defined:

$T : s \rightarrow o|_{\{a\}:*}$:   in transaction T, subject s can operate action(s) $\{a\}$ limited by condition $*$ on object o under access policies $\mathscr{P}(o,a) \subseteq \mathscr{P}(s,a)$;

$T : S \rightarrow o|_{\{a\}:*}$:   in transaction T, subset of subject S can operate action(s) $\{a\}$ limited by condition $*$ on object o under access policies $\mathscr{P}(o,a) \subseteq \mathscr{P}(S,a)$;

where $*$ is a condition to confine action(s) such as in a transaction T or in a contract C.   □

13.3.2   PRIVACY-PRESERVING MESSAGING PROTOCOL (PPMP)

Goal: For a given smart contract transaction T, one or a set of subject(s) $S \subseteq G$ can perform an action a, where $\mathscr{P}(S,a) \neq \mathscr{P}(\bar{S},a)$, in which the collective privilege (e.g., by colluding) of set $\bar{S}$ cannot satisfy the privilege given by subset S. This means that the data access policy $\mathscr{P}(S,a)$ can only be satisfied by the subgroup of participants in S. Thus, the *PPMP* protocol describes the data access privilege that only a subset of participants S can perform an action $a \in \{read, write, change\}$ in a transaction T. □

    Messages: In order to implement the access control privilege associated to an action in a smart contract, we use cryptographic approaches. Let's define three cryptographic enforcement operations:

$$c = \{Encryption(E), Decryption(D), Signature(Sig)\},$$

which can be applied to an action a in the smart contract transaction to enforce a security policy $\mathscr{P}$ for one or multiple subject(s).

    For an or multiple action(s) $\{a\}$ in a transaction T, a *PPMP* message is defined as:

$$PPMP(T, \{a\}, \mathscr{P}, c, s, o) \quad = \quad T : s \rightarrow o|_{\{a\}:\mathscr{P},c}; \tag{13.1}$$

$$or$$

$$PPMP(T, \{a\}, \mathscr{P}, c, S, o) \quad = \quad T : S \rightarrow o|_{\{a\}:\mathscr{P},c}. \tag{13.2}$$

    Based on the above definition, the *PPMP* protocol is actually the implementation of data access capabilities defined in Definition 13.5 by via conditions enforced by cryptographic operations $c = \{E, D, Sig\}$.

13.3.3   SMART CONTRACT PROTOCOLS

In this section, we present two smart contract protocols presented in the supply-chain example of Figure 13.2, namely, PPP (Public Parameters and Policies) contract and ppSCM (privacy-preserving Scheme) contract.

PPP Contract

Shown in Contract 13.1, *PPP* is a smart contract created by an initiator on the public blockchain. The initiator could be a TA who wants to negotiate attributes, global public parameters and policies with all other participants for using attribute-based encryption scheme and policy-based access control in a private blockchain. An attribute authority needs to call function *join* in the *PPP* first to join the negotiation and insert their attributes into the smart contract. The smart contract will form policies to be negotiated based on the inputs from all joined parties. The negotiation is a voting process on the collected policies and allow each joined party to vote at most once to select their preferred policy.

Contract 13.1 Pseudo-code of PPP Contract.

```
 1: struct AA = {...}                                ▷ Define Attribute Authority
 2: AAS = {AA₁,...,AAₙ}                              ▷ Address of each joined AA
 3: struct P = {...,count}                            ▷ Define Policy and their count
 4: PS = {P₁,...,Pₙ}                                 ▷ A set of policy to be negotiated
 5: function PPP(P p)
 6:     PS.push(new P(p))                            ▷ Initiator create a new policy

 7: function JOIN(Attribute attr)
 8:                                     ▷ Initiator grants a new AA to join the negotiation
 9:     require(isAllow(msg.sender)
10:     require(msg.sender ∉ AAS)
11:     AAS.push(new AA(msg.sender, attr))
12:     if (AA.attributes ∉ PS) then
13:         add AA.attributes to PS
14: function VOTE(P p)
15:     require((msg.sender ∈ AAS)
16:     require(not isVoted(addr'AA))
17:     PS.find(p).count⁺⁺
18: function GETPOLICIES( )
19:     require(msg.sender ∈ AAS)
20:     Policies = ∅
21:     for each p ∈ PS do
22:         Policies.push(p)
23:     return Policies
24: function GETPPP( )
25:     require(msg.sender ∈ AAS)
26:     return PS.findMaxVote()
```

Privacy-preserving scheme (ppSCM) contract

ppSCM contract is shown in Contract 13.2, which is a smart contract created by a supplier on a dedicated private blockchain. It uses the negotiated policy from *PPP* on public blockchain to create a contract for transactions involving supplier, buyer and carrier on the new created permission-based private blockchain. The permission to access the private blockchain is controlled by the access control policy.

   ppSCM will maintain purchase orders and invoices for the same buyer and supplier on a dedicated private blockchain. The data in purchase orders and invoices are

Contract 13.2 Pseudo-code of ppSCM Contract.

```
 1: address addr'Seller, addr'Buyer
 2: struct PO = {goods, quantity, number, price, safepay, shipment}    ▷ Purchase Order Data
 3: struct Shipment = {courier, price, safepay, payer, date}
 4: struct Invoice = {orderno, invoiceno}
 5: AccessPolicy = ∅
 6: Orders = ∅, Invoices = ∅
 7: orderseq = 0, invoiceseq = 0
 8: function PPSCM(address buyerAddr, Policy AccP)
 9:                                                      ▷ Initialize contract by a seller
10:     addr'Seller = sender
11:     addr'Buyder = buyerAddr
12:     AccessPolicy = AccP
13: function SENDORDER(string good, unit quantity)
14:                                                      ▷ Buyer send a PO to the seller
15:     require(msg.sender == addr'Buyer)
16:     Orders.push(new PO(good, quantity, orderseq++))
17:     OrderSent()
18: function SENDPRICE(PO po, unit priceP, unit priceS)
19:                                      ▷ Seller send prices (order and shipment) to buyer
20:     require(msg.sender == addr'Seller)
21:     require(isValid(po))
22:     po.price = priceP
23:     po.shipment.price = priceS
24:     PriceSent()
25: function SENDSAFEPAY(PO po)
26:                                                      ▷ Buyer send safe payment to seller
27:     require(msg.sender == addr'Buyer)
28:     require(isValid(po))
29:     SafepaySent()
30: function SENDINVOICE(PO po, unit date, address courier)
31:                                      ▷ Seller send invoice to Buyer and trigger the shipment
32:     require(msg.sender == addr'Seller)
33:     require(isValid(po))
34:     Invoices.push(new Invoice(po.number, invoiceseq++))
35:     po.shipment.date = date
36:     po.shipment.courier = courier
37:     InvoiceSent()
38: function DELIVERY(unit invoiceno, unit timestamp)
39:                                      ▷ Courier delivers the goods and trigger the payment
40:     require(isValid(Invoices[invoiceno]))
41:     po = PO[Invoices[invoiceno].orderno]
42:     require(po.shipment.courier == msg.sender)
43:     OrderDelivered()
```

protected by the selected access policy from *PPP*. Only the parties with appropriated attributes can query or update the data via transactions.

When a buyer would like to purchase products from a supplier, the supplier deploys *ppSCM* smart contract exclusively for the buyer's account. The buyer then put the purchase order on the supplier's *ppSCM* with product name and quantity by calling *sendOrder* function. Through an event, so-called *OrderSend*, the supplier could receive the order data and process it.

After received the purchase order, the supplier looks for the best shipping price on the carrier's smart contract. He then sends the order price and shipment price to the buyer by calling *sendPrice* and the buyer receives this through the event called *PriceSent*.

The buyer performs the safe payment of the grand total (order price + shipment price) through the smart contract in the public blockchain by the *SafepaySent()* event in the *ppSCM*. Theses coins goes to the smart contract account and waits there until the delivery.

After safe payment, the supplier sends the invoice with delivery date and some other data to the buyer by calling *sentInovice*. The buyer receives the invoice data through the event called *InvoiceSent*.

The carrier, after delivery the order to the buyer, marks the order as delivered on the *ppSCM* smart contract by calling *delivery* function. The event *OrderDelivery* then calls a smart contract in the public blockchain to payout the supplier for the order, and payout the carrier for the shipment.

13.3.4 POLICY-BASED DATA PRIVACY PROTECTION

PPP: Public Parameters and Policies

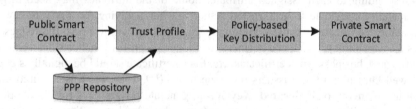

Figure 13.4: Trust and policy management procedure.

As shown the PoP example in Figure 13.2, we can abstract the In the PoP's trust and policy management procedure in Figure 13.4. For each private blockchain construction, we need to set up or choose a trust profile, i.e., either derived from a public smart contract to generate a PPP or reuse a previously establish PPP. A PPP is built based on the following data structure:

- D1: A set of Global Parameters (GP, see the *global parameter setup*) provides the global parameters that all ABE users need to use for key generation, encryption, and decryption.
- D2: An array of identities of authorities ($\{GID\}$) and their associate public parameters $\{MPK\}$ and/or federated public parameters $M\hat{P}K$, and

each parameter associated attributes $\{MPK\}, M\hat{P}K \rightarrow \{A\}$. The mapping between public parameters and attributes allow each private blockchain participant to select which public parameters to use in the ABE **Encrypt** procedure.

- D3: A set of policy examples that can be used for each of smart contract transaction during the private blockchain construction.

PPP is built using smart contracts over public blockchain (see Contract 13.1), and thus they are searchable. A public directory service can be used to store established PPP. For convenience, each PPP can be reused as a template for a new private blockchain construction. The *GID* is a public blockchain address, which is usually generated from a self-created public key. In D2, exposing *GID* will not reveal the real blockchain participant's identity. In many real business scenarios, suppliers may prefer to expose their real identity for easier key management procedure after the PPP establishment. In addition, when creating a new private blockchain, the participant can initiate an ***update smart contract*** to update the authority list and associated attributes, which can be implemented using the PPP creation smart contract (see Contract 13.1).

Private Key Distribution

Once a PPP is determined and trusted authorities are known, a private key distribution procedure is conducted as an off-chain procedure. The key distribution can be initiated by either a private blockchain participant or a trust authority. Using existing public key exchange protocols can allow the participants to derive private keys corresponding to each assigned attribute. Some of the attributes may need to get a capability certificate from a trusted authority when applying for a private key. A capability certificate is usually a digitally signed document to prove the requester has the capability to conduct a business function, e.g., professional certificates, bank certificates, business type certificates, etc. Each certificate should be digitally signed by well-known certificate issuers on requestor's *GID*. Then, the trusted authorities can use **KenGen** or **Federated KeyGen** to generate private keys for distribution. The key distribution is an off-chain procedure and can be done offline. Any existing public key or shared key-based key distribution schemes can be used, and the details are omitted in this chapter.

Data Object Encryption and Decryption

The data access protocol is specified in PPMP protocol. The data object operation diagram is presented in Figure 13.5. Both DApp (a web-based app) and smart contract (running within an Ethereum Virtual Machine (EVM)) run locally on a user's site, and the ABE encryption/decryption engine is also interfaced to the DApp locally. The encryption and decryption are performed between the DApp and the blockchain, and encryption/decryption engine.

In this solution, we consider the data object granularity is determined by access control policies. Using the same data access control policy provided in Section 13.2.3: $\mathcal{P}_1$=*The pricing and quantity can be accessed by the supplier and the buyer.* $\mathcal{P}_1$ is

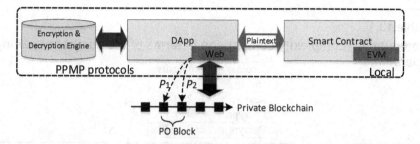

Figure 13.5: Data object encryption and decryption.

an example to specify the data protection when creating the PO. The PPMP message will be used by DApp, which runs an ABE encryption and decryption engine to create the PO blockchain block. Thus (13.2) can be written as:

$$PPMP(T, \{a\}, \mathscr{P}, c, s, o)$$
$$= T_{PO} : \{buyer, supplier\} \rightarrow price\&quantity|_{create:\mathscr{P}_1, E}.$$

PO should also contain an address information for shipment. If the data access policy is $\mathscr{P}_2$=*The shipment address information can be accessed by the buyer and the carrier.*, then the PPMP message can be:

$$PPMP(T, \{a\}, \mathscr{P}, c, s, o)$$
$$= T_{PO} : \{buyer, carrier\} \rightarrow shipping\ address|_{create:\mathscr{P}_2, E}.$$

The PO example presents two data access control policies $\mathscr{P}_1$ and $\mathscr{P}_2$ are involved, and thus on the blockchain, there should be at least two transactions corresponding to $\mathscr{P}_1$ and $\mathscr{P}_2$, respectively. The granularity of encrypted block on the blockchain is determined by using the same data access control policy without needing to create two different DEKs. Technically, we can combine $\mathscr{P}_1$ and $\mathscr{P}_2$ as one policy, however, there is no way to use one DEK to protect the data content to fulfill both of them. Other crypto actions such as decryption can be similarly created based on the PO example. We note that the crypto currency involved functions can be achieved by using public blockchain's payment channel approach [174]. Due to page limits, we do not provide details in this chapter.

13.4 PoP OPERATION AND PERFORMANCE ANALYSIS

13.4.1 COMPLEXITY ANALYSIS

Table 13.1 presents the computation overhead for the scheme measured by the number of paring operations. Table 13.2 summarizes the complexity of the algorithms of the presented scheme. Here, only computation performed on each individual attribute authority will be counted. Therefore, the complexity of (setup, federated setup) and (key generation, federated key generation) is the same. Thus, we only show Setup and KeyGen to represent these two.

Table 13.1

Computation complexity comparison in terms of the number of pairing operations

Schemes	Complexity
Setup	$2|U_i|$
KeyGen	$2|S|$
Encrypt	$5n\mathbf{E} + 1$
Decrypt	$2n\mathbf{P} + n\mathbf{E}$

$\mathbf{E}$ and $\mathbf{P}$ represents exponentiation and pairing respectively. U_i indicates the set of attributes managed by a certain attribute authority. S represents the set of attributes assigned to a user. n is the number of rows of the linear secret sharing matrix used in the encryption and decryption algorithm.

Complexity analysis of offloading overhead

Table 13.2

Computation complexity comparison in terms of the number of pairing operations

Schemes	DABE	device	edge node
Setup	$2|U_i|$	$2|U_i|$	0
KeyGen	$2|S|$	$2|S|$	0
Encrypt	$5n\mathbf{E} + 1$	$2n\mathbf{E}+1$	$3n\mathbf{E}$
Decrypt	$2n\mathbf{P} + n\mathbf{E}$	$n\mathbf{P} + n\mathbf{E}$	$n\mathbf{P}$

Table 13.2 compares the complexity of the original algorithms and the scheme with Offloaded encryption and decryption. From this summary, we can see that the during encryption, the edge node is able to do more than half of the computational work. As for the decryption, the edge node will share half number of the pairing operations. Since the main complexity is caused by exponentiation and pairing, we only count the number of these two operations. $\mathbf{E}$ and $\mathbf{P}$ represents exponentiation and pairing respectively. U_i indicates the set of attributes managed by a certain attribute authority. S represents the set of attributes assigned to a user. n is the number of rows of the linear secret sharing matrix used in the encryption and decryption algorithm.

13.4.2   COMPUTATION EVALUATION

The evaluation is based on Charm [12], i.e., a python library for pairing-based cryptography. The evaluation environment is a VM on top of Intel Xeon CPU E5-2650 v2 @ 2.60hZ. Two virtual CPU, 4GB RAM and 80GB hard disk are assigned for this VM. The performance evaluation of setup, keygen, encrypt, and decrypt is presented from Figure 13.6 to Figure 13.9. The computation overhead is medium. Moreover, considering supply-chain business functions' real-time need, these delays are negligible. However, considering IoT related business models with more stringent Realtime requirements, we presented an Offloaded version of the scheme.

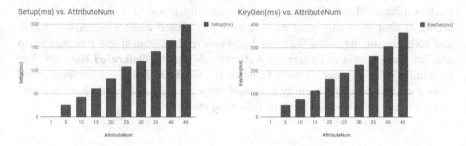

Figure 13.6: Setup time.   Figure 13.7: Key generation time.

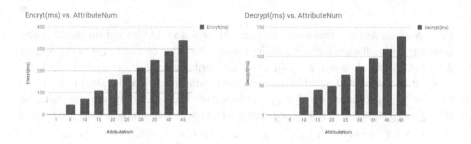

Figure 13.8: Encryption time.   Figure 13.9: Decryption time.

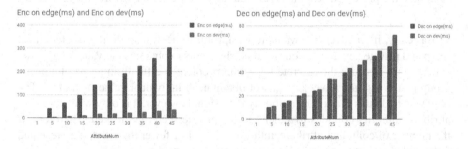

Figure 13.10: Encryption time.   Figure 13.11: Decryption time.

Computation Overhead with Offloading

Figures 13.10 and 13.11 compare the computation overhead between the edge node and the user's end. With offloading, most of the encryption work can be offload to the edge nodes and almost half of the decryption work can be eliminated from the user.

13.4.3 SECURITY ANALYSIS

The presented ABE scheme is based on Lewko's scheme from [137]. For interested readers, please refer to Lewko's work for security proof, in which the scheme is secure against both multiple (fewer than $n-1$) trusted authority collusion attack and collusion among users. We extend Lewko's scheme from single authority setup and key management to **Federated Authority Setup** and **Federated KeyGen**. The remaining work we need to do is to prove the federated setup and key generation algorithm does not cause security issue and break the collusion problem provided by Lewko's scheme. Then, we have the following theorem:

Theorem 13.1: Collusion Resistance

The presented scheme is resistant against attacks from colluded attribute authorities.

■

Proof Sketch. Assume that $n-1$ attribute authorities in AAS except for one attribute authority AA' colluded, each of which provides the value α_j and y_j ($AA_j \in AAS$ and $AA_j \neq AA'$). We denote the set of colluded attribute authorities as CAA. To compromise the presented scheme, the colluded authorities need to obtain the value of $\alpha = \sum_{j \in AAS} \alpha_j$ and $y = \sum_{j \in AAS} y_j$. However, they can only get the value of $\alpha' = \sum_{j \in CCA} \alpha_j$ and $y' = \sum_{j \in CCA} y_j$. Assume that they successfully calculate the value α and y with access to α', y', $e(g_1, g_1)^{\alpha_j}$ and $g_1^{y_j}$ ($AA_j = AA'$), then it is easy to see that the colluded attribute authorities solved well known computationally difficult problem discrete logarithm problem which brings up a contradiction. Therefore, we can prove the presented scheme is resistant against the colluded attribute authorities.

□

Basically, if an adversary wants to compromise the system during the federated setup and private key generation, what he/she wants to obtain is the value of $\sum_{j=1}^{n} \alpha_j$ and $\sum_{j=1}^{n} y_j$. Because the discrete logarithm problem is difficult in terms of a prime group that is big, the adversary cannot obtain each individual value α_j and y_j. The only way to do this is attribute authority collusion. However, it is only when all of the attribute authority colludes together can the private secret get leaked. Therefore, if the number of colluding attribute authority is $n-1$ or fewer than that, the presented federated algorithms are resistant against collusion attacks.

Security Analysis of the Scheme with Offloaded Encryption and Decryption

In the Offloaded encryption algorithm, both the message M and related information of the shared secret s is leaked to the edge node. In this way, the edge node is capable

Table 13.3

Blockchain feature comparison

Solution	Protect	Method	Compu.	Type	Access Policy
PoP	smart contract	off-chain execution & ABE	Medium	no limit	Yes
Hawk [132]	smart contract	off-chain execution & on-chain zkSNARK	heavy	no limit	No
Ekiden [54]	smart contract	off-chain Hardware Tee	low	no limit	No
Maxwell [156]	amount	on-chain	n/a	Bitcoin	No
ZeroCash [190]	identity	on-chain	n/a	Bitcoin	No

to help offload the computation overhead and also knows nothing about the encrypted message. In the decryption algorithm, the edge node is only responsible for the private key unrelated computation, i.e., $C_{1,x} \cdot e(H(GID), C_{3,x})$. Therefore, the private key is not leaked to the edge node, while the edge node can still help do the decryption. Therefore, we can prove easily that offloaded scheme is secure if the original scheme is secure.

13.4.4 COMPARATIVE STUDY

Several existing projects, e.g., Hyperledger [45], R3CEV's Corda [208], and the Gem Health network [176] provide private blockchain solution for business. The idea of cross-chain functionality is to enjoy the benefits from both public and private blockchains, in which solutions bent on delivering cross-chain functionality mean that many of the existing obstacles currently governing the exchange of value will gradually fade. In effect, cross-chain functionality could gather together the best features of blockchains [146], both private and public for the purposes of exchanging value across disconnected ecosystems. Ripple [44] has already made notable strides to this effect, with Inter-ledger already testing transactions across multiple ledgers simultaneously in different currencies. ZCash [105] provides privacy protection for Bitcoin [162] users. Hawk [132] and Ekiden [54] have been proposed using off-chain approaches to provide data privacy protection. However, none of existing solutions clearly addressed how to apply access control policies to enforce data privacy protection on transaction secrets.

Table 13.3 summarizes the main feature comparisons with existing major privacy-preserving solutions.

13.5 SUMMARY

In this chapter, we presented a blockchain solution on how to build private blockchains over public blockchains, called PoPs. A set of messaging and smart contract protocols were also presented to illustrate privacy-preserving functions of PoP. We use a supply-chain procurement procedure example to illustrate how PoP works.

Blockchain technologies for supply-chain and other business functions are emerging research and development areas. This presented work may lead to many research and development directions for the next step. First, more functional-rich policy-based access control solutions should be considered. The existing solution is based on Lewko's solution. Other features, such as attribute and user's revocation, should be considered; policy/attributes expiration should be also considered that allow more automatic policy-based access control features, etc. Second, the presented smart contracts only focus on PPP establishment and procurement. Other smart contracts such as cancellation/revocation of a contract should be also investigated. Third, we briefly discussed on how to use IoT device and how a bank can monitor blockchain-based transactions to allow them to decide business loan credibility of business parties.

References

1. Ethereum project. Available at https://www.ethereum.org.
2. Named Data Networking, 2015.
3. NCCOE discussion and impediments (to XACML and ABAC adoption). https://lists.oasis-open.org/archives/xacml/201604/msg00000.html, 2016.
4. Axiomatics. https://www.axiomatics.com/, 2017.
5. Home Healthcare Market worth 349.8 Billion USD by 2020. by Markets and Markets, available at http://www.marketsandmarkets.com/PressReleases/home-healthcare.asp, 2017.
6. Jericho's Enterspace. https://www.jerichosystems.com/products/decisioning.html, 2017.
7. New study: 63% of organizations will adopt distributed storage (SDS) by 2018. available at https://storpool.com/blog/63-percent-of-organizations-will-have-adopted-distributed-storage-by-2018, 2017.
8. Nextlabs' Dynamic Authorization. https://www.nextlabs.com/products/abac/, 2017.
9. Sia. https://sia.tech/, 2017.
10. XpressRules LLC. https://www.sbir.gov/sbirsearch/detail/1320237, 2017.
11. Zeutro. http://www.zeutro.com/, 2017.
12. "charm: A framework for rapidly prototyping cryptosystems". Available at https://pypi.org/project/charm-crypto/, accessed on Oct 2019.
13. Michel Abdalla, Dario Catalano, Alexander W Dent, John Malone-Lee, Gregory Neven, and Nigel P Smart. Identity-based encryption gone wild. In *International Colloquium on Automata, Languages, and Programming*, pages 300–311. Springer, Berlin, Heidelberg, 2006.
14. Andre Adelsbach and Ulrich Greveler. A broadcast encryption scheme with free-riders but unconditional security. In *International Conference on Digital Rights Management*, pages 246–257. Springer, 2005.
15. André Adelsbach, Ulrich Huber, and Ahmad-Reza Sadeghi. Property-based broadcast encryption for multi-level security policies. In *International Conference on Information Security and Cryptology*, pages 15–31. Springer, 2005.
16. Ruqayah R Al-Dahhan, Qi Shi, Gyu Myoung Lee, and Kashif Kifayat. Survey on revocation in ciphertext-policy attribute-based encryption. *Sensors*, 19(7):1695, 2019.
17. Fabio Angius, Mario Gerla, and Giovanni Pau. Bloogo: Bloom filter based gossip algorithm for wireless ndn. In *Proceedings of the 1st ACM Workshop on Emerging Name-Oriented Mobile Networking Design-Architecture, Algorithms, and Applications*, pages 25–30. ACM, 2012.
18. Somaya Arianfar, Teemu Koponen, Barath Raghavan, and Scott Shenker. On preserving privacy in content-oriented networks. In *Proceedings of the ACM SIGCOMM Workshop on Information-centric Networking*, pages 19–24, 2011.
19. G. Ateniese, R. Burns, R. Curtmola, J. Herring, L. Kissner, Z. Peterson, and D. Song. Provable data possession at untrusted stores. In *Proceedings of the 14th ACM Conference on Computer and Communications Security*, pages 598–609. ACM, 2007.

20. G. Ateniese, K. Fu, M. Green, and S. Hohenberger. Improved proxy re-encryption schemes with applications to secure distributed storage. *ACM Trans. Inf. Syst. Secur.*, 9(1):1–30, 2006.

21. B Balamurugan and P Venkata Krishna. Extensive survey on usage of attribute based encryption in cloud. *Journal of Emerging Technologies in Web Intelligence*, 6(3):263–272, 2014.

22. Feng Bao, Robert H Deng, Xuhua Ding, and Yanjiang Yang. Private query on encrypted data in multi-user settings. In *International Conference on Information Security Practice and Experience*, pages 71–85. Springer, 2008.

23. Amos Beimel. *Secure schemes for secret sharing and key distribution*. Technion-Israel Institute of technology, Faculty of computer science, 1996.

24. Mihir Bellare, Alexandra Boldyreva, and Adam O'Neill. Deterministic and efficiently searchable encryption. In *Annual International Cryptology Conference*, pages 535–552. Springer, 2007.

25. John Bethencourt, Amit Sahai, and Brent Waters. Ciphertext-policy attribute-based encryption. In *IEEE S&P'07*, pages 321–334, Oakland, CA, May 2007.

26. John Bethencourt, Amit Sahai, and Brent Waters. Ciphertext-policy attribute-based encryption. In *Security and Privacy, 2007. SP'07. IEEE Symposium on*, pages 321–334. IEEE, 2007.

27. Genci Bilali. Know your customer-or not. *U. Tol. L. Rev.*, 43:319, 2011.

28. Ratnabali Biswas, Kaushik Chowdhury, and Dharma P. Agrawal. Attribute allocation and retrieval scheme for large-scale sensor networks. *International Journal of Wireless Information Networks*, pages 303–315, 2006.

29. Andreas Bogner, Mathieu Chanson, and Arne Meeuw. A decentralised sharing app running a smart contract on the ethereum blockchain. In *Proceedings of the 6th International Conference on the Internet of Things*, pages 177–178. ACM, 2016.

30. D. Boneh, X. Boyen, and E.J. Goh. Hierarchical identity based encryption with constant size ciphertext. *Advances in Cryptology–EUROCRYPT 2005*, pages 440–456, 2005.

31. D. Boneh, G. Di Crescenzo, R. Ostrovsky, and G. Persiano. Public key encryption with keyword search. In *Eurocrypt*, pages 506–522. Springer, 2004.

32. D. Boneh and M. Franklin. Identity-based encryption from the Weil pairing. In *CRYPTO'01*, pages 213–229, Santa Barbara, CA, Aug. 2001.

33. D. Boneh, C. Gentry, and B. Waters. Collusion resistant broadcast encryption with short ciphertexts and private keys. In *Advances in Cryptology–CRYPTO 2005*, pages 258–275. Springer, 2005.

34. D. Boneh, A. Sahai, and B. Waters. Fully collusion resistant traitor tracing with short ciphertexts and private keys. *Annual International Conference on the Theory and Applications of Cryptographic Techniques*, pages 573–592. Springer, Berlin, Heidelberg, 2006.

35. D. Boneh and B. Waters. A fully collusion resistant broadcast, trace, and revoke system. *Proceedings of the 13th ACM conference on Computer and communications security*, pages 211–220, 2006.

36. D. Boneh and B. Waters. Conjunctive, subset, and range queries on encrypted data. In *Theory of Cryptography Conference*, pages 535–554. Springer, Berlin, Heidelberg, 2007.

37. Dan Boneh and Xavier Boyen. Efficient selective-id secure identity-based encryption without random oracles. In *International conference on the theory and applications of cryptographic techniques*, pages 223–238. Springer, 2004.

38. Dan Boneh, Xavier Boyen, and Hovav Shacham. Short group signatures. In *Annual International Cryptology Conference*, pages 41–55. Springer, 2004.
39. Dan Boneh and Matt Franklin. Identity-based encryption from the weil pairing. In *Annual International Cryptology Conference*, pages 213–229. Springer, 2001.
40. Dan Boneh and Michael Hamburg. Generalized identity based and broadcast encryption schemes. In *International Conference on the Theory and Application of Cryptology and Information Security*, pages 455–470. Springer, 2008.
41. Christoph Bösch, Pieter Hartel, Willem Jonker, and Andreas Peter. A survey of provably secure searchable encryption. *ACM Computing Surveys (CSUR)*, 47(2):18, 2015.
42. X. Boyen and B. Waters. Anonymous hierarchical identity-based encryption (without random oracles). 4117:290–307, 2006.
43. Xavier Boyen. Attribute-based functional encryption on lattices. In *Theory of Cryptography Conference*, pages 122–142. Springer, 2013.
44. Vitalik Buterin. Chain interoperability, 2016.
45. Christian Cachin. Architecture of the hyperledger blockchain fabric. In *Workshop on Distributed Cryptocurrencies and Consensus Ledgers*, volume 310, 2016. https://www.zurich.ibm.com/dccl/papers/cachindccl.pdf.
46. R. Canetti, T. Malkin, and K. Nissim. Efficient Communication-Storage Tradeoffs for Multicast Encryption, Advances in Cryptology-Eurocrypt'99. *Lecture Notes in Computer Science*, 1592:459–474, 1999.
47. G. Caronni, M. Waldvogel, D. Sun, and B. Plattner. Efficient security for large and dynamic multicast groups. *Proceedings of the IEEE 7th International Workshop on Enabling Technologies: Infrastructure for Collaborative Enterprises (WET ICE'98)*, 1998.
48. A. Carzaniga, M.J. Rutherford, and A.L. Wolf. A routing scheme for content-based networking. In *INFOCOM 2004*, pages 918–928.
49. I. Chang, R. Engel, D. Kandlur, D. Pendarakis, D. Saha, I.B.M.T.J.W.R. Center, and Y. Heights. Key management for secure Internet multicast using Boolean functionminimization techniques. *INFOCOM'99. Eighteenth Annual Joint Conference of the IEEE Computer and Communications Societies. Proceedings. IEEE*, 2, 1999.
50. M. Chase. Multi-authority attribute based encryption. *Theory of Cryptography Conference*, pages 515–534. Springer, Berlin, Heidelberg, 2007.
51. D. Chaum and E. Van Heyst. Group signatures. *Lecture Notes in Computer Science*, 547:257–265, 1991.
52. Jie Chen, Hoon Wei Lim, San Ling, Huaxiong Wang, and Khoa Nguyen. Revocable identity-based encryption from lattices. In *Information Security and Privacy*, pages 390–403. Springer, 2012.
53. Nanxi Chen, Mario Gerla, Dijiang Huang, and Xiaoyan Hong. Secure, selective group broadcast in vehicular networks using dynamic attribute based encryption. In *Proceedings of the 9th IFIP Annual Mediterranean Ad Hoc Networking Workshop (Med-Hoc-Net)*, 2010.
54. Raymond Cheng, Fan Zhang, Jernej Kos, Warren He, Nicholas Hynes, Noah Johnson, Ari Juels, Andrew Miller, and Dawn Song. Ekiden: A platform for confidentiality-preserving, trustworthy, and performant smart contract execution. *arXiv preprint arXiv:1804.05141*, 2018.
55. L. Cheung, J. Cooley, R. Khazan, and C. Newport. Collusion-resistant group key management using attribute-based encryption. Technical report, Cryptology ePrint Archive Report 2007/161, 2007. http://eprint.iacr.org.

56. L. Cheung and C. Newport. Provably secure ciphertext policy abe. In *CCS '07: Proceedings of the 14th ACM conference on Computer and communications security*, pages 456–465, New York, NY, USA, 2007. ACM.

57. Ling Cheung and Calvin Newport. Provably secure ciphertext policy ABE. In *SIGSAC*, pages 456–465, 2007.

58. R. Chow, P. Golle, M. Jakobsson, E. Shi, J. Staddon, R. Masuoka, and J. Molina. Controlling data in the cloud: outsourcing computation without outsourcing control. In *Proceedings of the 2009 ACM workshop on Cloud computing security*, pages 85–90. ACM, 2009.

59. Cisco. Cisco visual networking index: forecast and methodology, 2012–2017. 2013.

60. Clifford Cocks. An identity based encryption scheme based on quadratic residues. In *IMA International Conference on Cryptography and Coding*, pages 360–363. Springer, 2001.

61. IEEE Computer Society LAN MAN Standards Committee et al. Wireless lan medium access control (mac) and physical layer (phy) specifications. *ANSI/IEEE Std. 802.11-1999*, 1999.

62. T.M. Cover and J.A. Thomas. *Elements of information theory*. Wiley, 2006.

63. C. Cseh. Architecture of the dedicated short-range communications (DSRC) protocol. *IEEE Vehicular Technology Conference (VTC)*, 3:2095–2099, Ottawa, Ont., Canada, May 1998.

64. Reza Curtmola, Juan Garay, Seny Kamara, and Rafail Ostrovsky. Searchable symmetric encryption: Improved definitions and efficient constructions. *Journal of Computer Security*, 19(5):895–934, 2011.

65. Joan Daemen and Vincent Rijmen. *The design of Rijndael: AES-the advanced encryption standard*. Springer Science & Business Media, 2013.

66. C. Dannewitz, J. Golic, B. Ohlman, and B. Ahlgren. Secure naming for a network of information. In *INFOCOM 2010*, pages 1–6.

67. Angelo De Caro and Vincenzo Iovino. jpbc: Java pairing based cryptography. In *2011 IEEE symposium on computers and communications (ISCC)*, pages 850–855. IEEE, 2011.

68. Cécile Delerablée, Pascal Paillier, and David Pointcheval. Fully collusion secure dynamic broadcast encryption with constant-size ciphertexts or decryption keys. In *International Conference on Pairing-Based Cryptography*, pages 39–59. Springer, 2007.

69. Y. Desmedt and J. Quisquater. Public-key systems based on the difficulty of tampering (is there a difference between des and rsa?). In *Proceedings on Advances in cryptology—CRYPTO '86*, pages 111–117, London, UK, 1987. Springer-Verlag.

70. S. D. C. di Vimercati, S. Foresti, S. Jajodia, S. Paraboschi, and P. Samarati. Over-encryption: management of access control evolution on outsourced data. In *Proceedings of the 33rd international conference on very large data bases*, pages 123–134, Vienna, Austria, 2007.

71. Changyu Dong, Giovanni Russello, and Naranker Dulay. Shared and searchable encrypted data for untrusted servers. *Journal of Computer Security*, 19(3):367–397, 2011.

72. Qiuxiang Dong, Zhi Guan, and Zhong Chen. Attribute-based keyword search efficiency enhancement via an online/offline approach. In *Parallel and Distributed Systems (ICPADS), 2015 IEEE 21st International Conference on*, pages 298–305. IEEE, 2015.

73. Qiuxiang Dong, Zhi Guan, Liang Wu, and Zhong Chen. Fuzzy keyword search over encrypted data in the public key setting. In *International Conference on Web-Age Information Management*, pages 729–740. Springer, 2013.

74. Qiuxiang Dong and Dijiang Huang. Privacy-preserving matchmaking in geosocial networks with untrusted servers. In *Distributed Computing Systems (ICDCS), 2017 IEEE 37th International Conference on*, pages 2591–2592. IEEE, 2017.

75. Qiuxiang Dong, Dijiang Huang, Jim Luo, and Myong Kang. HIR-CP-ABE: Hierarchical Identity Revocable Ciphertext-Policy Attribute-Based Encryption for Secure and Flexible Data Sharing. Crypto eAchieve, available at https://eprint.iacr.org/2017/1101, 2017.

76. Qiuxiang Dong, Dijiang Huang, Jim Luo, and Myong Kang. ID-HABE: Incorporating ID-based Revocation, Delegation, and Authority Hierarchy into Attribute-Based Encryption. Crypto eAchieve, available at https://eprint.iacr.org/2017/1102, 2017.

77. Qiuxiang Dong, Dijiang Huang, Jim Luo, and Myong Kang. Achieving fine-grained access control with discretionary user revocation over cloud data. In *2018 IEEE Conference on Communications and Network Security (CNS)*, pages 1–9. IEEE, 2018.

78. Kennedy Edemacu, Hung Kook Park, Beakcheol Jang, and Jong Wook Kim. Privacy provision in collaborative ehealth with attribute-based encryption: Survey, challenges and future directions. *IEEE Access*, 7:89614–89636, 2019.

79. Keita Emura, Atsuko Miyaji, Akito Nomura, Kazumasa Omote, and Masakazu Soshi. A ciphertext-policy attribute-based encryption scheme with constant ciphertext length. In *International Conference on Information Security Practice and Experience*, pages 13–23. Springer, 2009.

80. C. Erway, A. Kupcu, C. Papamanthou, and R. Tamassia. Dynamic provable data possession. In *Proceedings of the 16th ACM conference on Computer and communications security*, pages 213–222. ACM, 2009.

81. A. Fiat and M. Naor. Broadcast Encryption, Advances in Cryptology-Crypto'93. *Lecture Notes in Computer Science*, 773:480–491, 1994.

82. Nikos Fotiou, Giannis F. Marias, and George C. Polyzos. Access control enforcement delegation for information-centric networking architectures. In *Proceedings of the second edition of the ICN workshop on Information-centric networking*, pages 85–90, 2012.

83. Nikos Fotiou, Pekka Nikander, Dirk Trossen, and GeorgeC. Polyzos. Developing Information Networking Further: From PSIRP to PURSUIT. In *Lecture Notes of the Institute for Computer Sciences, Social Informatics and Telecommunications Engineering*, pages 1–13. 2012.

84. David Mandell Freeman. Converting pairing-based cryptosystems from composite-order groups to prime-order groups. In *Annual International Conference on the Theory and Applications of Cryptographic Techniques*, pages 44–61. Springer, 2010.

85. J.A. Garay, J. Staddon, and A. Wool. Long-lived broadcast encryption. *Lecture Notes in Computer Science*, pages 333–352, 2000.

86. Gartner. Market Trends: Cloud-Based Security Services Market, Worldwide. Website is available at https://www.gartner.com/doc/, accessed Nov. 16, 2016.

87. C. Gentry. Practical identity-based encryption without random oracles. In *Eurocrypt*, volume 4004, pages 445–464. Springer, 2006.

88. C. Gentry. Fully homomorphic encryption using ideal lattices. In *Proceedings of the 41st annual ACM symposium on Theory of computing*, pages 169–178. ACM, 2009.

89. C. Gentry. Computing arbitrary functions of encrypted data. *Communications of the ACM*, 53(3):97–105, 2010.

90. M. Gerlach, A. Festag, T. Leinmuller, G. Goldacker, and C. Harsch. Security architecture for vehicular communication. In *Proceedings of the 5th International Workshop on Intelligent Transportation (WIT)*, Hamburg, Germany, March 2007.

91. M. Gerlach and F. FOKUS. Trust for Vehicular Applications. In *Proceedings of the Eighth International Symposium on Autonomous Decentralized Systems (ISADS)*, pages 295–304, 2007.

92. Simon Godik and Tim Moses. Oasis extensible access control markup language (xacml). *OASIS Committee Secification cs-xacml-specification-1.0*, 2002.

93. M.T. Goodrich, J.Z. Sun, and R. Tamassia. Efficient tree-based revocation in groups of low-state devices. *Lecture Notes in Computer Science*, pages 511–527, 2004.

94. V. Goyal, A. Jain, O. Pandey, and A. Sahai. Bounded ciphertext policy attribute based encryption. *Automata, Languages and Programming*, pages 579–591.

95. V. Goyal, O. Pandey, A. Sahai, and B. Waters. Attribute-based encryption for fine-grained access control of encrypted data. *Proceedings of the 13th ACM conference on Computer and communications security*, pages 89–98, 2006.

96. Matthew Green, Susan Hohenberger, and Brent Waters. Outsourcing the decryption of abe ciphertexts. In *USENIX Security Symposium*, volume 2011, 2011.

97. J. Guo, JP Baugh, and S. Wang. A Group Signature Based Secure and Privacy-Preserving Vehicular Communication Framework. In *Proceedings of Mobile Networking for Vehicular Environments (MOVE)*, pages 103–108, 2007.

98. Fei Han, Jing Qin, Huawei Zhao, and Jiankun Hu. A general transformation from kp-abe to searchable encryption. *Future Generation Computer Systems*, 30:107–115, 2014.

99. D.R. Hankerson, S.A. Vanstone, and A.J. Menezes. *Guide to Elliptic Curve Cryptography*. Springer, 2004.

100. Joan Hash, Pauline Bowen, Arnold Johnson, Carla Dancy Smith, and DI Steinberg. *An introductory resource guide for implementing the health insurance portability and accountability act (HIPAA) security rule*. US Department of Commerce, Technology Administration, National Institute of Standards and Technology, 2005.

101. J. Herranz, F. Laguillaumie, and C. Ràfols. Constant Size Ciphertexts in Threshold Attribute-Based Encryption. *Public Key Cryptography–PKC 2010*, pages 19–34, 2010.

102. HIPAA. Individuals' Right under HIPAA to Access their Health Information 45 CFR § 164.524. https://www.hhs.gov/hipaa/for-professionals/privacy/guidance/access/index.html, 2017.

103. Erik Hofmann, Urs Magnus Strewe, and Nicola Bosia. *Supply Chain Finance and Blockchain Technology: The Case of Reverse Securitisation*. Springer, 2017.

104. Susan Hohenberger and Brent Waters. Online/offline attribute-based encryption. In *International workshop on public key cryptography*, pages 293–310. Springer, 2014.

105. Daira Hopwood, Sean Bowe, Taylor Hornby, and Nathan Wilcox. Zcash protocol specification. Technical report, Tech. rep. 2016-1.10. Zerocoin Electric Coin Company, 2016.

106. F. Hsu and H. Chen. Secure file system services for web 2.0 applications. In *Proceedings of the 2009 ACM workshop on Cloud computing security*, pages 11–18. ACM, 2009.

107. Vincent C Hu, David Ferraiolo, Rick Kuhn, Adam Schnitzer, Kenneth Sandlin, Robert Miller, and Karen Scarfone. Guide to attribute based access control (abac) definition and considerations. *NIST Special Publication 800-162*, January 2014.

108. D. Huang, X. Zhang, M. Kang, and J. Luo. Mobicloud: A secure mobile cloud framework for pervasive mobile computing and communication. In *Proceedings of 5th IEEE International Symposium on Service-Oriented System Engineering*, 2010.

109. D. Huang, Z. Zhou, and Z. Yan. Gradual Identity Exposure Using Attribute-Based Encryption. In *Social Computing (SocialCom), 2010 IEEE Second International Conference on*, pages 881–888. IEEE, 2010.

110. Dijiang Huang. Pseudonym-based cryptography for anonymous communications in mobile ad hoc networks. *International Journal of Security and Networks*, 2(3-4):272–283, 2007.

111. Dijiang Huang, Satyajayant Misra, Mayank Verma, and Guoliang Xue. Pacp: An efficient pseudonymous authentication-based conditional privacy protocol for vanets. *IEEE Transactions on Intelligent Transportation Systems*, 12(3):736–746, 2011.

112. Dijiang Huang and Mayank Verma. Aspe: Attribute-based secure policy enforcement in vehicular ad hoc networks. *Ad Hoc Networks*, 7(8):1526–1535, 2009.

113. Dijiang Huang and Zhijie Wang. Enabling comparable data access control for lightweight mobile devices in clouds, July 11 2017. US Patent 9,705,850.

114. Dijiang Huang and Zhijie Wang. Enabling comparable data access control for lightweight mobile devices in clouds, September 21 2017. US Patent App. 15/617,035.

115. Dijiang Huang, Tianyi Xing, and Huijun Wu. Mobile cloud computing service models: a user-centric approach. *IEEE network*, 27(5):6–11, 2013.

116. Dijiang Huang and Zhibin Zhou. Methods, systems, and apparatuses for optimal group key management for secure multicast communication, September 16 2014. US Patent 8,837,738.

117. Dijiang Huang, Zhibin Zhou, and Zhu Yan. Gradual Identity Exposure Using Attribute-Based Encryption. In *SocialCom*, pages 881–888, 2010.

118. John Hughes and Eve Maler. Security assertion markup language (saml) v2. 0 technical overview. *OASIS SSTC Working Draft sstc-saml-tech-overview-2.0-draft-08*, pages 29–38, 2005.

119. Yong Ho Hwang and Pil Joong Lee. Public key encryption with conjunctive keyword search and its extension to a multi-user system. In *International conference on pairing-based cryptography*, pages 2–22. Springer, 2007.

120. IETF PIIX Working Group. Public-Key Infrastructure (X.509) (PKIX), available at http://www.ietf.org/html.charters/pkix-charter.html. Internet RFCs.

121. ITE. Traffic Management Data Dictionary (TMDD) and Message Sets for External Traffic Management Center Communications (MS/ETMCC). Access on October 2019, available at https://www.ite.org/technical-resources/standards/tmdd/.

122. Praveen Jayachandran. The difference between public and private blockchain. available at https://www.ibm.com/blogs/blockchain/2017/05/the-difference-between-public-and-private-blockchain/, 2017.

123. Brent Waters John Bethencourt, Amit Sahai. Ciphertext-Policy Attribute-Based Encryption. http://acsc.csl.sri.com/cpabe/, 2006.

124. M. Kallahalla, E. Riedel, R. Swaminathan, Q. Wang, and K. Fu. Plutus: Scalable secure file sharing on untrusted storage. In *FAST '03: Proceedings of the 2nd USENIX Conference on File and Storage Technologies*, pages 29–42, 2003.

125. Apu Kapadia, Patrick P Tsang, and Sean W Smith. Attribute-based publishing with hidden credentials and hidden policies. In *NDSS*, volume 7, pages 179–192, 2007.

126. Arijit Karati, Ruhul Amin, and GP Biswas. Provably secure threshold-based abe scheme without bilinear map. *Arabian Journal for Science and Engineering*, 41(8):3201–3213, 2016.

127. J. Katz, A. Sahai, and B. Waters. Predicate encryption supporting disjunctions, polynomial equations, and inner products. In *EUROCRYPT'08: Proceedings of the theory and applications of cryptographic techniques 27th annual international conference on Advances in cryptology*, pages 146–162, Berlin, Heidelberg, 2008. Springer-Verlag.

128. E. Keller, J. Szefer, J. Rexford, and R.B. Lee. NoHype: virtualized cloud infrastructure without the virtualization. In *Proceedings of the 37th annual international symposium on Computer architecture*, pages 350–361. ACM, 2010.

129. Rami Khalil and Arthur Gervais. Revive: Rebalancing off-blockchain payment networks. In *Proceedings of the 2017 ACM SIGSAC Conference on Computer and Communications Security*, pages 439–453. ACM, 2017.

130. Andreas Klein, Christian Mannweiler, Joerg Schneider, and Hans D Schotten. Access schemes for mobile cloud computing. In *2010 eleventh international conference on mobile data management*, pages 387–392. IEEE, 2010.

131. Teemu Koponen, Mohit Chawla, Byung-Gon Chun, Andrey Ermolinskiy, Kye Hyun Kim, Scott Shenker, and Ion Stoica. A data-oriented (and beyond) network architecture. In *SIGCOMM*, pages 181–192, 2007.

132. Ahmed Kosba, Andrew Miller, Elaine Shi, Zikai Wen, and Charalampos Papamanthou. Hawk: The blockchain model of cryptography and privacy-preserving smart contracts. In *Security and Privacy (SP), 2016 IEEE Symposium on*, pages 839–858. IEEE, 2016.

133. Praveen Kumar, PJA Alphonse, et al. Attribute based encryption in cloud computing: A survey, gap analysis, and future directions. *Journal of Network and Computer Applications*, 108:37–52, 2018.

134. Eemil Lagerspetz and Sasu Tarkoma. Mobile search and the cloud: The benefits of offloading. In *2011 IEEE international conference on pervasive computing and communications workshops (PERCOM workshops)*, pages 117–122. IEEE, 2011.

135. Cheng-Chi Lee, Pei-Shan Chung, and Min-Shiang Hwang. A survey on attribute-based encryption schemes of access control in cloud environments. *IJ Network Security*, 15(4):231–240, 2013.

136. Allison Lewko, Amit Sahai, and Brent Waters. Revocation systems with very small private keys. In *Security and Privacy (SP), 2010 IEEE Symposium on*, pages 273–285. IEEE, 2010.

137. Allison Lewko and Brent Waters. Decentralizing attribute-based encryption. In *Annual International Conference on the Theory and Applications of Cryptographic Techniques*, pages 568–588. Springer, 2011.

138. B. Li, A. Prabhu Verleker, D. Huang, Z. Wang, and Y. Zhu. Attribute-based access control for icn naming scheme. In *Proceedings of the IEEE Conference on Communications and Network Security*, CNS, pages 391–399, 2014.

139. Bing Li, Dijiang Huang, Zhijie Wang, and Yan Zhu. Attribute-based access control for icn naming scheme. *IEEE Transactions on Dependable and Secure Computing*, 15(2):194–206, 2016.

140. Bing Li, Ashwin Prabhu Verleker, Dijiang Huang, Zhijie Wang, and Yan Zhu. Attribute-based access control for icn naming scheme. In *Proceedings of the 17th ACM Conference on Computer and Communications Security*. IEEE, 2014.

141. Bing Li, Zhijie Wang, and Dijiang Huang. An efficient and anonymous attribute-based group setup scheme. In *GLOBECOM*, pages 861–866, 2013.

142. Chao Li, Bo Lang, and Jinmiao Wang. Outsourced kp-abe with enhanced security. In *International Conference on Trusted Systems*, pages 36–50. Springer, 2014.

143. J. Li, Q. Wang, C. Wang, N. Cao, K. Ren, and W. Lou. Fuzzy keyword search over encrypted data in cloud computing. In *INFOCOM, 2010 Proceedings IEEE*, pages 1–5. IEEE.

144. Ming Li, Shucheng Yu, Ning Cao, and Wenjing Lou. Authorized private keyword search over encrypted data in cloud computing. In *2011 31st International Conference on Distributed Computing Systems*, pages 383–392. IEEE, 2011.

145. PBC Library. Curve Param generation. 2015.

146. Joe Liebkind. Public vs Private Blockchains: Challenges and Gaps. available at https://www.investopedia.com/news/public-vs-private-blockchains-challenges-and-gaps/, 2018.

147. Xiaodong Lin, Xiaoting Sun, Pin-Han Ho, and Xuemin Shen. Gsis: A secure and privacy-preserving protocol for vehicular communications. *IEEE Transactions on vehicular technology*, 56(6):3442–3456, 2007.

148. Chi-Wei Liu, Wei-Fu Hsien, Chou Chen Yang, and Min-Shiang Hwang. A survey of attribute-based access control with user revocation in cloud data storage. *IJ Network Security*, 18(5):900–916, 2016.

149. Yuan Liu, Licheng Wang, Lixiang Li, and Xixi Yan. Secure and efficient multi-authority attribute-based encryption scheme from lattices. *IEEE Access*, 7:3665–3674, 2018.

150. D. Lubicz and T. Sirvent. Attribute-based broadcast encryption scheme made efficient. *Progress in Cryptology–AFRICACRYPT 2008*, pages 325–342.

151. Jim Luo, Qiuxiang Dong, Dijiang Huang, and Myong Kang. Attribute based encryption for information sharing on tactical mobile networks. In *MILCOM 2018-2018 IEEE Military Communications Conference (MILCOM)*, pages 1–9. IEEE, 2018.

152. B. Lynn. *On the implementation of pairing-based cryptosystems*. PhD thesis, Stanford Univesity, http://crypto.stanford.edu/pbc/thesis.pdf, 2007.

153. B. Lynn. PBC Library The Pairing-Based Cryptography Library. In *http://crypto.stanford.edu/pbc/*, Accessed Oct 2019.

154. Ben Lynn. The pairing-based cryptography library, 2006.

155. U. Maurer and Y. Yacobi. Non-interactive public-key cryptography. *Advances in Cryptology–EUROCRYPT'91 (LNCS 547)*, pages 498–507, 1991.

156. Gregory MaxwelL. Project Gmaxwell. Access on October 2019, available at https://github.com/gmaxwell.

157. E.J. McCluskey. Minimization of Boolean functions. *Bell System Technical Journal*, 35(5):1417–1444, 1956.

158. Antti P Miettinen and Jukka K Nurminen. Energy efficiency of mobile clients in cloud computing. *HotCloud*, 10(4-4):19, 2010.

159. S. Mittra. Iolus: A framework for scalable secure multicasting. *ACM SIGCOMM Computer Communication Review*, 27(4):277–288, 1997.

160. A. Miyaji, M. Nakabayashi, and S. Takano. New explicit conditions of elliptic curve traces for FR-reduction. *IEICE Transactions on Fundamentals of Electronics, Communications and Computer Sciences*, 84(5):1234–1243, 2001.

161. MJ Moyer, JR Rao, and P. Rohatgi. A survey of security issues in multicast communications. *Network, IEEE*, 13(6):12–23, 1999.

162. Satoshi Nakamoto. Bitcoin: A peer-to-peer electronic cash system. 2008.

163. D. Naor, M. Naor, and J. Lotspiech. Revocation and tracing schemes for stateless receivers. *Lecture Notes in Computer Science*, pages 41–62, 2001.

164. M. Naor and B. Pinkas. Efficient trace and revoke schemes. In *Financial cryptography: 4th international conference, FC 2000, Anguilla, British West Indies, February 20-24, 2000: proceedings*, page 1. Springer Verlag, 2001.

165. T. Nishide, K. Yoneyama, and K. Ohta. Attribute-based encryption with partially hidden encryptor-specified access structures. *Applied Cryptography and Network Security*, 5037:111, 2008.

166. Takashi Nishide, Kazuki Yoneyama, and Kazuo Ohta. Attribute-based encryption with partially hidden encryptor-specified access structures. In *Proceedings of the 6th international conference on Applied cryptography and network security*, pages 111–129, 2008.

167. Justin O'Connell. What Are the Use Cases for Private Blockchains? The Experts Weigh In. available at `https://bitcoinmagazine.com/articles/what-are-the-use-cases-for-private-blockchains-the-experts-weigh-in-1466440884/`, 2016.

168. R. Ostrovsky and B. Waters. Attribute-based encryption with non-monotonic access structures. In *Proceedings of the 14th ACM conference on Computer and communications security*, pages 195–203. ACM New York, NY, USA, 2007.

169. P. Papadimitratos, L. Buttyan, J-P. Hubaux, F. Kargl, A. Kung, and M. Raya. Architecture for Secure and Private Vehicular Communications. In *Proccedings of the 7th International Conference on ITS Telecommunications*, pages 1–6. Sophia Antipolis, France, 2007.

170. B. Parno and A. Perrig. Challenges in securing vehicular networks. In *Workshop on hot topics in networks (HotNets-IV)*, pages 1–6. College Park, MD USA, 2005.

171. A. Perrig, D. Song, and J. Tygar. ELK, A New Protocol for Efficient Large-Group Key Distribution. *IEEE SYMPOSIUM ON SECURITY AND PRIVACY*, pages 247–262, 2001.

172. A. Perrig and JD Tygar. *Secure Broadcast Communication in Wired and Wireless Networks*. Springer, 2003.

173. M. Pirretti, P. Traynor, P. McDaniel, and B. Waters. Secure attribute-based systems. In *CCS '06: Proceedings of the 13th ACM conference on Computer and communications security*, pages 99–112, New York, NY, 2006. ACM.

174. Joseph Poon and Thaddeus Dryja. The bitcoin lightning network: Scalable off-chain instant payments. *draft version 0.5*, 9:14, 2016.

175. R. Poovendran and JS Baras. An information-theoretic approach for design and analysis ofrooted-tree-based multicast key management schemes. *IEEE Transactions on Information Theory*, 47(7):2824–2834, 2001.

176. J Prisco. The blockchain for healthcare: Gem launches gem health network with philips blockchain lab. *Bitcoin Magazine*, 2016.

177. Ioannis Psaras, Wei Koong Chai, and George Pavlou. Probabilistic in-network caching for information-centric networks. In *Proceedings of the Sigcomm workshop on Information-centric networking*, pages 55–60, 2012.

178. A. Ramachandran, Z. Zhou, and D. Huang. Computing Cryptographic Algorithms in Portable and Embedded Devices. *Portable Information Devices, 2007. PORTABLE07. IEEE International Conference on*, 25-29:1–7, 2007.

179. M. Raya and J.P. Hubaux. Efficient secure aggregation in VANETs. *Proceedings of the 3rd international workshop on Vehicular ad hoc networks*, pages 67–75, 2006.

180. M. Raya and J.P. Hubaux. Securing vehicular ad hoc networks. *Journal of Computer Security*, 15(1):39–68, 2007.

181. M. Raya, P. Papadimitratos, V.D. Gligor, J.P. Hubaux, and S. EPFL. On Data-Centric Trust Establishment in Ephemeral Ad Hoc Networks. In *Proceedings of IEEE Infocom*, 2008.

182. Mariana Raykova, Binh Vo, Steven M Bellovin, and Tal Malkin. Secure anonymous database search. In *Proceedings of the 2009 ACM workshop on Cloud computing security*, pages 115–126. ACM, 2009.

183. T. Ristenpart, E. Tromer, H. Shacham, and S. Savage. Hey, you, get off of my cloud: exploring information leakage in third-party compute clouds. In *Proceedings of the 16th ACM conference on Computer and communications security*, pages 199–212. ACM, 2009.

184. S Roy and M Chuah. Secure data retrieval based on ciphertext policy attribute-based encryption (cp-abe) system for the dtns. *Lehigh CSE Tech. Rep.*, 2009.

185. SAE. ISP-vehicle location referencing standard. SAE Standard J1746. Access on October 2019, available at https://www.sae.org/standards/content/j1746_199912/, 2001.

186. A. Sahai and B. Waters. Fuzzy Identity-Based Encryption. *Advances in Cryptology–Eurocrypt*, 3494:457–473.

187. K. Sampigethaya, L. Huang, M. Li, R. Poovendran, K. Matsuura, and K. Sezaki. CARAVAN: providing location privacy for VANET. In *Proceedings of Embedded Security in Cars (ESCAR)*, 2005.

188. Ravi S Sandhu, Edward J Coyne, Hal L Feinstein, and Charles E Youman. Role-based access control models. *Computer*, 29(2):38–47, 1996.

189. T. Sasao. Bounds on the average number of products in the minimum sum-of-products expressions for multiple-value input two-valued output functions. *Computers, IEEE Transactions on*, 40(5):645–651, May 1991.

190. Eli Ben Sasson, Alessandro Chiesa, Christina Garman, Matthew Green, Ian Miers, Eran Tromer, and Madars Virza. Zerocash: Decentralized anonymous payments from bitcoin. In *Security and Privacy (SP), 2014 IEEE Symposium on*, pages 459–474. IEEE, 2014.

191. Paul Allan Schott. *Reference guide to anti-money laundering and combating the financing of terrorism*. World Bank Publications, 2006.

192. Daniel Servos and Sylvia L Osborn. Current research and open problems in attribute-based access control. *ACM Computing Surveys (CSUR)*, 49(4):65, 2017.

193. Siamak F Shahandashti and Reihaneh Safavi-Naini. Threshold attribute-based signatures and their application to anonymous credential systems. In *International Conference on Cryptology in Africa*, pages 198–216. Springer, 2009.

194. A. Shamir. How to share a secret. *Communications of the ACM*, 22(11):612–613, 1979.

195. Adi Shamir. Identity-based cryptosystems and signature schemes. In *Workshop on the theory and application of cryptographic techniques*, pages 47–53. Springer, 1984.

196. Guo Shanqing and Zeng Yingpei. Attribute-based signature scheme. In *2008 International Conference on Information Security and Assurance (ISA 2008)*, pages 509–511. IEEE, 2008.

197. A.T. Sherman and D.A. McGrew. Key Establishment in Large Dynamic Groups Using One-Way Function Trees. *IEEE Transactions on Software Engineering*, pages 444–458, 2003.

198. Jie Shi, Junzuo Lai, Yingjiu Li, Robert H Deng, and Jian Weng. Authorized keyword search on encrypted data. In *European Symposium on Research in Computer Security*, pages 419–435. Springer, 2014.

199. Victor Shoup. Lower bounds for discrete logarithms and related problems. In *Eurocrypt*, pages 256–266, 1997.

200. Rasal Shraddha and Tidke Bharat. Enhancing flexibility for abe through the use of cipher policy scheme with multiple mediators. In *Proceedings of the 3rd International Conference on Frontiers of Intelligent Computing: Theory and Applications (FICTA) 2014*, pages 457–464. Springer, 2015.

201. Sapna Singh. A Trust Based Approach For Secure Access Control In Information Centric Network. *International Journal of Information and Network Security (IJINS)*, 1:97–104, 2012.

202. J. Snoeyink, S. Suri, and G. Varghese. A lower bound for multicast key distribution. *Computer Networks*, 47(3):429–441, 2005.

203. D.X. Song, D. Wagner, and A. Perrig. Practical techniques for searches on encrypted data. *sp*, page 0044, 2000.

204. Wenhai Sun, Shucheng Yu, Wenjing Lou, Y Thomas Hou, and Hui Li. Protecting your right: Attribute-based keyword search with fine-grained owner-enforced search authorization in the cloud. In *IEEE INFOCOM 2014-IEEE Conference on Computer Communications*, pages 226–234. IEEE, 2014.

205. Yi Sun, Seyed K. Fayaz, Yang Guo, Vyas Sekar, Yun Jin, Mohamed Ali Kaafar, and Steve Uhlig. Trace-Driven Analysis of ICN Caching Algorithms on Video-on-Demand Workloads. In *Proceedings of the 10th ACM International Conference on Emerging Networking Experiments and Technologies*, pages 363–376, 2014.

206. H. Tanaka. A realization scheme for the identity-based cryptosystem. In *Advances in Cryptology¡ªCRYPTO*, volume 293, pages 341–349, 1987.

207. S. Tsujii and T. Itoh. An ID-based cryptosystem based on the discrete logarithm problem. *IEEE J. SELECTED AREAS COMMUN.*, 7(4):467–473, 1989.

208. Eric Wall and Gustaf Malm. Using blockchain technology and smart contracts to create a distributed securities depository. 2016.

209. D. Wallner, E. Harder, and R. Agee. Key Management for Multicast: Issues and Architectures RFC 2627. *IETF, June*, 1999.

210. C. Wang, Q. Wang, K. Ren, and W. Lou. Privacy-preserving public auditing for data storage security in cloud computing. In *INFOCOM, 2010 Proceedings IEEE*, pages 1–9. IEEE, 2010.

211. Peishun Wang, Huaxiong Wang, and Josef Pieprzyk. Common secure index for conjunctive keyword-based retrieval over encrypted data. In *Workshop on secure data management*, pages 108–123. Springer, 2007.

212. Peishun Wang, Huaxiong Wang, and Josef Pieprzyk. An efficient scheme of common secure indices for conjunctive keyword-based retrieval on encrypted data. In *International workshop on information security applications*, pages 145–159. Springer, 2008.

213. Peishun Wang, Huaxiong Wang, and Josef Pieprzyk. Keyword field-free conjunctive keyword searches on encrypted data and extension for dynamic groups. In *International conference on cryptology and network security*, pages 178–195. Springer, 2008.

214. Peishun Wang, Huaxiong Wang, and Josef Pieprzyk. Threshold privacy preserving keyword searches. In *International Conference on Current Trends in Theory and Practice of Computer Science*, pages 646–658. Springer, 2008.

215. W. Wang, Z. Li, R. Owens, and B. Bhargava. Secure and efficient access to outsourced data. In *Proceedings of the 2009 ACM workshop on Cloud computing security*, pages 55–66. ACM, 2009.

216. Weijia Wang, Zhijie Wang, Bing Li, Qiuxiang Dong, and Dijiang Huang. IR-CP-ABE: Identity Revocable Ciphertext-Policy Attribute-Based Encryption for Flexible Secure

Group-Based Communication. Crypto eAchieve, available at `https://eprint.iacr.org/2017/1100`, 2017.

217. Zhijie Wang and Dijiang Huang. Privacy-preserving mobile crowd sensing in ad hoc networks. *Ad Hoc Networks*, 73:14–26, 2018.

218. Zhijie Wang, Dijiang Huang, Huijun Wu, Bing Li, and Yuli Deng. Towards distributed privacy-preserving mobile access control. In *2014 IEEE Global Communications Conference*, pages 582–587. IEEE, 2014.

219. Zhijie Wang, Dijiang Huang, Yan Zhu, Bing Li, and Chun-Jen Chung. Efficient attribute-based comparable data access control. *IEEE transactions on computers*, 64(12):3430–3443, 2015.

220. B. Waters. Ciphertext-policy attribute-based encryption: An expressive, efficient, and provably secure realization. *ePrint report*, 290, 2008.

221. Brent Waters. Efficient identity-based encryption without random oracles. In *Annual International Conference on the Theory and Applications of Cryptographic Techniques*, pages 114–127. Springer, 2005.

222. Puwen Wei, Xiaoyun Wang, and Yuliang Zheng. Public key encryption without random oracle made truly practical. In *International Conference on Information and Communications Security*, pages 107–120. Springer, 2009.

223. Wikipedia. United states traffic rules and regulations. Access on October 2019, available at `https://en.wikipedia.org/wiki/Traffic`.

224. C.K. Wong, M. Gouda, and SS Lam. Secure group communications using key graphs. *Networking, IEEE/ACM Transactions on*, 8(1):16–30, 2000.

225. Shota Yamada, Nuttapong Attrapadung, Goichiro Hanaoka, and Noboru Kunihiro. A framework and compact constructions for non-monotonic attribute-based encryption. In *International Workshop on Public Key Cryptography*, pages 275–292. Springer, 2014.

226. Z. Yan, H. Hongxin, A. Gail-Joon, H. Dijiang, and W. Shanbiao. Towards temporal access control in cloud computing. In *INFOCOM*, pages 2576 –2580, march 2012.

227. Yanjiang Yang, Haibing Lu, and Jian Weng. Multi-user private keyword search for cloud computing. In *2011 IEEE Third International Conference on Cloud Computing Technology and Science*, pages 264–271. IEEE, 2011.

228. Y.R. Yang, X.S. Li, X.B. Zhang, and S.S. Lam. Reliable group rekeying: a performance analysis. In *Proceedings of the 2001 conference on Applications, technologies, architectures, and protocols for computer communications*, pages 27–38. ACM New York, NY, USA, 2001.

229. S. Yu, K. Ren, and W. Lou. Attribute-based content distribution with hidden policy. In *Secure Network Protocols, 2008. NPSec 2008. 4th Workshop on*, pages 39–44. IEEE, 2005.

230. S. Yu, K. Ren, and W. Lou. Attribute-based on-demand multicast group setup with membership anonymity. In *SecureComm '08: Proceedings of the 4th international conference on Security and privacy in communication netowrks*, pages 1–6, New York, NY, USA, 2008. ACM.

231. S. Yu, C. Wang, K. Ren, and W. Lou. Achieving Secure, Scalable, and Fine-grained Data Access Control in Cloud Computing. *INFOCOM'1010. Conference of the IEEE Computer and Communications Societies. Proceedings. IEEE*, pages 1–9, 2010.

232. S. Yu, C. Wang, K. Ren, and W. Lou. Attribute based data sharing with attribute revocation. In *Proceedings of the 5th ACM Symposium on Information, Computer and Communications Security*, pages 261–270. ACM, 2010.

233. Shucheng Yu, Kui Ren, and Wenjing Lou. Attribute-based on-demand multicast group setup with membership anonymity. In *Proceedings of the 4th International Conference on Security and Privacy in Communication Netowrks*, pages 18:1–18:6, 2008.

234. Yu-Ting Yu, Chris Tandiono, Xiao Li, You Lu, MY Sanadidi, and Mario Gerla. Ican: Information-centric context-aware ad-hoc network. In *2014 International Conference on Computing, Networking and Communications (ICNC)*, pages 578–582. IEEE, 2014.

235. Eric Yuan and Jin Tong. Attributed based access control (abac) for web services. In *IEEE International Conference on Web Services (ICWS'05)*. IEEE, 2005.

236. A. Yun, C. Shi, and Y. Kim. On protecting integrity and confidentiality of cryptographic file system for outsourced storage. In *Proceedings of the 2009 ACM workshop on Cloud computing security*, pages 67–76. ACM, 2009.

237. Jiang Zhang, Zhenfeng Zhang, and Aijun Ge. Ciphertext policy attribute-based encryption from lattices. In *Proceedings of the 7th ACM Symposium on Information, Computer and Communications Security*, pages 16–17. ACM, 2012.

238. Jian Zhao, Haiying Gao, and Junqi Zhang. Attribute-based encryption for circuits on lattices. *Tsinghua Science and Technology*, 19(5):463–469, 2014.

239. Qingji Zheng, Shouhuai Xu, and Giuseppe Ateniese. Vabks: verifiable attribute-based keyword search over outsourced encrypted data. In *IEEE INFOCOM 2014-IEEE Conference on Computer Communications*, pages 522–530. IEEE, 2014.

240. Zhibin Zhou and Dijiang Huang. On efficient ciphertext-policy attribute based encryption and broadcast encryption. In *Proceedings of the 17th ACM conference on Computer and communications security*, pages 753–755. ACM, 2010.

241. Zhibin Zhou and Dijiang Huang. An optimal key distribution scheme for secure multicast group communication. In *2010 Proceedings IEEE INFOCOM*, pages 1–5. IEEE, 2010.

242. Zhibin Zhou and Dijiang Huang. Efficient and secure data storage operations for mobile cloud computing. *IACR Cryptology ePrint Archive*, 2011:185, 2011.

243. Zhibin Zhou and Dijiang Huang. Efficient and secure data storage operations for mobile cloud computing. In *2012 8th International Conference on Network and Service Management (CNSM) and 2012 Workshop on Systems Virtualiztion Management (SVM)*, pages 37–45. IEEE, 2012.

244. Zhibin Zhou and Dijiang Huang. Gradual identity exposure using attribute–based encryption. *International Journal of Information Privacy, Security and Integrity*, 1(2):278–297, 2012.

245. Zhibin Zhou, Dijiang Huang, and Zhijie Wang. Efficient privacy-preserving ciphertext-policy attribute based-encryption and broadcast encryption. *Computers, IEEE Transactions on*, 64(1):126–138, 2013.

246. Yan Zhu, Hongxin Hu, Gail-Joon Ahn, Mengyang Yu, and Hongjia Zhao. Comparison-based encryption for fine-grained access control in clouds. In *Proceedings of the Second ACM Conference on Data and Application Security and Privacy*, pages 105–116. ACM, 2012.

247. Yan Zhu, Dijiang Huang, Chang-Jyun Hu, and Xin Wang. From rbac to abac: constructing flexible data access control for cloud storage services. *IEEE Transactions on Services Computing*, 8(4):601–616, 2014.

248. Yan Zhu, Di Ma, Chang-Jun Hu, and Dijiang Huang. How to use attribute-based encryption to implement role-based access control in the cloud. In *Proceedings of the 2013 International Workshop on Security in Cloud Computing*, pages 33–40. ACM, 2013.

249. Yan Zhu, Di Ma, Dijiang Huang, and Changjun Hu. Enabling secure location-based services in mobile cloud computing. In *Proceedings of the second ACM SIGCOMM workshop on Mobile cloud computing*, pages 27–32. ACM, 2013.

250. X. Zou, Y.S. Dai, and E. Bertino. A Practical and Flexible Key Management Mechanism For Trusted Collaborative Computing. *INFOCOM 2008. The 27th Conference on Computer Communications. IEEE*, pages 538–546, 2008.

Index